Asphaltenes

Fundamentals and Applications

Asphaltenes

Fundamentals and Applications

Edited by

Eric Y. Sheu

Texaco Research and Development Center
Beacon, New York

and

Oliver C. Mullins

Schlumberger-Doll Research
Ridgefield, Connecticut

Springer Science+Business Media, LLC

Library of Congress Cataloging in Publication Data

Asphaltenes: fundamentals and applications / edited by Eric Y. Sheu and Oliver C. Mullins.
p. cm.
"Incorporating the proceedings of an international symposium on the Characterization of Petroleum Colloids, held at the 24th Annual Meeting of the Fine Particle Society, held August 24–25, 1993, in Chicago, Illinois"—T.p. verso.
Includes bibliographical references (p. –) and index.
ISBN 978-1-4757-9295-9 ISBN 978-1-4757-9293-5 (eBook)
DOI 10.1007/978-1-4757-9293-5
1. Asphaltene—Congresses. I. Sheu, Eric Y. II. Mullins, Oliver C. III. Fine Particle Society, Meeting (24th: 1993: Chicago, Ill.)
TP692.4.A8A743 1995 95-43746
665.5'388—dc20 CIP

Incorporating the proceedings of an international symposium on the Characterization of Petroleum Colloids, held at the 24th Annual Meeting of the Fine Particle Society, held August 24–25, 1993, in Chicago, Illinois

ISBN 978-1-4757-9295-9

Originally published by Plenum Press, New York in 1995
Softcover reprint of the hardcover 1st edition 1995

10 9 8 7 6 5 4 3 2 1

Dedicated to our wives Julie M. Sheu and Cynthia K. Mullins for their support and forbearance with our many, extended stays at National Laboratories.

PREFACE

Asphaltenes have traditionally been viewed as being extremely complex, thus very hard to characterize. In addition, certain fundamental properties of asphaltenes have previously been inaccessible to study by traditional macroscopic methods, further limiting understanding of asphaltenes. These limitations inhibited development of descriptions regarding the microscopic structure and solution dynamics of asphaltenes. However, a variety of more recent studies have implied that asphaltenes share many chemical properties with the smaller, more tractable components of crude oils. Recent measurements have indicated that asphaltene molecular weights are not as large as previously thought, perhaps in the range of 600 to 1000 amu. In addition, new experimental methods applied to asphaltene chemical structures have been quite revealing, yielding a broad understanding. Consequently, the ability to relate chemical structure with physical and chemical properties can be developed and extended to the understanding of important commercial properties of asphaltenes.

This book treats significant new developments in the fundamentals and applications of asphaltenes. In the first section of the book, new experimental methods are described that characterize asphaltene structures from the molecular to colloidal length scale. The colloidal properties are understandable in terms of asphaltene chemical structures, especially with regard to the heteroatom impact on bonding. However, quantitative measurements of the enthalpy of asphaltene self-association still need to be determined. In the second section of this book, the fundamental understanding of asphaltenes is related directly to asphaltene utilization. Fundamental characteristics of asphaltene and asphalt impact various applications such as paving, coating, encapsulation, and oil well production.

The editors are grateful to the contributors for their diligence and hard work. We would also like to thank Plenum Publishing Company and Patricia Vann for their assistance and dedication in the publication of this book. O.C.M. would like to thank his parents, Professors June B. Mullins and William W. Mullins, for introducing him to the great satisfaction of academic and scientific research.

Eric Y. Sheu
Oliver C. Mullins

CONTENTS

Chapter I

COLLOIDAL PROPERTIES OF ASPHALTENES IN ORGANIC SOLVENTS

Eric Y. Sheu and David A. Storm

Texaco Research and Development
P.O. Box 509, Beacon, New York 12508

I. INTRODUCTION

Asphaltene is a component of the petroleum liquids defined operationally. It represents the most refractory (or the heaviest) fraction. A generally accepted definition of asphaltenes is by their solvent solubility. The fraction insoluble in heptane (or pentane) but soluble in toluene is called asphaltene [1]. This fraction usually has high apparent molecular weight (often measured by vapor pressure osmometry, VPO). Because it is very refractory, asphaltenes cannot be refined with currently technology. Thus, in many vacuum residua, asphaltene is the main component. In addition to the non-refinerable nature, asphaltene has been known to initiate wellbore plugging, pipeline deposition; hinder the refining yields, and initiate coking, et cetera. Such hinderance on production and processing have made asphaltene one of the most focused materials in petroleum research. The ultimate goal is to either separate asphaltene from the petroleum liquids before entering the refining processes or "upgrade" it to a less refractory (or lighter) fraction. For both cases, one needs to understand the fundamental chemistry of asphaltene.

The relevant physical chemical parameters for asphaltene separation (from crude oil or vacuum resids) are polarities, solubility and temperature, if one uses the solvent fractionation method. Other parameters may be encountered, if additives are used to enhance asphaltene precipitation during separation process or if one wants to upgrade asphaltene. No matter how one chooses to treat asphaltenes, an understanding of the fundamental parameters, such as the average molecular weight and the molecular structure, is always needed. For example, if one wants to upgrade asphaltenes, then, the task is to evaluate the energy needed, in order to "break" asphaltenes into lower molecular weight products (or less refractory fractions), so that one can refine them with the current technology. Obviously, the first parameter needed is the apparent molecular weight by which one evaluates how good the upgrading process is. The next parameter is - how much energy is required ? This depends on the structure of asphaltenes and their physical and chemical properties in the petroleum liquids. For example, if asphaltenes in crude oils or other petroleum liquids (such as vacuum residua) self-assemble to form agglomerates (i.e, they are dispersed as aggregates, instead of as molecules), then, one may need only little energy to break these agglomerates into their

Asphaltenes: Fundamentals and Applications, Edited by E. Y. Sheu and O. C. Mullins
Plenum Press, New York, 1995

respective molecular forms. For such systems, the observed high molecular weight is an apparent one, due to agglomeration. From the economic point of view, this is a better case for asphaltene upgrading than for separation. On the other hand, if asphaltenes are dispersed as their molecular forms of very high average molecular weights, then, much higher energy is needed to break the chemical bonds and reduce the molecular weights. In this case, asphaltene separation may be a better choice. In order to differentiate these two dispersed structures, the key parameter is clearly the true molecular weight of asphaltene.

There has been much work dealing with the asphaltene molecular weight determination. The techniques used are well documented [2]. The molecular weights reported span a wide range from a few hundred to several million daltons. These results naturally lead to the speculation of asphaltene self-association. The concept of asphaltene self-association has a long history. To the best of our knowledge, the colloidal picture of asphaltene was first proposed by Nellensteyn [3] and tested by Mack to interpret the rheological data [4]. Unfortunately, there was no enough scientific evidence provided to support this view. In addition, the data analysis procedure was suspectable. Due to this, the colloidal view did not draw much attention at the time. In 1940, Pfeiffer and Saal [5] proposed a detailed model for a asphaltenes/maltenes/resin complex. It was the first time the term micelle was adopted. The main points of this work were (1) asphaltenes form the micellar core and are peptized by maltenes (or resins) to form a very diffused surface boundary (2) asphaltene micelles may further agglomerate to form clusters depending on the aromatic ratio between the asphaltene core and the peptides, and (3) the average molecular weight of an asphaltene micelle is between 8,000 and 14,000, subjected to a certain degree of over-estimate. All three points brought up in this work are crucial for fundamental understanding of asphaltene chemistry and for petroleum operations. In this chapter we will investigate these three points using more sophisticated techniques and theories. Focus will be on the colloidal properties of asphaltenes, in terms of their micellization, micellar structure, as well as the energies involved in the micellization.

II. OVERVIEW

A. Asphaltene Micellization

Micellization is a phenomenon describing the self-association process of the surface active materials in aqueous solutions. These surface active materials, known as surfactant, tend to self-assemble into colloidal-like particles for better suspension in the solutions. This phenomenon occurs only when the surfactant concentration exceeds a threshold known as critical micelle concentration (CMC). Although micellization represents a self-association phenomenon in general, it is conventionally limited to the self-association initiated by the hydrophobic-hydrophilic imbalance. As a result, it is used only for the a limited class of surface active agents. This highly constrained definition of micellization is certainly not suitable for describing the asphaltene self-association phenomena. The driving force that leads to micellization of asphaltenes is not yet understood. What has been know thus far is that asphaltene self-association is definitely not due to hydrophobic-hydrophilic imbalance. In fact, the only common ground between the asphaltene and the conventional surfactant systems is that both of them self-associate. Due to this, we shall relax all the constraints that limit the definition of micellization and use the term "micellization" to describe the asphaltene self-association phenomenon.

Experimental evidence of micellization often relies on the quantitative determination of the CMC. For conventional surfactant systems, the solvents are aqueous. It is relatively easy to determine the CMC and observe the micellization phenomenon directly. There are many known methods for CMC determination [6]. Among them, the surface tension, conductivity measurement, and the equivalent osmotic pressure are the popular methods. In the case of asphaltene solution, the solvent is hydrophobic, which makes it much more difficult to detect the CMC and observe the micellization phenomenon experimentally. For example, the equivalent conductance or the conductivity methods may not apply because of the low conductivity of the solvents. Conceptually, there are at least three methods for the oil-continuous systems: (1) static or dynamic surface tension (2) microcalorimetry measurement and (3) direct size measurement using X-ray or neutron scattering techniques.

The surface tension method provides evidence similar to the conventional surfactant systems. It is based on the Gibbs adsorption isotherm. The measurement is usually performed as a function of concentration. The micelle starts to form when the surface tension curve shows a distinguishable breaking point on the surface tension plot as a function of the logarithm of concentration.

For most conventional surfactant systems, one often sees an approximately linear decrease of the surface tension, and exhibits a break when concentration exceeds the CMC. After CMC, the surface tension remains fairly constant. However, there is conceptual difficulty associated with this technique when dealing with an asphaltene system. This is to define the dual nature of an asphaltene molecule. For a surfactant molecule, there is no confusion, because it contains a hydrophobic tail(s) and a hydrophilic head(s). These two portions show vast hydrophobicity contrast. Thus, when surfactant is added to the aqueous solution, the hydrophobic tails stick out of the surface, resulting in rapid reduction of the surface tension. However, in asphaltene, the origin of the micellization energy is not known, thus, the dual nature is hard to define. A simple straight forward way to resolve this problem is to assume the aliphatic chain and the polynuclear aromatics of an asphaltene molecule represent the solvent-loving and the solvent-hating (or vice versa) portions. Based on this, one needs to find a solvent with a surface tension much higher than that of a saturated chain solvent (such as alkyl chain) in order to observe the decreasing of surface tension when asphaltene alkyl chains are adsorbed to the surface. This argument was tested by Sheu et al. using pyridine as the solvent, which has the right surface tension [7].

The microcalorimetry method requires a good energy resolution, since the micellization energy may be quite low. Anderson et al. successfully applied this method for CMC determination [8] with the reported values on the same order as obtained by Sheu et al. The size measurement using X-ray or neutron scattering is in principle the most direct and *in situ* method. Unfortunately, there is not yet any data available up to date.

B. Asphaltene Micellar Structure

A hypothesized micellar structure of asphaltene was proposed by Pfeiffer and Saal [5] (see Figure 1). The polynuclear aromatics form a dense core and peptized by resins or maltenes. Based on this hypothetical picture, resins or maltenes are necessary in order to suspend the asphaltene core. This structure of asphaltene is an asphaltene-resin complex rather than an asphaltene micelle. In fact, asphaltenes can form suspended micelles without being peptinized by resins or maltenes, at least in many organic solvents [7].

The true asphaltene micellar structure was first proposed by Dickie and Yen [9] based on the X-ray scattering data. The model Dickie and Yen proposed contains many details. It was proposed for the neat asphaltene phase, but the model was found to be applicable to the suspension systems as well. Based on this model, asphaltenes have strong self-association propensity, forming globular micelles as the elementary units. Within a unit, the polynuclear aromatics of the asphaltene molecules stack to form local discotic structures. However, the vast difference in molecular structure among each individual asphaltene molecule prevents these agglomerates from forming a discotic phase, which requires the monomers to have an

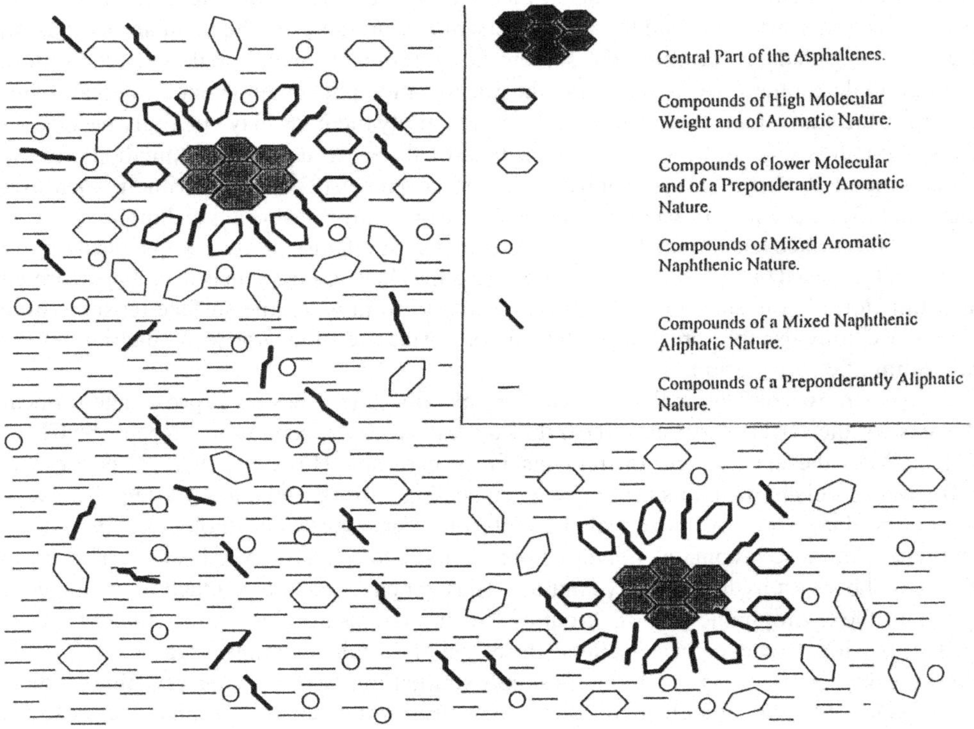

Figure 1. The model for asphaltene-resin complexes proposed by Pfeiffer and Saal [5].

identical molecular structure. Instead, the asphaltenes randomly stack to form globular micelles as proposed in this model. Dickie and Yen also proposed that the elementary micellar units can further cluster into larger particles similar to flocs. This structural model appears to be quite accurate, from the experimental data shown thus far. In the following we will show our data for asphaltene in toluene from the dilute regime (~ 1.0 wt%) to the very concentrated regime (~ 80 wt%). We found the structure gradually evolves from dispersed micelles into clustered micelles with a intra-fractal structure, which more or less follows the Yen's model.

C. Micellization Mechanism

Although the self-association of asphaltene in organic solvents has been proposed, speculated, and observed for quite some time, the micellization mechanism was not explored until recently [10]. The true mechanism is not yet understood.

As we have mentioned earlier, asphaltene self-association is driven by free energies different from the hydrophobic energy. For some time, hydrogen bonding energy and Van der Waal energy were speculated. However, they were later found to be not as relevant to the asphaltene micellization processes. In late 1980's, the involvement of the intra-molecular dipole moments and the dynamic charges were first investigated by Marusk et al. [10]. Later, Shiffert et al. [11] introduced the concept of donor and acceptor. Based on these two works we investigated the charge behavior of asphaltene, both statically and dynamically. From our results, we speculated that the micellization energy is related to the charge transfer energy. Detailed description of this study will be given in section V.

III. ASPHALTENE MICELLIZATION AND THE CRITICAL MICELLE CONCENTRATION

A. Asphaltene Extraction and the Surface Tension Measurement

Asphaltenes used in the experiments presented in this chapter are extracted from the Ratawi (Neutral Zone) vacuum residue (VR) by the solvent fractionation method. The Ratawi VR was dissolved in n-heptane with a 1 gram VR to 40 ml heptane (HPLC grade) ratio. The mixture was gently stirred at room temperature with the beakers covered with watch glass to prevent heptane evaporation. The stirring lasted for ~ 16 hours. The mixture was then filtered using the Whatman no. 5 filter paper. The insoluble fraction (asphaltene) was dried under a stream of nitrogen to remove the residual heptane. A constant weight was obtained after several days of drying. A mass balance and elemental analysis were performed for both soluble and insoluble fractions to ensure complete solvent removal. Table I gives the elemental analysis of the extracted asphaltene. The hydrogen to carbon ratio (H/C) was found to be 1.22.

In the critical micellar concentration (CMC) study, we used the surface tension method. Pyridine was selected as the solvent based on the principle described in section II. The Krüss surface tensiometer was used for this study. The measurements were performed as a function of the asphaltene concentration by gradually adding a small amount of 1 wt% asphaltene/pyridine solution to a 10 ml pure pyridine in the surface tension vessel to increase the asphaltene concentration. Figure 2 shows the Gibbsian plot of the surface tension data as a function of the asphaltene concentration. A clear break (i.e., CMC) was observed at ~ 0.05 wt% [12].

One notes from this plot that the surface tension does not reduce much (3 dynes/cm) before reaching its CMC, indicating a low adsorption energy of asphaltenes to the surface. In other word, asphaltenes are not very surface active, as far as pyridine is concerned. However, the observed CMC is low which means asphaltenes exhibit strong self-association propensity. This dilemma leads to an inferential conclusion that the energy governing the surface activity may not be the energy that induces the self-association. This is a very important conclusion. If this is indeed the case, the CMC value so obtained may be quantitative only to the first order of approximation, but may not be very accurate. This may be explained as follows.

When asphaltenes are first introduced to the pure pyridine solvent, the solvent-hating portion of the asphaltene molecules prefer to adsorb onto the surface, in order to reduce the surface tension energy. This is similar to a surfactant system where the hydrophobic tails stick out of the water surface to minimize the tail-water contact. However, the difference between the asphaltene systems and the surfactant systems is the solvent-loving part of the molecules. In the asphaltene case, the solvent-loving part of the asphaltene molecules may not have as great affinity to the solvent as to the solvent-loving portion of another asphaltene

Table I. Elemental analysis for Ratawi asphaltene

Element	Percent of element in asphaltene
C	82.7
H	8.4
N	0.92
S	7.73
Ni (ppm)	145
V (ppm)	308

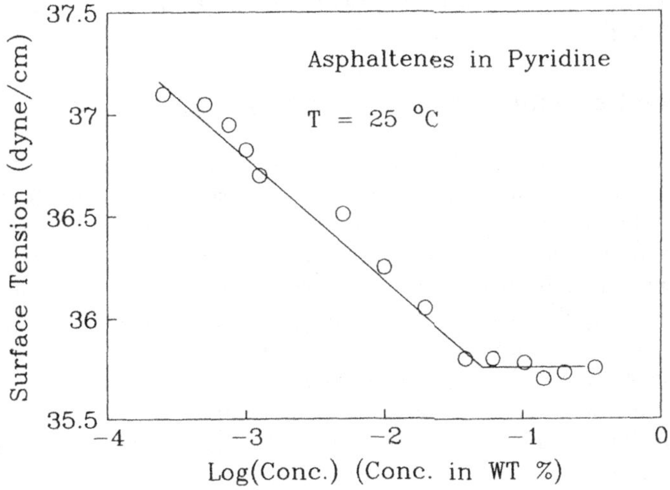

Figure 2. Surface tension of Ratawi asphaltene in pyridine as a function of asphaltene concentration. A clear break was observed signifying the onset of self-association (the critical micelle concentration).

molecule. This is because the affinity between the solvent-loving parts of the asphaltene molecules and the solvent depends on the solubility energy, which is governed by the like-dislike principle. Based on this principle, the interactions between the solvent-loving portion of two asphaltenes molecules hold a certain degree of advantage, since they are structurally similar. Moreover, the interactions can also be based on other principles. They may involve interactions such as, strong van der Waal, π-π electron interactions, or even charge transfer. As a result, self-association in the bulk and adsorption to the surface may proceed simultaneously. This is to say that the CMC may not be well defined and the break observed in surface tension is nothing but an indication that the surface adsorption process is terminated. The micellization process may have started at much lower concentration.

As we have pointed out, this is a scenario we believe to be reasonable. Follow up research is definitely necessary to understand the correlation between the bulk and the surface activities when the concentration is below the CMC. Additionally, it is clear the origin of the micellization energy of asphaltene needs to be investigated further.

The CMC measured using the microcalorimetry was conducted by Anderson et al. [8]. In that work, the micellization energy was measured directly. However, it suffers from the same problem described previously. The energy curves were not like a delta function indicating that the micellization can not be well defined by the asphaltene concentration.

An accurate method to examine the entire process is to apply the small angle X-ray scattering technique which is able to detect the formation of micelles as a function of the monomers concentration. The techniques can probe the formation of macromolecules *in situ,* and thus the onset of the micellization and the degree of micellization (i.e., the partition coefficient of micelles and monomers). The only experimental difficulty is the X-ray source. To the best of our knowledge, a synchrotron X-ray source may be needed, in order to provide reasonable scattering statistics.

B. Gibbs Adsorption Isotherm and Molecular Weight

Although the surface tension method may not accurately determine CMC, it however, provides the evolution of the asphaltene concentration at the surface sublayer. One thus can use the surface tension data to estimate the average molecular weight of asphaltene using the Gibbs equation. Taylor [13] had used our data to estimate the molecular weight according to the following procedure. First, the surface concentration of asphaltenes, Γ, was computed using the Gibbs equation and the surface tension data,

$$\Gamma = -\frac{1}{RT}\frac{d\gamma}{d\ lnc} \tag{1}$$

where γ is the surface tension, R the ideal gas constant, T the temperature, and c the concentration. Γ was obtained from the slope just before the break point. From the Γ values calculated, the surface area per adsorbed asphaltene molecule, A_{mol}, can be obtained by $A_{mol} = 1/(N_A\Gamma)$ where N_A is the Avogadro's number. Taylor then used this A_{mol} value, the asphaltene molecular density (~ 1.1 g/cm^3), and an assumed molecular geometry at the surface to calculate the molecular weight. He obtained ~4770 Daltons for the Ratawi asphaltene. This molecular weight turns out to be comparable to the apparent molecular weight of many reported values. However, there were problems associated with the procedure as we pointed out later [14]. The A_{mol} obtained should represent the area in which only one asphaltene molecule can be found, instead of the area completely occupied by one asphaltene molecule, as assumed by Taylor. This is because the adsorbed asphaltenes do not have a compact geometric packing. There are interactions between molecules and between molecules and solvent. These interactions keep the adsorbed asphaltene molecules in a two-dimensional equilibrium with an inter-molecular distance larger than the molecular diameter. The inter-molecular interaction and the solvation, which affect A_{mol}, can be quantitatively estimated [15]. We followed the Taylor procedure and took into account these effects to obtain an average molecular of 870-1000 Daltons for Ratawi asphaltene. We believe this is a more reasonable molecular weight range. Recently, Hunt [16] used mass spectrometry to measure the molecular weight of asphaltene and obtained a 500-1000 Daltons molecular weight range which is consistent with our result.

C. Dynamic Surface Tension and Association-Dissociation Kinetics

One important property of a micellar system is that their micellar association/dissociation is a thermally reversible process, provided the surfactants do not degrade. In studying asphaltene micelles, it is necessary to examine if asphaltene micelles have the same property. This is a crucial step toward adopting the micellar theories for analysis of asphaltene systems. If the self-association process of asphaltenes is not a reversible process, one needs to be careful in using surfactant theories for interpretation of asphaltene data.

Experimentally, one can examine the reversibility using dynamic surface tension measurement. We prepared a 0.01 wt% (below CMC) and a 0.5 wt % (above CMC) asphaltene/pyridine solutions, sealed in a pressure cell, and heated to 250°C for 4 hours. The intention of this step was to dissociate the micelles. After four hours, the cell was quenched to room temperature and the dynamic surface tension was measured. If the samples degraded at 250°C, the asphaltene molecular configurations would change. They would in turn change the surface sublayer concentration Γ, and thus the surface tension. This is to say one should be able to detect the degradation by the surface tension if it does occur.

The dynamic surface tension technique was used to investigate how the asphaltene concentration at the surface sublayer changes as the system evolves toward equilibrium. This includes the desorption process and the micellization process (for the 0.5 wt % case). In the dynamic surface tension measurement, we did not employ the bubble method (commonly used for dynamic surface tension measurement on surfactant system which has equilibrium

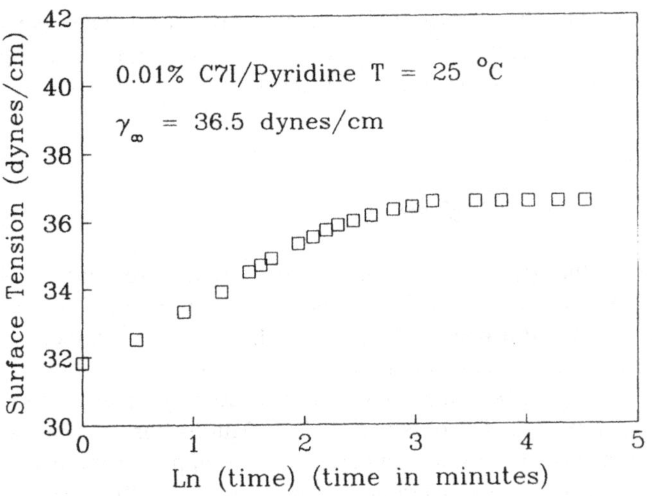

Figure 3. Dynamic surface tension for the quenched asphaltene/pyridine solution at 25 °C with concentration below CMC. The kinetic process toward equilibrium is much slower than surfactant systems.

kinetics in milli-seconds to seconds). This is because we anticipated very slow equilibrium kinetics. The reason was that each asphaltene molecule is structurally different, which should slow down the adsorption/desorption processes as well as the self-assembling process when forming micelles. Our dynamic surface tension data proved this point.

Figure 3 and Figure 4 show the dynamic surface tensions for the 0.01 wt % and the 0.5 wt% asphaltene/pyridine solutions, after heating for four hours at 250°C, and quenching to room temperature. The surface tension increases slowly as a function of time, indicating a gradual asphaltene desorption from the surface sublayer. If the systems are reversible, the equilibrium surface tension should coincide with the value obtained before heating as shown in Figure 2. The equilibrium surface tensions obtained after quenching were 36.5 and 35.8 dyne/cm respectively, consistent with the concentration dependent measurements (see Figure 2). These results prove that the systems did not degrade and the micellar association/dissociation are thermodynamically reversible.

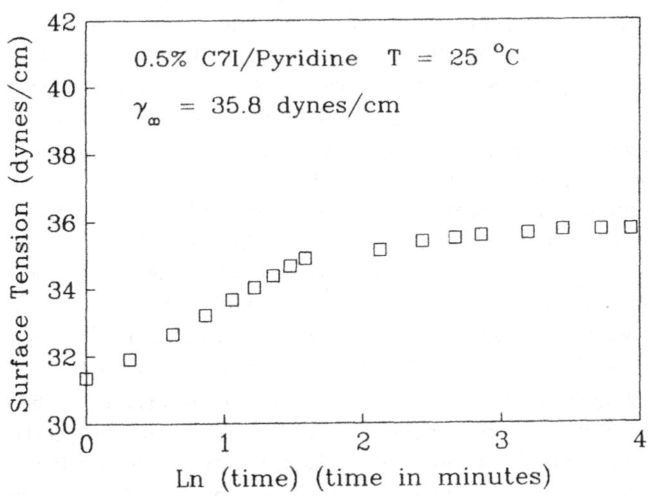

Figure 4. Same plot as Figure 3 at 0.5 wt % (above CMC).

As we expected, the equilibrium kinetics is very slow compared to the surfactant systems. In order to characterize the kinetic process from the dynamic surface tension measurements, we adopted a simple relation,

$$\gamma(t) = \gamma_{\infty} [1 - e^{-Kt}] \tag{2}$$

where γ_{∞} is the asymptotic surface tension, and k represents the characteristic rate constant of the equilibrium process. This relation will be adequate, if there is only one kinetic process or if the desorption process and the subsequent micellization process (only for the 0.5% case which is above CMC) are two processes in a sequential order. In other words, the micellization process does not occur until the desorption is nearly completed. If these two processes occur simultaneously, Eq.(2) cannot be used to extract the rate constants for the desorption and the micellization from the dynamic surface tension data. Fortunately, the asphaltene systems we dealt with appear to have two distinctive kinetic processes, one for desorption and one for micellization.

Figure 5 and Figure 6 show the results of the analyses using Eq.(2). As one can see, the 0.01% sample (below CMC) exhibits only one slope, signifying that the equilibrium kinetics involves only the desorption process. The kinetics constant k was found to be 0.18 minutes^{-1}. On the other hand, the 0.5% case showed two distinctive slopes. The initial slope represents the desorption kinetics process with a 0.36 minutes^{-1} kinetic constant, much faster than the 0.01% case. This faster kinetic rate is due to the asphaltene concentration. The higher initial concentration of asphaltenes at the sublayer (this reflects on its lower initial surface tension as shown in figure 4) significantly lowers the adsorption energy (or easier to desorb). This is to say the activation energy for desorption to occur is lower compared to the 0.01 wt % case. Thereby, it results in a faster desorption kinetic process. The second slope observed corresponds to the micellization process. The kinetic constant was found to be about 0.12 minute^{-1}, a much slower process. This indicates that the structure of each asphaltene molecule is very different and the system is very complex. Thus, it requires much longer time for the monomers to aggregate, pack into micelles, and reach the new equilibrium. This slow micellization kinetic process suggests that any asphaltene characterization studies should take into account the kinetic effect. In particular, when temperature, concentration, or

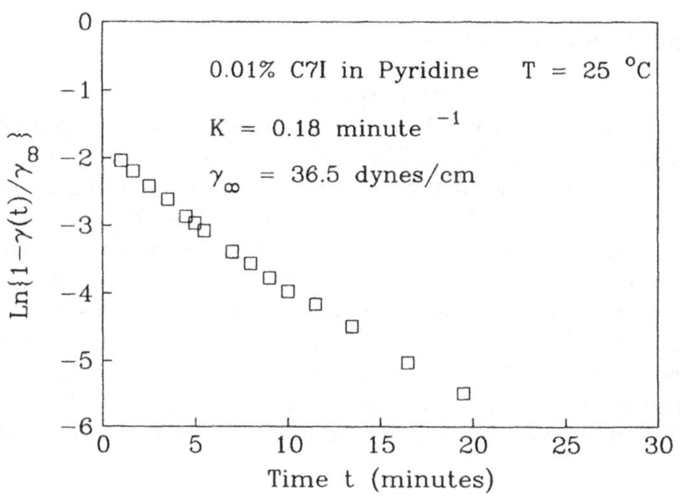

Figure 5. Analysis of Figure 3 data using Eq.(2). A linear behavior was obtained representing the desorption kinetic process.

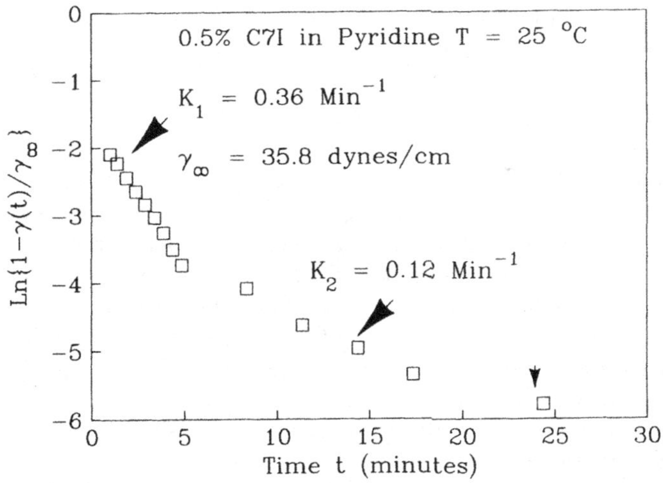

Figure 6. Same plot as Figure 5. Two linear regions can be identified. The first one corresponds to the desorption process while the second slope shows the micellization kinetic process.

pressure is varied, one should always age the system for enough long period of time before experiment is performed.

IV. ASPHALTENE MICELLAR STRUCTURE

The next topic to study, after learning that asphaltenes form micelles with slow kinetic process and the micellization is a reversible process, is naturally the structure of asphaltene micelles, and their thermodynamic properties. Here we will present two characterization approaches. One is the rheological method, representing a macroscopic approach and the other is the small angle scattering technique which is a microscopic technique.

A. Rheological Measurements

In our viscosity study, Ratawi asphaltene and toluene was used to prepare the asphaltene micelle solutions. The solvent extraction method, same as described previously, was used to fractionate asphaltenes. The asphaltene solutions were prepared by simply dissolving the solid asphaltenes in toluene at room temperature using a vortex mixer. The solutions were then aged for 24 hours before measurement. The viscometer used was a Brookfield low shear rate viscometer with a 0-1000 centipoise working range and a 0.1% accuracy of the full scale. The cell was a cuvette cell, immersed in a temperature bath. The open end of the cuvette cell was capped to prevent solvent evaporation. Two experiments were performed. One was to measure the viscosity as a function of the asphaltene concentration and the other as a function of temperature. The solutions were Newtonian for concentration up to 70 wt% in a temperature range from -30 to 60°C. Thus, the non-Newtonian effect will not be discussed here. In the following we will analyze and discuss our viscosity study.

A-1. Shape and Asphaltene-Solvent Interactions. Figure 7 shows the relative viscosity η_r (= η/η_s, where η is the solution viscosity and η_s the solvent viscosity) of the asphaltene

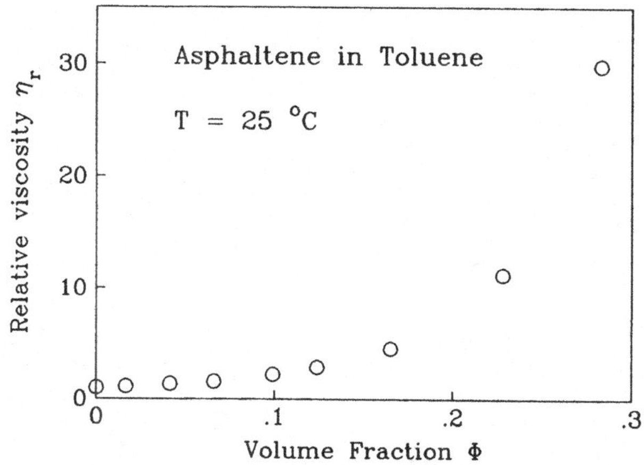

Figure 7. Relative viscosity of Ratawi asphaltene/toluene solution as a function of asphaltene volume fraction.

solution as a function of the asphaltene volume fraction ϕ [15]. The curve is similar to a typical colloidal viscosity curve. In the very dilute regime, the viscosity increases linearly with asphaltene concentration (the Einstein region) and deviates from it when concentration further increases. Apparently, the Einstein equation,

$$\eta_r = (1 + 2.5\phi) \tag{3}$$

is applicable only in the dilute regime. However, the fundamental question here is: are asphaltene micelles spherical? To resolve this question, we adopted a simple equation developed by Pal and Rhodes [17]. This equation is certainly not rigorous but it can provide a rough idea about the shape of the suspended particle using low shear viscosity data. The Pal and Rhode equation reads

$$\eta_r = (1 + K\phi)^{-2.5} \tag{4}$$

where K represents the solvation-particle interactions. When $K = 1$, no solvation occurs. The exponent in Eq.(4) is 2.5 which is for a spherical particle. If one uses this equation to fit the experimental data or plot the data according to

$$\eta^{\alpha} = 1 + K\phi \tag{5}$$

with α being adjusted until a linear curve is obtained. In this way, α and K (the slope) can be determined uniquely. Figure 8 is the Pal and Rhodes plot. The most appropriate exponent for the curve to show linearity was indeed 2.5. This indicates that the asphaltene particles dispersed in toluene are hydrodynamically spherical. The K value so obtained was ~ 2.8 which is much higher than a hard sphere ($K = 1$). This is understandable, because we do not anticipate asphaltene micelles to have smooth surfaces (direct evidence can be found from the Porod analysis of the small angle scattering data to be discussed in the B subsection). Therefore, when the micelles diffuse, the surrounding solvent molecules are dragged as if they are part of the micelles. This can be seen as a solvation phenomenon.

The degree of solvation can be quantitatively determined by an empirical formula proposed by Eiler [18]

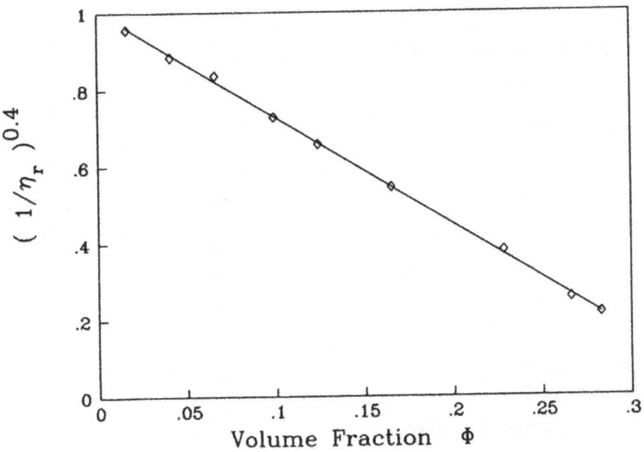

Figure 8. The Pal and Rhodes analysis, Eq.(4), suggests a spherical hydrodynamic geometry of the asphaltene micelles (see text for details).

$$\frac{(\eta_r^{1/2} - 1)}{\phi} = \frac{[\eta]}{2} + \frac{1}{\phi_m}(\eta_r^{1/2} - 1) \tag{6}$$

In Eq.(6), ϕ_m is the maximum packing volume fraction where the viscosity diverges. To determine ϕ_m, we argue from the statistical mechanical theory point of view. For many colloidal systems, or even simply liquids, their microscopic structures show that the particles often pack according to a face center cubic structure with respect to the nearest neighbor particles or according to a random packing, In these two cases, ϕ_m is 0.74 and 0.68 respectively. If one uses 0.74 for ϕ_m, and plots the viscosity data according to Eq.(6), one can obtained the experimental ϕ_m which is lower than 0.74. Apparently, the solvation accounts for their difference. Figure 9 is an Eiler plot from which the experimental ϕ_m was found to be 0.43. From this value and 0.74 for non-solvated case, we found the solvation layer to be about 20% in thickness. This suggests that the asphaltene micelles do pack loosely and have very rough surfaces.

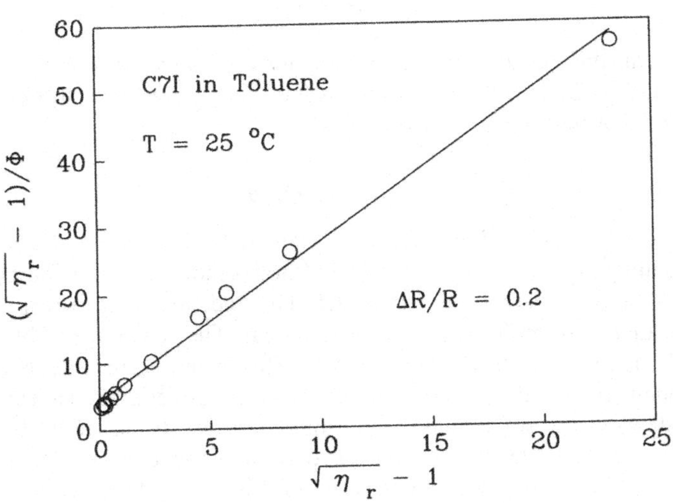

Figure 9. The Eiler plot, Eq.(6), shows a 20% solvated layer (see section IV-A1).

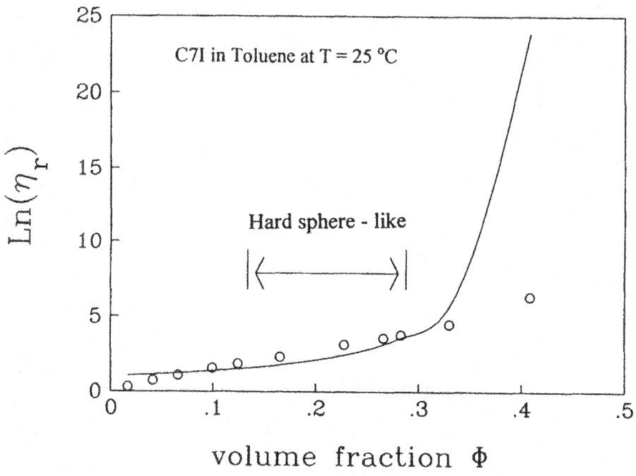

Figure 10. Comparison with a hard sphere model (the Campbell-Forgacs plot [18]).

Up to this point, the equations used for the viscosity analysis were all developed for hard sphere systems. In order to justify our analysis, we compared our relative viscosity data to a theoretical hard sphere curve developed by Campbell and Forgacs [19] (see Figure 10). As one can see, the data follow the theoretical curve fairly well until ~ 0.32 volume fraction. This means that we can safely use the equations developed for a hard sphere system, as long as we maintain the volume fraction below ~ 0.32.

A-2. Concentrated Regime and Micelle-Micelle Interactions. As indicated previously, the asphaltene system can be treated as a hard sphere system, provided the solvation effect is properly taken into account. However, the interactions between particles should be accounted for when volume fraction is above the Einstein regime (~ 0.1). Here we adopted the Grimson-Barker theory [20] to evaluate the inter-particle interactions. Grimson and Barker derived a viscosity equation for analyzing the low shear viscosity of an interacting system. They assumed a general inter-particle potential of the following form

$$V(r) = \Lambda\left[\frac{\sigma}{r-\sigma}\right]^n \tag{7}$$

where Λ is a coupling constant representing the amplitude of the potential, σ is the particle diameter, r the inter-particle distance, and n is a parameter characterizing the dependence of the potential as a function of r. Based on this potential, Grimson and Barker derived a simple analytical equation,

$$\eta_r = (1+2.5\phi) + \Lambda\left(\frac{\phi}{\phi_m}\right)^{\frac{n+1}{3}}\left[1 - \left(\frac{\phi}{\phi_m}\right)^{1/3}\right]^{-n} \tag{8}$$

Figure 11 shows the Grimson-Barker analysis for the asphaltene/toluene systems. This potential form can explain the data fairly well for volume fraction up to ~ 0.25. The n obtained was about 2 with Λ ~ 0.6. This means that the interaction is Columbic. Λ is a coupling constant qualitatively representing the interaction strength.

The above simple analyses suggest that one can apply Eq.(4)-(8) to analyze the viscosity data of an asphaltene system and characterize its physical properties to a fairly good extent.

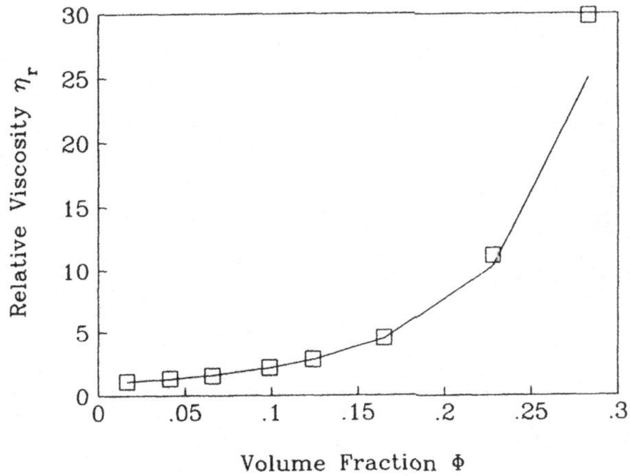

Figure 11. The Grimson-Barker plot, Eq.(8) [19], shows the inter-micellar interaction potential is describable by Eq.(7).

A-3. Temperature Effect and Glass Transition. One can also investigate the thermal effect on the viscosity behavior of an asphaltene system to determine the physical parameter such as the glass transition temperature. Figure 12 is an Arrhenius plot of the measured viscosity as a function of temperature. For both systems shown, strong non-linear behavior was observed. It indicates the participation of non-linear interactions, in addition to the Arrhenius kinetic energy. In order to understand the origin of this interaction, we adopted the argument from the mode-mode coupling theory. This theory is based on the coupling of two phenomena observed at two concentration extremes. One is at the dilute regime where the particle can freely diffuse without inter-particle interactions. In this case, the autocorrelation of the density fluctuation shows an exponential decay. This is to say that the probability of finding a particle at position r decays exponentially as a function of time. This regime is called the diffusion mode. In this concentration regime, the Einstein equation is applicable. On the other hand, when the system is very close to crystallization, the particle movements are confined within a cell. The particles can only fluctuate within the cell. This mode is called the fluctuation mode. In this extreme, only fluctuation characteristic shows, which is again an exponential form.

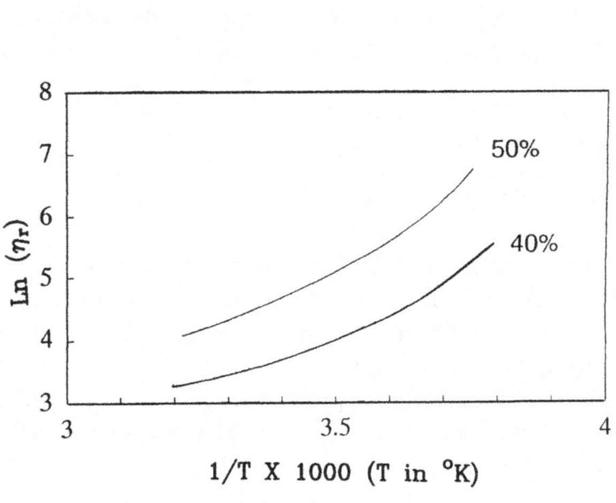

Figure 12. Arrhenius plot for 40 and 50 wt% Ratawi asphaltene in toluene. The non-linearity reveals that the inter-micellar interactions have modified the thermal rheological behavior.

However, when system concentration is between these two extremes, the particles are somewhat localized within a loosely defined cell, and a particle may diffuse out of the cell after a certain number of fluctuations within the cell. To theoretically describe such a phenomenon, one needs to couple the diffusion and the fluctuation modes with a residence time of the particle in the cell depending on a function known as the memory function. Since the coupling of these two modes is through the memory function and the fact that the diffusion frequency and the fluctuation frequency are very different, the resulting transport property is expected to be very non-linear as a function of the logarithm of temperature. In 1985, Das and Mazenko [21] adopted this mode-mode coupling theory to derive a simple formula for the temperature-viscosity relation for the glass phase transition,

$$\eta_r = \left| \frac{T}{T_g} - 1 \right|^{\mu} \tag{9}$$

where $\mu = 1.8$ for $T > T_g$ and T_g is the glass transition temperature. Figure 13 shows the temperature effect on viscosity according to Eq.(9). The μ values obtained 1.82 for volume fraction $\phi = 0.31$ and 1.25 for $\phi = 0.28$. Apparently, the $\phi = 0.31$ case undergoes a glass-like transition while the $\phi = 0.28$ case seems to behave like a viscous liquids with strong inter-particle interactions. The glass transition temperature obtained for $\phi = 0.31$ case was -10°C.

B. Small Angle Neutron Scattering

B-1. Theory. After the macroscopic characterization using viscosity data, we turned to the microscopic techniques for further investigation of the asphaltene micellar systems. Small angle neutron scattering (SANS) technique was chosen for this purpose. Small angle scattering technique makes use of a fundamental concept that the interaction between a probe (such as neutron or X-ray) and a colloidal particle should reflect the structural characteristics of the particle as well as its surroundings. This makes the small angle scattering technique important for structural study of colloidal systems. The way small angle scattering determines the colloidal structure is through the following mechanism. In a scattering process, the incident wave is created by X-ray, electric field (laser

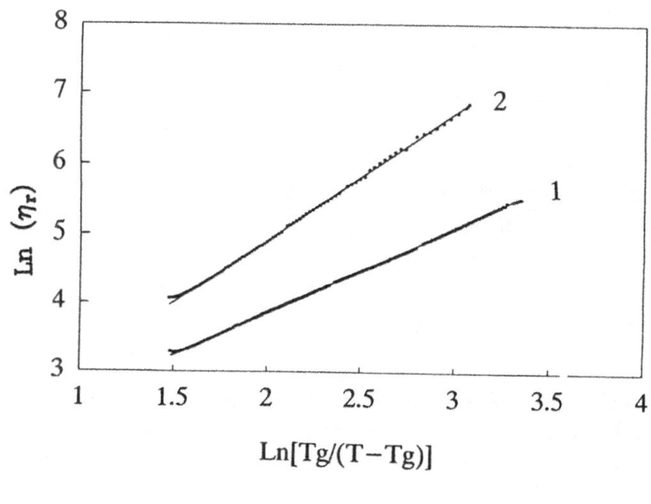

Figure 13. An analysis of the thermal rheological data using the Das and Mazenko equation [20], Eq.(9). A 1.82 slope was obtained for μ indicating the onset of a glass-like transition at T_g ~ 254 K.

scattering) or particles such as neutron. The interactions between the incident wave (in this case the neutron) and the particles (in this case the asphaltene micelles) depend on the intra-particle as well as the inter-particle structures of the particles. Thus, different particles and inter-particle structures result in different spatial scattering patterns of the scattered wave. These scattering patterns (it is the differential cross section per unit volume of the system) carry the structural information of the particles. In principle, the particle structure and the inter-particle structure can be rigorously determined through analysis of the scattering pattern. This scattering pattern is usually defined in a space characterized by the Bragg wave vector,

$$\vec{Q} = \vec{k}_i - \vec{k}_f \tag{10}$$

where k_i and k_f are the momentum transfer of the incident and the scattered wave respectively. When the scattering angle is small, the interactions between the incident wave and the scatterers are predominately elastic or quasielastic. In this case, $|k_i| = |k_f| = 2\pi/\lambda$, where λ is the wavelength. Consequently, the scattering vector Q has a magnitude of

$$Q = |\vec{Q}| = \frac{4\pi}{\lambda} n \sin(\frac{\theta}{2}) \tag{11}$$

where n is the index of refraction of the system and θ is the scattering angle. The index of refraction of water for laser lights is 1.33. For some saturated alkyl chain oil it is about 1.4. In the case of X-ray and neutron, n is very close to unity. This is why n is often omitted in the formulation of X-ray or neutron scattering.

The scattering intensity as a function of angle θ, and thus Q, provides the structural information of the scatterers in the spatial space through the Fourier transform of the intensity spectrum [22]. The resolution of the experiment in θ naturally propagates to the spatial space. Thus, the maximum resolution in the spatial space depends on the maximum Q value achievable in an experiment. As a general rule, if Q_m is the maximum achievable Q value, the spatial resolution will be

$$\Delta R = \frac{\pi}{Q_m} \tag{12}$$

As mentioned earlier, the pattern of the scattered wave in the Q space represents the differential scattering cross section, provided the energy transfer between the incident wave and the scattered wave is negligible. This is usually true when the scattering angle is small (or when the incident wave length is long see Eq.(11)). Under this condition, it is quantitatively accurate to apply the static approximation concept. The differential cross section $d\Sigma/d\Omega$ (or the scattering intensity function) in this case can be written as,

$$\frac{d\Sigma}{d\Omega} = \frac{1}{V} < |\sum_{\alpha=1}^{N} \sum_{\alpha'=1}^{N} b_\alpha b_{\alpha'} \exp[i\vec{Q}\cdot(\vec{r}_\alpha - \vec{r}_{\alpha'})]| > \tag{13}$$

where b_α is the bound scattering length of the α-th nucleus (in the case of neutron scattering) in the scatterer, and r_α is the position vector of the α-th nucleus. The bracket represents an average over all possible configurations. For simplicity, Eq.(13) can be rewritten as

$$\frac{d\Sigma}{d\Omega} = \frac{1}{V} < |\sum_{\alpha=1}^{N} b_\alpha \exp(i\vec{Q}\cdot\vec{r}_\alpha)|^2 > \tag{14}$$

When the scattering angle is small, the bound state scattering length b_α depends on the species of the nuclei and their corresponding spin states, it is thus more advantageous to separate b_α into the coherent part and the incoherent part. The incoherent part is Q independent. Experimentally, it contributes to a constant background, and is generally subtracted before data analysis proceeds. Mathematically, these two parts can be expressed as

$$(\frac{d\Sigma}{d\Omega})^{inc} = \frac{1}{V} \sum_{\alpha=1}^{N} (b_\alpha^{inc})^2 \tag{15}$$

and

$$(\frac{d\Sigma}{d\Omega})^{coh} = \frac{1}{V} < |\sum_{\alpha=1}^{N} b_\alpha^{coh} \exp(i\vec{Q}\cdot\vec{r}_\alpha)|^2 > \tag{16}$$

In what follows, $d\Sigma/d\Omega$ will be used for the coherent scattering cross section only, unless it is denoted.

For most colloidal systems, the colloidal particles are in a relatively well defined structure distinguishable from the surrounding solvent molecules. It is thus convenient to mathematically treat each colloid as a scattering center. From this view point, the scattering volume can be divided into M cells with the center of mass of the colloid being the center of each cell. Eq.(16) in this case becomes

$$(\frac{d\Sigma}{d\Omega}) = \frac{1}{V} < |\sum_{i=1}^{M} \exp(i\vec{Q}\cdot\vec{r}_i) \sum_{cellj} b_{ij} \exp(i\vec{Q}\cdot\vec{x}_{ij})|^2 > \tag{17}$$

where $\vec{x}_{ij}$ is the position vector for the j-th nucleus in the i-th cell. In Eq.(17), the second summation is evaluated for the i-th particle. Since it is an intra-particle term, we thus define the form factor as

$$F_i(\vec{Q}) = \sum_{cellj} b_{ij} \exp(i\vec{Q}\cdot\vec{x}_{ij}) \tag{18}$$

At this point we introduce a scattering length density for the i-th particle $\rho_i(r)$, defined as

$$\rho_i(\vec{r}) = \sum_{j} b_{ij}\delta(\vec{r}-\vec{x}_{ij}) \tag{19}$$

and assume the solvent molecules are much smaller than the colloids, so that the solvent can be treated as a continuous media with a constant scattering length density ρ_s. With the $\rho_i(r)$ of Eq.(19) and the assumption for ρ_s, one can write $F_i(Q)$ as

$$F_i(\vec{Q}) = \int_{particle\ i} d^3r\ [\rho_i(\vec{r})-\rho_s]\exp(i\vec{Q}\cdot\vec{r}) + \int_{cell\ i} d^3\ \rho_s \exp(i\vec{Q}\cdot\vec{r}) \tag{20}$$

The second term of Eq.(20) is approximately a delta function at $Q = 0$. Thus, for $Q > 0$ we have

$$F_i(\vec{Q}) = \int_{particle\ i} d^3r\ [\rho_i(\vec{r})-\rho_s]\exp(i\vec{Q}\cdot\vec{r}) \tag{21}$$

Obviously, $F_i(Q)$ depends only on the structure of the colloids, if the particles are uniform. Substitute Eq.(18) and (21) into Eq.(17), one immediately obtains, assuming the particles are monodisperse,

$$\frac{d\Sigma}{d\Omega}(Q) = \frac{N_p}{V}|F(Q)|^2\frac{1}{N_p}<\sum_{i=1}^{N_p}\sum_{i'=1}^{N_p}\exp[i\vec{Q}\cdot(\vec{r}_i-\vec{r}_{i'})]> \tag{22}$$

Define $P(Q) = |F(Q)|^2$, n_p = number density of the particles= N_p/V, and the inter-particle correlation factor (known as inter-particle structure factor) as

$$S(Q) = \frac{1}{N_p}<\sum_{i=1}^{N_p}\sum_{i'=1}^{N_p}\exp[i\vec{Q}\cdot(\vec{r}_i-\vec{r}_{i'})]> \tag{23}$$

then, the differential cross section $d\Sigma/d\Omega$ can be simplified as

$$\frac{d\Sigma}{d\Omega}(Q) = n_pP(Q)S(Q) \tag{24}$$

In the case of polydisperse systems, Eq.(24) becomes

$$\frac{d\Sigma}{d\Omega}(Q) = \frac{N}{V}<P(Q)>\tilde{S}(Q) \tag{25}$$

where

$$<P(Q)> = <|F(Q)|^2> = \sum_{p=1}^{n}\frac{N_p}{N}|F_p(Q)|^2,\quad (N = \sum_{p=1}^{n}N_p) \tag{26}$$

and $\tilde{S}(Q)$ is an effective one-component inter-particle structure factor,

$$\tilde{S}(Q) = 1 + \beta(Q)[S(Q)-1],\quad \beta(Q) = \frac{|<F(\vec{Q})>^2|}{<|F(\vec{Q})|^2>} \tag{27}$$

From Eq.(22) to (27) one can see that $P(Q)$ is governed by the structure of the particle while $S(Q)$ is governed by the inter-particle correlations. For various particle geometries, $P(Q)$ has analytical expressions [22] and can be easily computed. For example, $P(Q)$ for a spherical particle of radius R is

$$P(Q) = |F(Q)|^2,\quad F(Q) = V(\rho-\rho_s)\frac{3j_1(QR)}{QR} \tag{28}$$

where ρ is the average scattering length density of the colloid, and $j_1(x)$ is the spherical Bessel function of the first kind. For a triaxial ellipsoid of semiaxes a, b, and c, the form factor $P(Q)$ is

$$P(Q) = \int_0^1 dx \int_0^1 \Phi^2\left\{Q[a^2\cos^2(\frac{1}{2}\pi x) + b^2\sin^2(\frac{1}{2}\pi x)(1-y^2) + c^2y^2]^{\frac{1}{2}}\right\}dy \tag{29}$$

For a hollow cylinder of height H, outer radius R_1 and inner radius R_2, $P(Q)$ has the following form

$$P(Q) = \int_0^1 \Phi[Q,\sqrt{R_1(1-\mu^2)},\sqrt{R_2(1-\mu^2)}]\frac{\sin(\frac{QH\mu}{2})}{QH\mu}d\mu$$

$$\Phi(Q,R_1,R_2) = \frac{1}{1-(\frac{R_2}{R_1})^2}[2J_1(QR_1)-(\frac{R_2}{R_1})^2 2J_1(QR_2)] \tag{30}$$

For other geometries the reader can refer to reference [23].

Comparing with $P(Q)$, it is much more difficult to compute $S(Q)$ because the basic inter-particle correlation function can not be presumed or determined trivially. For a well dispersed colloidal system, $S(Q)$ can be calculated through the statistical mechanical theory by solving the Ornstein-Zernike equation [24]. Several colloidal structures factors were determined using this theory [25-28]. The $S(Q)$ so computed is generally quite accurate. However, in the case of asphaltene, the $S(Q)$ is either negligible (for concentration < 10%) or should be treated differently from the Ornstein-Zernike approach which is only suitable for a well dispersed system. In the cases discussed here, $S(Q)$ will be neglected in the dilution regime and be constructed according to a fractal theory when the asphaltene concentration is high.

B-2. Experiment. The SANS experiments for the asphaltene systems reported here were performed on the spectrometers at two national laboratories, namely, Argonne National Laboratory (ANL) and National Institute of Standards and Technology (NIST). At ANL, neutrons are generated from a spallation source by accelerating protons to 450 MeV and bombarding a target composed of 77.5% of ^{235}U. The generated neutrons have a wave length (λ) distribution very close to a Maxwellian distribution from ~ 4 to 12 Å, even after being moderated by a liquid hydrogen cold source. The range of the wave lengths used was from ~ 0.5 to 14 Å. These generated neurons were filtered by a MgO single crystal to remove the fast neutrons. They are then collimated by two consecutive sollers collimators before entering the sample cell. The sample cell is made of quartz of various path lengths depending on the neutron beam transmission. The cell holder was designed to take 7 samples for automatic measurement in a temperature controlled environment. The scattered neutrons were recorded by a two dimensional position sensitive detector of 20 x 20 x 3.2 cm^3 sensitive area. Since a wide spectrum of neutron wave lengths were used, the

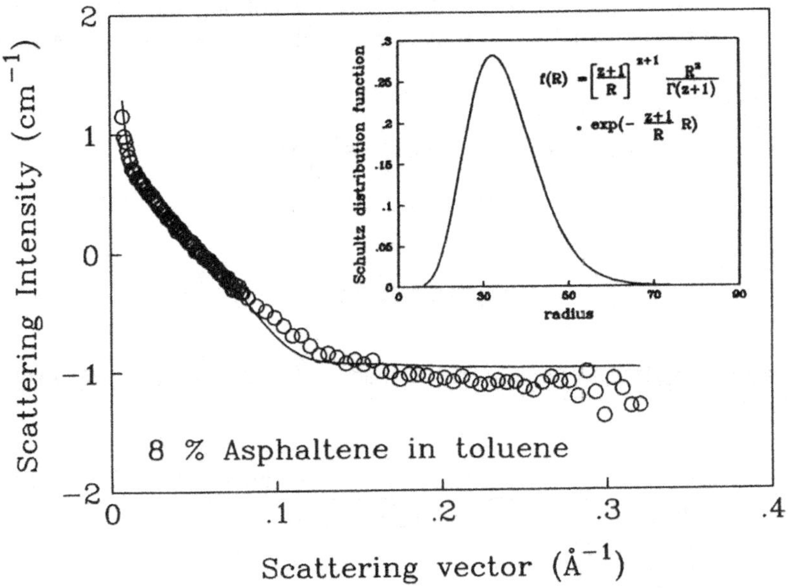

Figure 14. A typical small angle neutron scattering spectrum, I(Q), as a function of the scattering vector Q, Eq.(11). The solid line is the calculated I(Q) using a spherical model with a size distribution according the Schultz distribution function. The inset shows the particle size distribution extracted.

Q range was largely determined by the neutron wave length distribution rather than the sample to detector distance. As a result the geometric configuration was fixed at a relative short sample-to-detector distance. However, the Q range is fairly wide (from 0.008 to 0.35 $Å^{-1}$).

The SANS spectrometer at NIST was the NG7 spectrometer. At NIST, the neutrons were produced from a nuclear reactor of 25 MW power. The neutrons produced were moderated by a cold source before entering a guide pipe. The neutrons were then selected according to their wave length using a chopper. The neutrons can be selected according to the resolution, $\Delta\lambda/\lambda$. For the asphaltene experiment, $\Delta\lambda/\lambda = 0.1$ was selected. The Q range here can be adjusted by the sample-to-detector distance, the stopper size, and the number of guides for the neutrons. We chose a configuration to cover the small Q range. This allowed us to examine a large length scale. The Q range chosen was from 0.002 to 0.005 $Å^{-1}$. This configuration was suitable for the fractal study of asphaltene solutions at high concentrations. The details of this work will be discussed in the following (B-4).

B-3. Micellar Formation, Growth and Polydispersity. The SANS study on the asphaltene micellar formation, growth and polydispersity were conducted at the ANL spectrometer. Figure 14 shows a typical SANS spectrum along with the analysis which is based on a spherical micellar structure with a Schultz distribution for the micelle size. In an earlier work [22] we had proposed a self-consistent method to judge the goodness of a presumed structural and a size distribution function. Combining this method and the data fitting, one can find the most appropriate micellar structure and the size distribution. Using this methodology, we found the spherical structural model with a Schultz distribution in size was the most appropriate model to describe our asphaltene scattering data.

Table II tabulates the parameters extracted from such an analysis. One can see that the micelles do not grow, so does the polydispersity. This can be explained as follows.

Table II. The extracted parameters from SANS analysis

Conc. (wt%)	Toluene/pyridine*	Radius (Å)	Polydispersity (%)
8	100/0	32.4 ± 0.4	1.76
8	85/15	31.3 ± 0.3	19.2
8	70/30	30.4 ± 0.3	20.4
8	55/45	30.8 ± 0.3	20.5
8	30/70	30.1 ± 0.3	21.7
4	80/20	30.8 ± 0.4	18.6
4	60/40	31.6 ± 0.3	19.7
4	40/60	31.1 ± 0.3	20.4
4	20/80	30.7 ± 0.4	21.3
2	80/20	31.6 ± 0.4	19.0
2	60/40	31.8 ± 0.3	19.4
2	40/60	31.3 ± 0.4	19.6
2	20/80	30.8 ± 0.3	20.2
1	80/20	31.6 ± 0.4	19.7
1	60/40	31.5 ± 0.4	19.0
1	40/60	32.0 ± 0.5	19.7
1	20/80	31.0 ± 0.4	19.0

Asphaltene micelles were made of monomers of very different structure. Thus, the micelles so constructed are expected to have a very unique geometric packing. When more asphaltene monomers are added, unless they can fit to an existing crumple surface, it is much easier for the added monomer to form new micelles. Obviously, the restriction of micellar growth is mainly due to the packing constraints near the micellar surface. The other possible mechanism is to break the existing micelles and re-construct bigger micelles. However, this is very unlikely because (1) enough energy is needed to break the micelles, and (2) the natural structural distribution of the monomer likely set a limit for the number of monomers to pack into a free energy-efficient micelle. By adding more asphaltenes apparently does not provide enough thermodynamic impact for micellar to break. And the added monomers do not change the basic structural distribution of the monomers. Thus, the added asphaltenes will continue to form micelles. One can see this effect from the scattering intensity (see Figure 2 in ref. 29).

The analysis also showed that different ratios of toluene/pyridine mixtures do not introduce much effect on the micellar size, growth, or polydispersity either. Again, it is because the solvation energy varied by mixing toluene and pyridine was not high enough to overcome the packing energy and initiate the re-distribution of the micelles. One can see clearly from the scattering intensity versus concentration plot (Figure 2 in ref. 29) that the relation becomes nonlinear when the asphaltene concentration exceeds a threshold concentration. The threshold concentration was ~ 8%. Although no clear explanation was given in this reference, we speculate that agglomerates start to form via clustering of the elementary micelles. These clusters exhibit an intra-cluster fractal geometry to be described below.

B-4. Fractal Geometry of Dense Asphaltene Colloids. The asphaltenes used for the fractal study was also from the Ratawi vacuum residue and the asphaltene extraction followed the same procedure as described in II-A. The SANS experiment was performed at NIST for concentration from 5 to 80 wt% in deuterated toluene at room temperature. The scattering intensity distributions were analyzed using a fractal model developed primarily for a surfactant-protein complex by Chen and Teixeira [30].

To build this fractal model for SANS data analysis, the elementary micelles were assumed monodisperse while the clusters were assumed polydisperse with each cluster made of k identical elementary micelles (the unit particles). Under these assumptions, the scattering intensity of a single k cluster will be proportional to the following interference factor

$$\left\langle \left| \sum_{m=1}^{k} A \exp[i\vec{Q}\cdot\vec{r}_m] \right|^2 \right\rangle \tag{31}$$

where A is the scattering amplitude of a micelle and $\vec{r}_m$ is the position vector of its center of mass. This quantity is equal to A^2k^2 at $Q = 0$ when the unit particles in a cluster are identical. If the system consists of many clusters, the total scattering intensity from the system would consist of contributions from: (1) the number of unit particles in a cluster that contributes to the A^2 factor, (2) the interference between the elementary particles within a cluster (the intra-cluster structure factor, $S_k(Q)$), and (3) the interference between clusters. For simplicity, we further assumed that the inter-cluster interferences are negligible based on a random arrangement of the clusters. This assumption seems unreasonable for a dense system. However, it can be justified, even for dense concentration, in the following argument. It is known thermodynamically that in a polydisperse cluster system, there are many more small clusters than large clusters. But, the scattering intensity at the small Q range are dominated by the contribution from the large clusters. As a result, the scattering intensity is largely contributed by the "dilute" large clusters which are "dispersed" in a sea of small clusters. Thus, from the scattering intensity point of view, it is equivalent to a dilute system, and the assumption of negligible inter-cluster interference is justified. Using this assumption, the scattering intensity $I(Q)$ will be dependent on the number of unit particles in a cluster and the intra-cluster structure factor only,

$$I(Q) = A^2 \sum_{k=0}^{\infty} N(k)\, k^2\, S_k(Q) \tag{32}$$

where N(k) represents the number of clusters that contain k unit particles. Since the scattering intensity should reduce to Ak^2 at $Q = 0$, $S_k(Q)$ should be formulated such that it is normalized to unity at $Q = 0$. A proper normalized $S_k(Q)$ was given by Chen and Teixeira [30] for a fractal cluster consisting of k elementary particles with a fractal dimension D_f. It has the following form,

$$S_k(Q) = \frac{\sin[(D_f-1)\tan^{-1}(QR_k)]}{(D_f-1)QR_k(1+Q^2R_k^2)^{\frac{D_f-1}{2}}} \tag{33}$$

where R_k is the radius of gyration of the k-cluster, and it is related to the radius of gyration of the elementary particle R_1 by

$$R_k = R_1\, k^{\frac{1}{D_f}} \tag{34}$$

With $S_k(Q)$ formulated, the next step is to give an appropriate form for $N(k)$. Following Stauffer [31], $N(k)$ can be written in the following form,

$$N(k) \propto k^{-\tau}\, e^{-\frac{k}{s}} \tag{35}$$

where S is a measure of the average number of the unit particles in a cluster, and τ is an index parameter representing the degree of polydispersity of the cluster sizes for a given S. A smaller τ value gives a broader cluster size distribution. τ is an important parameter. It not only indicates the polydispersity, but also reveals how a cluster is formed (i..e, the clustering mechanism) [32-35]. Since $k = 1$ represents the elementary particle, $N(k)$ should be normalized accordingly,

$$N(k) = \frac{S^{\tau-2}\, k^{-\tau}}{\Gamma\left(2-\tau,\frac{1}{S}\right)} \exp\left[-\frac{k}{S}\right] \tag{36}$$

where $\Gamma(a,b)$ is the incomplete Euler Gamma function. Substitute $S_k(Q)$ and $N(k)$ in Eq.(32) by Eq.(33) and (36) and replace the summation by an integration from 1 to infinity (assuming the cluster size has a continuous distribution, or the unit particle is small enough), one gets [36]

$$I(x,S) = \frac{S}{\Gamma(2-\pi)}\left[F(3-\tau,x)\,[1+x^2]^{\frac{-D_f(3-\tau)}{2}} + G(2-\tau,x)\left[\frac{x}{h}\right]^{-D_f}\right] \tag{37}$$

where $x = Q\xi$ with ξ being the correlation length defined as

$$\xi = hR_1 S^{\frac{1}{D_f}} \tag{38}$$

$\Gamma(x)$ is gamma function, $F(a,x) = \Gamma(a)\text{-}\Gamma(a,u)$ with

$$u = \left[\frac{h^2(1+x^2)}{x^2}\right]^{\frac{D_f}{2}} \tag{39}$$

and

$$h = \sqrt{\frac{D_f(D_f+1)}{6}} \tag{40}$$

The function $G(a,x)$ is

$$G(a,x) = \sin\left[\frac{(D_f-1)\pi}{2}\right] \frac{\Gamma\left(a,(\frac{x}{h})^{D_f}\right)}{D_f-1} \tag{41}$$

Eq.(37) is the equation used in this work for SANS analysis. S, D_f, and τ are the adjustable parameters used in the data fitting.

One can see from this model that it is subject to some limitation because the unit particles are assumed to be monodisperse. Aside from this limitation, this model is flexible in the sense that the fractal dimension and the polydispersity of the cluster τ are taken into account explicitly. In this way, these two parameters can be extracted simultaneously in the data analysis and allow one to study the fractal arrangement of the clusters and how the clusters are formed.

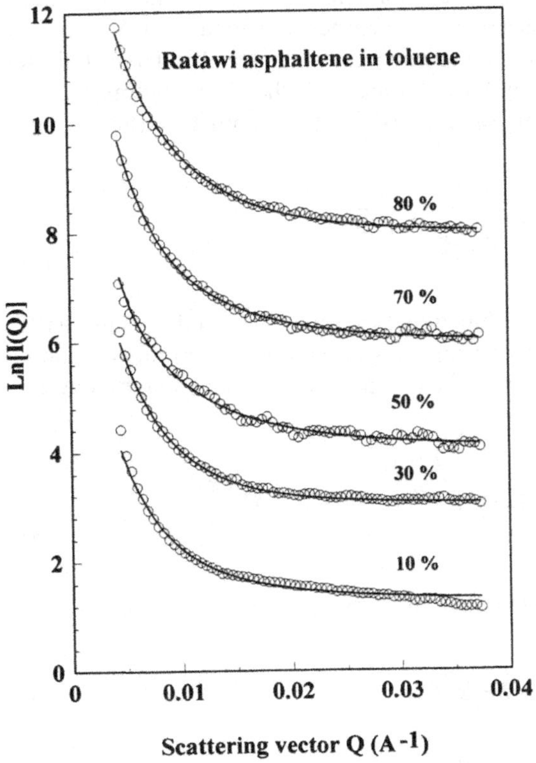

Figure 15. Small angle neutron scattering spectra for Ratawi/toluene systems at high concentrations. The spectra were shifted by a constant value for better comparison.

Figure 15 shows $I(Q)$ versus Q plots for concentration from 10 to 80 %. The curves were shifted (by adding an arbitrary constant to each scattering intensity to avoid overlap) for better presentation. Symbols are the experimental data and the solid lines are the calculated curves using the fractal model described above. Throughout the course of the data analysis, we consistently check the convergence of the fitting by giving different starting values of D_f, τ, and S. An unique convergence was found for all the cases studied here.

As one can see from Figure 15, the agreement between data and the calculated $I(Q)$ is excellent, except for the 10% case at very low Q region. This is due to the limited applicability of the model. As we mentioned earlier, this model is formulated based on an assumption that the system consists of clusters formed by the linkage of a group of monodisperse unit particles. The "form" of linkage depends on the interactions between the unit particles. For different types of interactions, the system can result in percolation, flocculation, clustering, and even sedimentation. These phenomena will be reflected in the fractal dimension D_f, the cluster polydispersity τ, and the average number of unit particles in a cluster S. In the case of 10% solution, most of the micelles (the unit particle in this model) remain dispersed. There are practically no clusters in the solution. It is thus expected to have D_f close to a three dimensional packing (i.e., $D_f \sim 3.0$) and $S \sim 1$. Since there are no clusters, the calculated curve based on the existence of clusters are expected to deviate from the data at this limit. This is particular true for the small Q region where $I(Q)$ is largely contributed by the large particles. Because of the model limitation the extracted parameter values may not be physically significant in this extreme case. We will discuss this when the extracted parameters are discussed.

Figure 16 gives S and D_f as a function of the asphaltene concentration. S remains very low until ~ 40%. Within this concentration range the asphaltene micelles (the unit particles

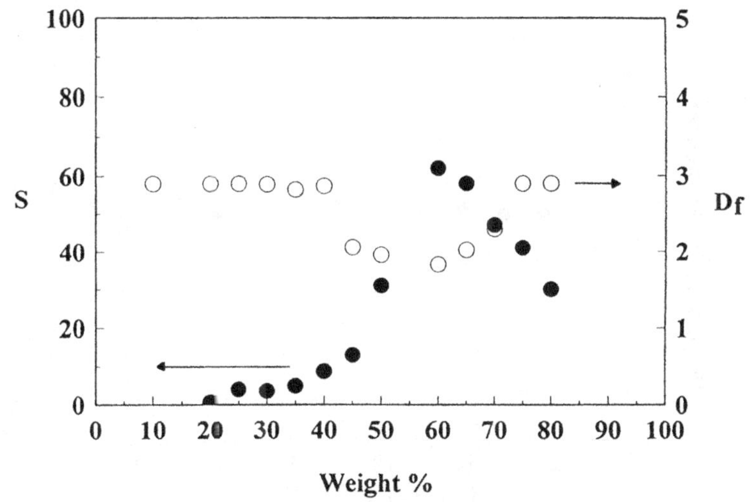

Figure 16. The number of unit micelles per cluster, S, and the fractal dimension D_f obtained from the analysis using Eq.(37).

that form clusters) do not seem to form big clusters. As concentration goes beyond 40%, big clusters start to form and the S parameter increases dramatically. It reaches a maximum value of ~ 60 at 60 %. In the same time, the fractal dimension decreases from ~ 3 to ~ 1.8. The values are comparable to that obtained from ESR work (D_f = 1.65 - 1.68) [37]. The way S behaves implies that the micelles form fractal clusters rapidly when the concentration reaches a threshold (in this case 40 wt %). From inter-micellar interaction point of view, the micelles stay dispersed until ~ 40 % means that the entropic energy of the micelles dominates the system free energy up to this concentration. Under this circumstance, more and more small clusters or unit particles will be created when the asphaltene concentration increases. When the concentration is greater than 40 %, the inter-micellar interactions overcome the entropic energy and start to bring the micelles together to form clusters.

In addition to the rapid increase of S for asphaltene greater than ~ 40 %, a more striking feature was observed for concentration greater than 60%. This is (1) S saturates at ~ 60 % and starts to decrease as concentration is further increased, and (2) D_f increases correspondingly while τ more or less remains constant.

Naturally, one would expect S to continuously increases via forming bigger clusters if the inter-micellar interactions continue to dominate. If this is the case, one would expect to see an increasing S and the D_f would level off with a value more or less the same as that at the 60% concentration. This is because the mechanism of the micellar clustering remains the same. In other words, the clusters, regardless of the increasing size, have the same self-similar structure (i.e., the same fractal dimension). On the other hand, if the entropic energy takes over at about 60 wt % again, then both S and D_f values will remain roughly constant. This is because an increasing S will become entropically unfavorable. Therefore, the added asphaltenes will form more clusters instead of making the existing clusters to grow. However, what we observed was decreasing S with increasing D_f, like going back to the dilute cases. To explain this behavior, we note that 60 weight % of asphaltene in toluene corresponds approximately to 0.53 asphaltene volume fraction. Since D_f is equal to 1.8 at this volume fraction, it means the space occupied by the clusters is in fact larger than 0.53. It will be about 0.8 (0.53 x 3/1.8, where 3 represents the 3 dimensional space filling), very close to the maximum packing of a soft-core system. This estimate is rough but is good to

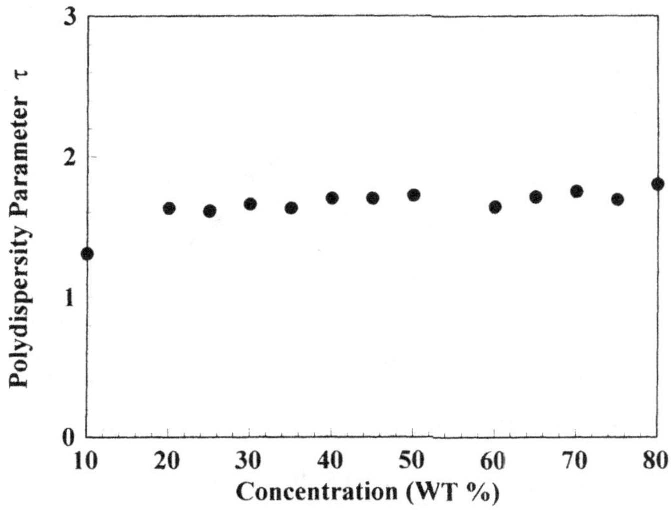

Figure 17. The τ values extracted as a function of asphaltene concentration.The τ values obtained were ~ 1.5 - 1.7 suggesting the clustering process is reaction limited.

the first order of approximation. Based on this, the system already reaches its maximum packing fraction at 60 wt%. Thus, further concentration increase would either precipitate the asphaltenes, or inverse the phase (i.e., toluene becomes the dispersed phase while asphaltenes become the solvent). Here we attribute the decline of S and the increase of D_f to a phase inversion phenomenon, which is commonly observed in surfactant systems [38]. In the inverse phase, toluene molecules form dispersed droplets within asphaltene matrix and become the scattering centers from the incident neutron point of view. Thus, neutrons "see" these toluene droplets in an asphaltene media of different scattering length density. If this is the scenario, S will decreases to 1 when phase is completely reversed and the toluene droplets are perfectly dispersed. Unfortunately, we did not observe this extreme phenomenon for concentration up to 80%. We obtained $S = 30$ at 80 % with $D_f = 2.9$.

The other parameter worth discussing is the polydispersity parameter τ. Figure 17 shows its behavior as a function of asphaltene concentration. One immediately notices that τ remains more or less constant (1.7 ± 0.1) except for the 10 % case (1.31 ± 0.1). This explains the argument we gave above for the deviation between the measured and the calculated I(Q) observed in Figure 15 for the 10 % at the small Q region. Other than the 10% case, the τ values obtained are between 1.6 to 1.7.

The value of τ provides information about polydispersity, which in turn tells us how a cluster is formed [32]. It is known that many thermodynamic equilibrium leads to a particular form of polydispersity [39], depending on the inter-particle interactions. In our case here, the interactions between micelles (the unit particles) lead the system to the formation of fractal clusters. This is analogous to an aggregation (micellization) phenomenon where the resulting polydispersity of the aggregates (micelles) can be expressed in terms of an index parameter. This index parameter carries the information about how the aggregation was formed. From the formulation given here (see Eq.(5)) τ serves as an index parameter to represent the system polydispersity. It therefore contains information about the mechanism of the cluster formation [32-35].

It has been known that the reaction limited aggregation (RLA) process leads to a τ value of 1.5. The asphaltene clustering observed here suggests that the asphaltene clustering process is likely triggered by the RLA mechanism. This is an important conclusion. In general, fractal clustering can arise from percolation, diffusion limited aggregation (DLA), and RLA etc. Percolation is a physical process which does not need a well defined activation energy for the process to occur. DLA is another physical process with its aggregation kinetic process limited by the diffusion of the unit particles (in this case the asphaltene micelles). RLA controls the aggregation process similar to a reaction. It has a well defined activation energy. Thus, one expects the process to occur rapidly as the energy given goes beyond the activation energy. This is likely the rapid S increase we observed at ~ 40 %. Similarly, if one applies enough thermal energy to overcome the activation energy the clusters can be dissociated. In fact, a previous study [29] showed the clusters form at 25°C and dissociate at 167°C. Within this temperature range the asphaltene micelles (the unit particles) remain intact. This indicates that the activation for clustering is much less than asphaltene micellization energy. In other words, the micellization and the micellar clustering have very different mechanisms.

Using the fractal analysis to characterize a dense asphaltene system is advantageous, because the asphaltene molecules, micelles or even clusters do not have a well defined shape or packing mechanism. They are complex in structures and shapes. Therefore, the fractal analysis is an appropriate approach, which provides the parameters that are related to the overall structural properties such as D_f, S, and τ obtained here. Moreover, fractal object represents a special class of spatial arrangement, which allows one to quantify the geometric arrangement through the fractal dimension and the polydispersity index τ. Thus, information like how the clusters are formed can be deduced from such an analysis. In fact, the work of Janardhan and Mansoori [40] demonstrated that this approach can even predict the amount of deposition as well as the size distribution.

V. ASPHALTENE MICELLIZATION MECHANISM—DIELECTRIC RELAXATION AND CONDUCTIVITY MEASUREMENT

Up to this point we have learned that asphaltenes behave like surfactant molecules in polar solvents such as toluene and pyridine. They form micelles and the micelles further agglomerate to form bigger clusters with intra-cluster fractal geometry. However, we still do not understand what is the main energy that initiates the micellization. Three groups, including ours, reported their findings regarding this point [41]. All three groups pointed to the charges carried by asphaltene molecules. We know charges exist in asphaltenes, mainly associated with the heteroatoms such as sulfur, nitrogen, vanadium, nickel, etc. In the following we will report our dielectric work which lead us to believe that the self-association is controlled by the localized charges.

The dielectric relaxation study was carried out in order to understand the dynamics of the localized charges in asphaltene micelles, if there are any. The experiment was performed on a HP 4192A, Low Frequency Impedance Analyzer. The instrument has a working frequency range from 5 Hz to 13 MHz. The sample cell used was a cylindrical cup filled with ~ 40 cm^3 of the sample. A liquid immersion-type four-electrode cell from Rosemont Analytical was used to measure the capacitance and the dissipation factor as a function of the electric field frequency. From the frequency dependence the dielectric storage and loss spectra were obtained. The cell consists of four electrodes of 2.25 cm^2 (a total area = 9 cm^2) with 1 mm spacing between plates. The sensitive areas of the electrodes were coated with platinum black to create rough surfaces for preventing local polarizability. The measurement was made for temperatures from 10°C to 60°C. A water circulator was used to controller the temperature with ± 0.1°C accuracy. Since the working limit of the electrodes is about 7 MHz, the analysis was performed for frequency up to 7 MHz only.

A dielectric relaxation spectrum shows how fast the permanent or induced dipoles in the system respond to the applied field frequency and amplitude. Through the response spectrum (phase shifts and amplitude) as a function of the applied frequency, one can readily deduce the microscopic properties of the system, such as the sizes of the dipoles and their dynamics. Since the applied alternating field is a wave function, the response spectrum has a complex form as a function of frequency. The real part stands for the storage spectrum while the imaginary part represents the loss spectrum. If the dipoles objects are monodisperse in size with identical dipole strength, then, the relaxation spectrum should follow the so called Debye relaxation form,

$$\epsilon(\omega) = \epsilon_\infty + \frac{\Delta\epsilon}{1 + i\omega\tau} \tag{42}$$

where $\varepsilon(\omega)$ is the dielectric permittivity at frequency ω, and ε_∞ is the dielectric permittivity at high frequency, $\Delta\varepsilon$ represents the difference between the zero frequency permittivity ε_o and ε_∞, and τ is the relaxation time. The Debye relaxation is usually observed for the case when there is no frequency spread in relaxation, or the relaxation of each individual particle (dipole) is identical. This situation often occurs at dilute concentration or at high temperature where the system approaches the ideal condition. For most complex systems like micellar solutions, the dipole strength from each micelle may be different. In this case the observed dielectric response represents an average of multiple mode relaxations. To characterize such a system, Cole and Cole [42] derived a simple equation, well known as Cole-Cole equation. The Cole-Cole equation reads

$$\epsilon(\omega) = \epsilon_\infty + \frac{\Delta\epsilon}{1+(i\omega\tau)^{1-\beta}} \tag{43}$$

where $(1-\beta)$ is the stretch exponent, representing the effect from the multiple-mode relaxations or the effect of inter-dipole interactions. When $\beta = 0$, it reduces to the Debye relaxation. Since $\varepsilon(\omega)$ is a complex function, it can be divided into a real part and an imaginary part, assuming the conductivity loss is negligible. For asphaltene/toluene systems, this assumption is reasonable, because toluene has very low conductivity.

In principle, one can analyze the storage data (the real part) or the loss data (the imaginary part) to extract structural and dynamic information. However, if it is a dispersed system and the dipoles are the dispersed phase (in our case, the asphaltene micelles), the loss data often shows a well defined peak, representing the average relaxation mode. In this case analyzing the loss spectrum is more accurate, since the peaked spectrum is more sensitive. Because our systems all exhibit a pronounced peak, we therefore selected the loss spectrum for data analysis. From Eq.(43), the analytical form of the dielectric loss can be explicitly expressed,

$$\epsilon''(\omega) = \frac{\Delta\epsilon(\omega\tau)^{1-\beta}\cos(\frac{2\pi\beta}{2})}{1-2(\omega\tau)^{1-\beta}\sin(\frac{\pi\beta}{2})+(\omega\tau)^{2-2\beta}} \tag{44}$$

Eq.(44) was the dielectric loss equation we used to fit the experimental data, in which $\Delta\varepsilon$, β, and τ were taken as adjustable parameters. Experimentally, $\varepsilon''(\omega)$ was obtained by

$$\epsilon''(\omega) = \frac{C(\omega)d}{\epsilon_o A D(\omega)} \tag{45}$$

where $C(\omega)$ are $D(\omega)$ are the measured capacitance and the dissipation factor, d (= 0.1 cm) is the distance between electrode plates, $\varepsilon_o = 8.854 \times 10^{-14}$ Farad/cm is the dielectric constant of the vacuum, and A (= 9 cm^2) is the total electrode area.

Using Eq.(44) for analysis allows us to extract the average relaxation time τ, the frequency spread (represented by the stretch exponent) and the dipole strength ($\Delta\varepsilon$).. From τ, one can calculate the dipole moment weighted particle size using the Stokes-Einstein equation for the rotational diffusion:

$$\frac{1}{\tau} = \frac{k_B T}{4\pi\eta R_d^3} \tag{46}$$

where k_B is the Boltzmann constant, T the temperature, η the effective viscosity, and R_d the dipole moment weighted particle radius.

Figure 18 shows the dielectric loss spectrum for 20% asphaltene in toluene at 50°C, as a function of ω. The solid line is the Cole-Cole fit. The agreement is reasonably good. For most of the samples we studied, the fitting quality is reasonable, except for a high concentration case (50% at 30°C), and a low temperature case (20% at 10°C), where a second peak (or a shoulder) becomes apparent (Figure 19a and Figure 19b). If one were to fit such a two-peak spectrum, a summation of two Cole-Cole equations is necessary. In this case, six free parameters would be involved. This way, the fitting may be ambiguous, because of possible non-unique convergence. We thus decided to fit all our data with a single Cole-Cole equation. The fit may not be satisfactory for the double peak cases, but the major features are qualitatively reasonable. Table III tabulates the extracted $\Delta\varepsilon$, $(1-\beta)$, τ values, and the dipole radii, R_d, calculated according Eq.(46).

The results show that $\Delta\varepsilon$ remains fairly constant for all the conditions studied. This is expected, because the asphaltene micelles are not expected to experience fundamental structural changes under our experimental conditions (temperature or concentration) to alter $\Delta\varepsilon$. The stretch exponent $(1-\beta)$ obtained increases gradually from ~0.5 to ~0.8 (Figure 20), with increasing temperature. This is also expected, because the system should become more and more ideal as temperature increases. If this is the case, the relaxation process should gradually evolve toward the Debye relaxation process ($(1-\beta)=1$). However, within the

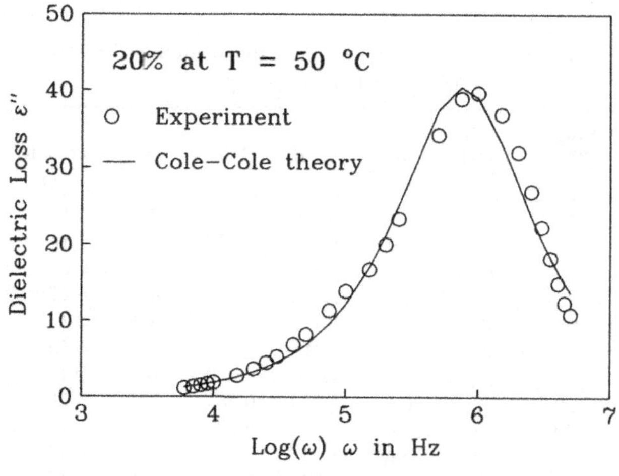

Figure 18. The dielectric loss $\varepsilon''(\omega)$ as a function of the frequency of the applied field. The solid line is the calculated curve using the Cole-Cole equation, Eq.(44).

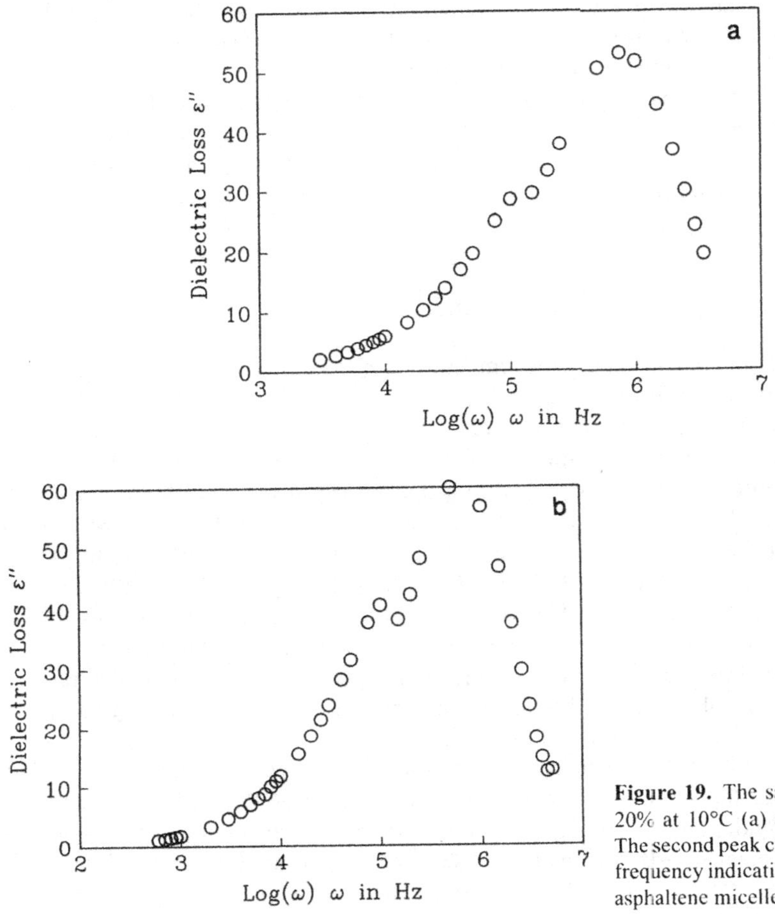

Figure 19. The same plot as Fig. 1 for 20% at 10°C (a) and 50% at 30°C (b). The second peak clearly shows at a lower frequency indicating the clustering of the asphaltene micelles.

Table III. The extracted parameter from the ε''(ω) fit using the Cole-Cole equation

Temp. (°C)	Conc. (wt%)	Δε	τ (μs)	1-β	R_d (Å)	η (Poise)
10	20	57.5	1.785	0.596	30.3	0.20*
20	20	57.5	1.629	0.659	30.8	0.18
30	30	59.9	1.512	0.677	30.4	0.18
30	50	55.3	2.631	0.505	32.7	0.25*
35	20	58.5	1.466	0.730	30.8	0.17
35	30	57.0	1.435	0.720	30.0	0.18
50	20	56.6	1.317	0.778	33.0	0.13
50	30	55.7	1.369	0.720	32.6	0.14
50	50	6.8	1.879	0.580	32.2	0.20
60	20	56.0	1.250	0.800	33.7	0.12
60	30	56.6	1.365	0.733	33.7	0.13
60	50	56.2	1.749	0.621	33.5	0.17

* There are two peaks (or shoulder) on the dielectric spectrum.

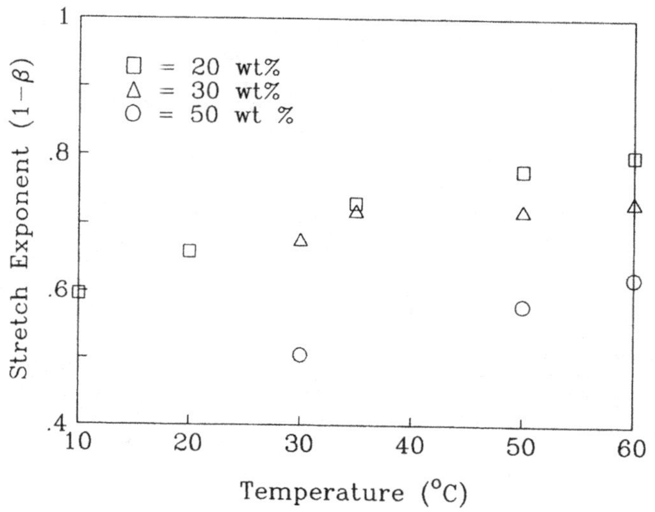

Figure 20. The stretch exponent as a function of temperature for three concentrations: (□) 20 %, (Δ) 30% and (O) 50%.

temperature range we studied, the (1 – β) values were still below unity, although it does increase. This indicates that the systems are still far from ideal. This is because of the relative high concentrations (20 to 50%) with significant inter-micellar interactions. Polydispersity may be the other factor. These two effects seem to become more profound as temperature decreases. Figure 21 shows the average relaxation time τ. It decreases as a function of temperature and increases substantially for the 50% case. Again, this is due to the inter-micellar interaction that slows down the relaxation time, because the relaxation is restricted by the neighboring micelles.

The radii, R_d, obtained for single peaked cases using Eq.(46) (see Table III) are consistent with the results obtained from our previous small angle neutron scattering studies

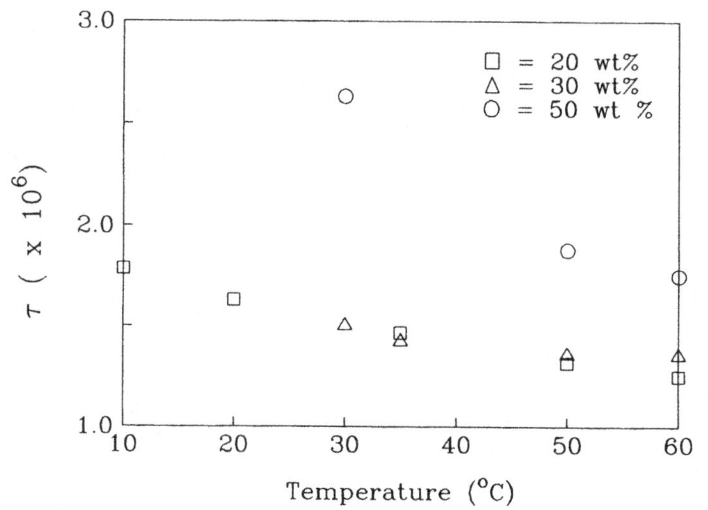

Figure 21. The mean dielectric relaxation time τ as a function of asphaltene concentration and temperature.

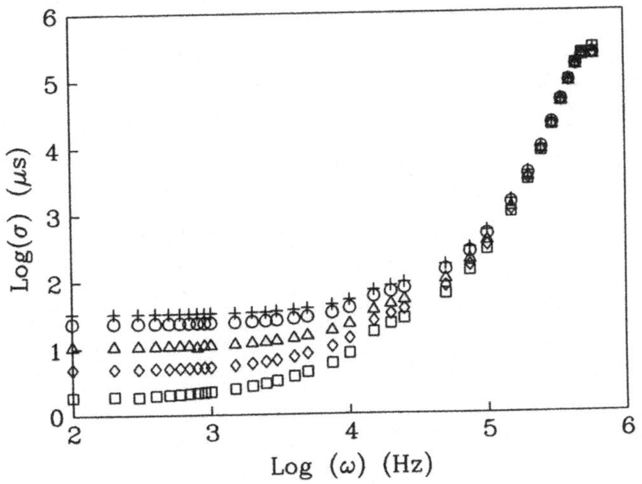

Figure 22. The conductivity as a function of frequency of the applied field (□: 10°C; ◇: 20°C; △: 35°C; O: 50°C; +: 60°C). At high frequency the conductivity becomes temperature independent, indicating that the micelles do not swell when temperature increases from 10 to 60°C.

(see Table II) [43]. The average particle sizes obtained approximately constant. One has to note that the errors of the calculated particle sizes may come from the effect of inter-particle interactions, as well as the viscosity measurements. These two error sources can account of approximately 5 to 10% error. Thus, one should not try to get a trend from the calculated average size. Instead, an approximately constant size should be concluded at this point. As for the double peaked cases (Figure 19), even the radii obtained are consistent with the scattering data, we do not intend to make any conclusion.

An immediate conclusion from the above results is that there definitely exist dipoles within the asphaltene micelles, and the dipole moments are polydisperse ($\beta > 0$). If these dipole moments are static, the conductivity within the micelles will exhibit purely thermodynamical behavior. In this case, the conductivity for frequency greater than the threshold value (i.e, greater than the peak frequency in the loss spectrum) should depend on temperature, according to the Arrhenius law. However, our previous work (ref. 41, 43, see Figure 22) showed that the conductivity is nearly temperature independent for ω above this frequency. On the other hand, the static conductivity exhibits Arrhenius behavior (see Figure 23). This suggests that most of

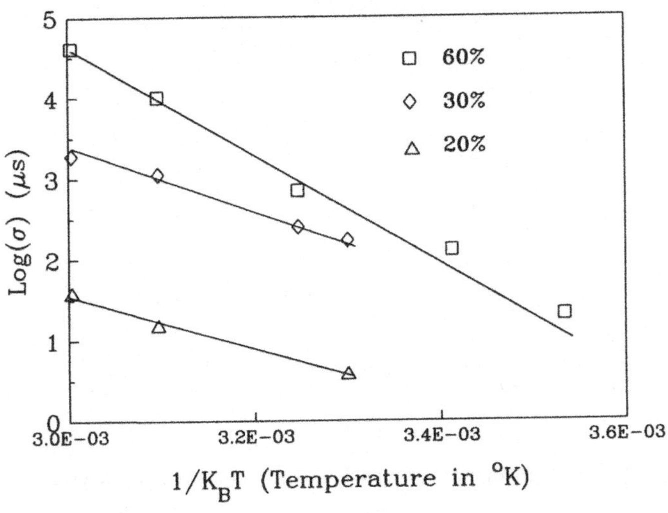

Figure 23. Temperature dependence of the conductivity in an Arrhenius plot. The 60 wt% shows different slope representing a different thermal mechanism.

the dipoles relaxations and their dynamics are dominated by the induced dipoles. More importantly, the charges actually exchange rapidly within the asphaltene micelles. This result is the conclusion we like to draw at this point. It directly indicates that the micellization is relevant to the charge exchange. In the following, we will give our explanation/speculation about the anomalous behavior for 50% concentration and the second peak observed.

In Figure 20 and 21, the 50% case shows a distinctive difference from the lower concentration cases for both $(1-\beta)$ and τ. Our interpretation is that the system grows into a self-similar structure via clustering (see the fractal analysis in VI-B-4). Under this situation, a dipole (a micellar unit) in the cluster will experience a caging effect during the dipole rotation precess. This is to say that the micelles in the cluster will be somewhat restricted by the neighboring micelles, when it responds to the applied alternating field via rotation. As a result, the clustering dipoles add up to become an integrated dipole representing the cluster. In this case, the dipole response to the dielectric relaxation depends on the rotational diffusion constant of the cluster, which is definitely slower. One thus anticipates to see a peak at the lower frequency.

The appearance of the second peak occurs at ~ 105 kHz for the high concentration and low temperature case. Strictly speaking, this second peak may result from (1) the clustering of the asphaltene micelles, which forms a self-similar structure (i.e., fractal structure); (2) the percolation in which the inter-micelles potential is low for a certain physical properties (to be described below), or (3) the networking effect (in which the chemical bonding or structural intercalation sets in). In many oil continuous microemulsions containing water droplets, the system easily undergoes percolation, if the volume fraction of the water droplet exceeds the percolation threshold. The driving force for percolation to occur is the attractive interaction, due to different Hamakar constant between water and oil. One simple way to judge the percolation is through the conductivity behavior [44]. If the conductivity suddenly rises at a threshold concentration, it is likely that the system percolates at that concentration. At this concentration, the potential for ions movement is essentially zero, allowing ions to move freely between droplets. This is particularly true for water containing microemulsions. If this is the case, the conductivity near the percolation threshold should behave like [45]

$$\begin{aligned} \sigma &\sim (\phi_c - \phi)^{-s}, \qquad \phi < \phi_c \\ \sigma &\sim (\phi - \phi_c)^{-\mu}, \qquad \phi < \phi_c \end{aligned} \tag{47}$$

where, σ is the conductivity, ϕ_c is the percolation volume fraction, ϕ is the volume fraction, and s and μ are the critical exponents. s and μ should be ~0.7 and ~1.9 respectively for a percolating system. Our conductivity data do not show such behavior (Figure 24) The slopes obtained differ from 0.7 and 1.9. Thus, the percolation process is unlikely. The networking process is also not possible, because the viscosity remains Newtonian, while a network-like structure will show non-Newtonian behavior. In fact, our viscosity work [46] showed the network forming for asphaltene in de-asphalted oil at $T > 350°C$, and the viscosity becomes shear-thinning. Below this temperature, it is Newtonian. We thus believe that our systems are viscous liquids, likely undergoing clustering process with increasing inter-micellar correlation strength.

VI. TWO-PHASE EQUILIBRIUM AND ASPHALTENE INTERFACIAL PROPERTIES

At this point we know that asphaltenes form micelles with the association energy likely related to the charges carried by the heteroatoms. If we parameterize this self-association propensity by a parameter parallel to the overall molecular polarity (the definition of

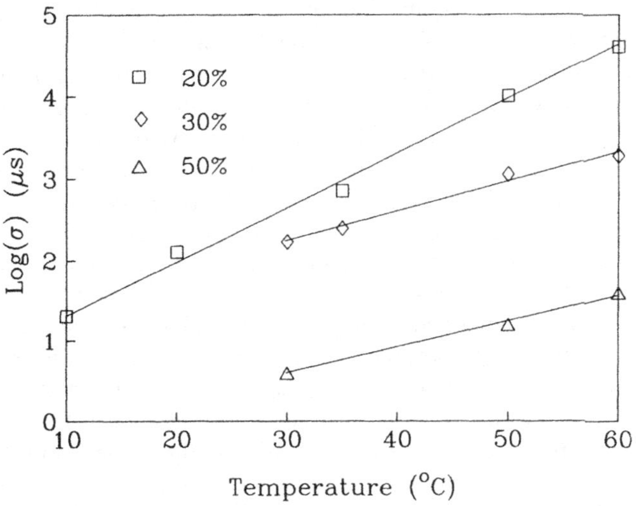

Figure 24. The conductivity plot according to Eq.(6) using temperature as the measuring parameter: (□) 20 %, (◊) 30% and (△) 50%. The curves, although show linear behavior, do not give the right slope as suggested in Eq.(47) Thus, the second peak should not arise from percolation process.

asphaltenes using solvent extraction can be seen as a polarity separation method), we can then modify this polarity to see if we can control the asphaltene "surface activity". To be precise, it should be the self-association propensity instead of the surface activity, because the association is not necessarily dominated by the surface activity. However, in seeking ways to make use of asphaltene, we still focus on its surface activity, assuming that it somewhat reflects the self-association. Thus, it is rational to examine the interfacial activity of asphaltenes by mixing an asphaltene containing organic solution with an aqueous solution of different pH values. To probe how asphaltene molecules react to this environment and characterize their interfacial activity, we measured the dynamic interfacial tension (DIFT). Again, we used the Krüss surface tensiometer and the Wilhelmy plate method.

Prior to the oil-aqueous phase equilibrium experiment and the DIFT measurement, the asphaltenes were in toluene at the desired concentration and aged for at least 24 hours before being equilibrated with the aqueous solutions.

To make the measurement consistent, we slightly modified the interfacial tension measurement procedure recommended by the manufacturer, in terms of the control of the fluid level and the plate submergence level. This modified method gave a better control of the zeroing and the plate wetting line. This procedure was as follows: (1) the instrument was zeroed with the platinum plate submerged in a constant volume light (oil) phase at a marked oil level; (2) the plate was then removed and cleaned; (3) an equal volume of the heavy phase as that of the light phase for calibration was prepared and its surface tensions measured according to the procedure recommended by the manufacturer without the zeroing step, and (4) the light phase was added (dropwise) to the heavy phase until the plate was submerged to the previously marked level. Because the light phase was added drop-wise, the true zero time information can not be rigorously defined. Thus, the zero time reported here should only be used as a reference. A 20 to 30 seconds of deviation from the real zero time is expected.

It is important to maintain the equal volume of the heavy phase and the light phase. When unequal volumes of them were used, appreciable error may be introduced. This is true even though the volume of the light phase was kept constant so that the plate submerged level remain the same. To prevent toluene from evaporation, we cover the vessel with a Teflon cover to minimize the evaporation. Since a hairline was drawn to indicate the light phase level, we checked the level, at zero time and at the end of each measurement. No observable evaporation was found.

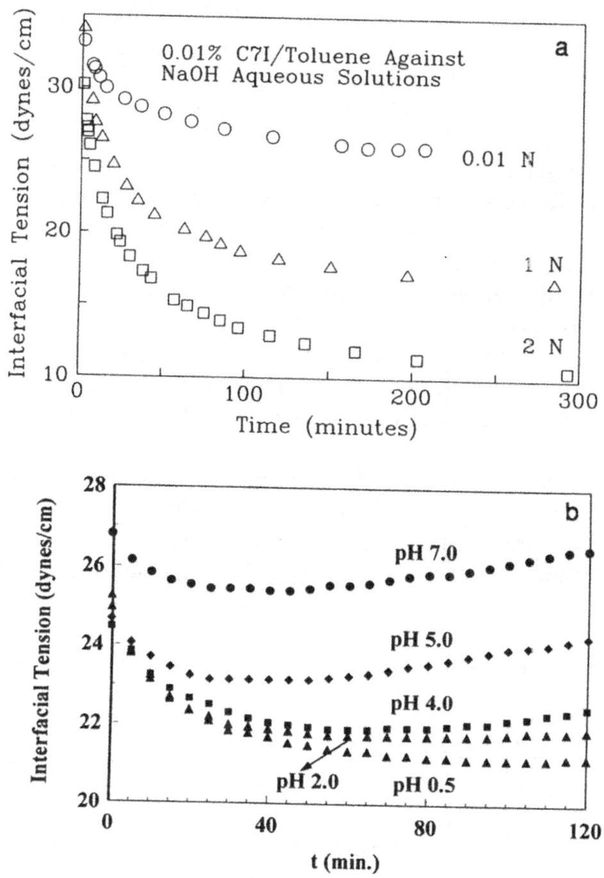

Figure 25. **(a)** Typical dynamic interfacial tension as a function of time for Asphaltene/caustic solution at different normalities. **(b)** Typical dynamic interfacial tension for asphaltene/acid interface.

Using this measuring method, we systematically examined the asphaltene interfacial activity as a function of the aqueous pH values from very basic to very acidic lower phase. Figure 25a and Figure 25b demonstrate the fundamental difference of the DIFT behavior of asphaltenes at the oil/caustic and at the oil/acid solutions. While DIFT shows a simple decaying function for the caustic solutions, the acidic cases seem to exhibit two kinetics processes. One is the initial process which reduces the interfacial tension to a minimum point within a period of ~ 40 minutes. During this period, asphaltenes continuously adsorb to the interface. Following this process is the second process which brings the interfacial tension back up, signifying either some chemical reactions are occurring which modify the interfacial tension, or asphaltenes start to diffuse back to the bulk. Since there are two processes in the acid cases, we adopted two diffusion controlled processes to describe the data. The equations we used was [47]

$$\gamma(t) = \gamma_o e^{-\sqrt{\beta t}} + \gamma_\infty \tag{48}$$

where $\gamma(t)$ is the time-dependent interfacial tension, γ_o is a proportional constant, γ_∞ is the equilibrium interfacial tension, and β is a kinetic parameter that characterizes the DIFT reduction rate. Its physical meaning is equivalent to a diffusion constant or an adsorption rate constant. In the case of desorption, Eq.(48) can be expressed as

$$\gamma(t) = \gamma_o e^{\sqrt{\alpha t}} + \gamma_\infty \tag{49}$$

where the α parameter represents the desorption rate constant. It is important to note that both α and β are quantities equivalent to the diffusion constant and the desorption constant respectively. But they are not exactly the same as the diffusion constant or the desorption constant. This is because the system examined here is not at the equilibrium state while the diffusion constant or the Gibbs isotherm are defined based on an equilibrium state.

In Table IV we summarize the equivalent diffusion constants α and β for all the systems studied. As one can see from Table IV, higher asphaltene concentration always reduces the equilibrium interfacial tension (γ_∞). This is true for both caustic and acidic cases. However, γ_∞ is generally much lower for the caustic cases than the acidic cases.

Based on information summarized in Table IV, one may look for stronger caustic solutions to see if an emulsion can be made using asphaltenes as the emulsifier. However, it was not the case we observed. Instead, we found that when we lowered the acidic solution to pH = 1, a 5 % asphaltene solution can form emulsion and uptake an equal volume of the aqueous phase. The emulsions thus formed appear to be stable for more than 60 days at 25°C. They are water-in-oil emulsions according to the conductivity measurement. Figure 26 demonstrates this point. Structurally, asphaltenes appear to serve as the surfactant component in these emulsions. Their spatial arrangements are such that the hydrophilic water droplets

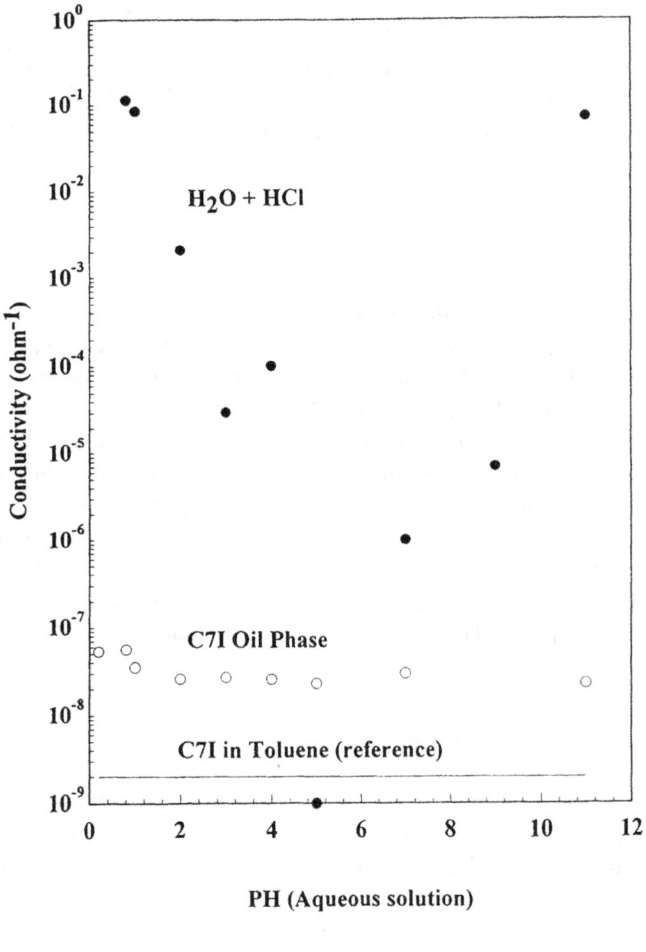

Figure 26. Conductivity as a function of pH. For pH > 7, the solution was prepared by water and NaOH. The relative low conductivity of the C7I (asphaltene solution) compared to the aqueous phase (the filled circles) indicates that the emulsions are water-in-oil.

Table IV. Extracted parameter values from adsorption/desorption study

C (wt%)	pH	α (min^{-1})	β (min^{-1})	γ_∞ (dyne/cm)
1e-5		—	0.039	15.8
0.01	12	—	0.020	24.5
	+	—	0.043	16.0
	++	—	0.042	9.9
0.025	2	—	0.021	22.0
	3	—	0.021	23.5
	4	—	0.019	25.8
	5	—	0.018	27.8
	7	—	0.022	30.0
	+	—	0.048	13.8
0.05	2	—	0.018	21.0
	3	—	0.024	23.0
	4	—	0.019	24.5
	5	—	0.022	26.0
	7	—	0.021	29.0
	+	—	0.055	10.1
0.1	0.5	0.0	0.063	17.0
	2	0.076	0.071	20.6
	4	0.072	0.085	21.4
	5	0.080	0.094	22.5
	7	0.099	0.114	25.0
	+	—	0.217	8.0

+[NaOH] = 1N, ++ [NaOH] = 2N.

can be prevented from contacting the hydrophobic solvent. Detailed structure is not yet known at this point.

Since this is the first emulsion made using asphaltenes as the surfactant, there is a need to characterize the emulsions. In the following section we shall present the our viscosity experiment for characterization of the base treated asphaltene solutions.

To explain the asphaltene interfacial activity, we separate the data into two series: the acid and the base series. The main feature of the acid cases is their peculiar DIFT behavior toward equilibrium. The base cases, on the other hand, show simple asphaltene adsorption. Thus, the issue to be discussed is how the asphaltene solution is altered after asphaltene adsorption.

We shall discuss the acid cases in the following and the case of base in the next section where the rheological properties of the based treated asphaltene solutions are investigated.

Figure 27 to 29 show the dynamic interfacial tension (DIFT) at three asphaltene concentrations as a function of time. In each figure, the $\gamma(t)$ curves were arbitrarily shifted by adding a constant background to avoid overlapping of the curves. Based on these three figures, one sees the interfacial tension decreases initially for time t up to ~ 40 minutes. When t > 40 minutes, the interfacial tension may rise again (at high asphaltene concentration and high pH) or may level off and reach the new equilibrium state when asphaltene concentration is low (except the pH = 2 case at 0.05 wt% asphaltene). The data for t < 40 minutes were interpreted by Eq.(48) and the data after t > 40 minutes by Eq.(49) if the interfacial tension rises again.

Figure 30 to 32 show the plots of Figure 27 to 29 according to Eq.(48) and (49). A linearly decaying behavior was obtained for all cases in the t < 40 minutes time period. If

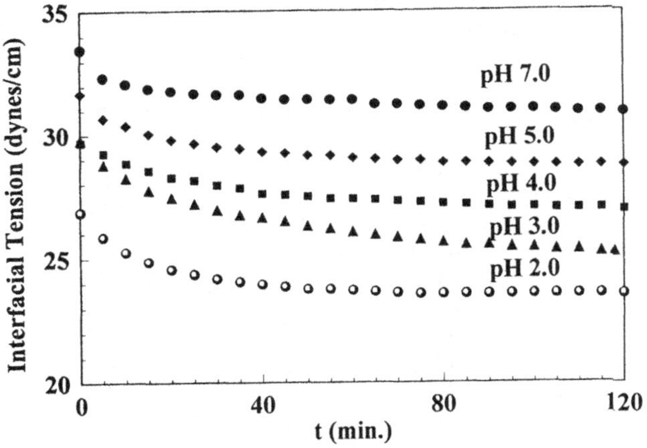

Figure 27. Interfacial tension $\gamma(t)$ versus t for 0.025 wt% asphaltene/toluene solution equilibrated with HCl solutions of various pH values.

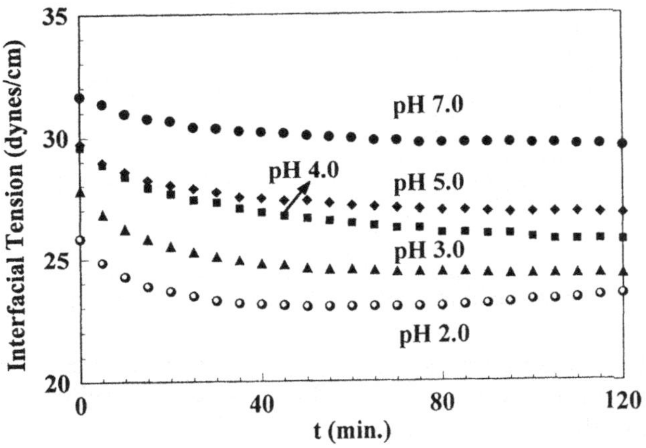

Figure 28. Interfacial tension $\gamma(t)$ versus t for 0.05 wt% asphaltene/toluene solution equilibrated with HCl solutions of various pH values.

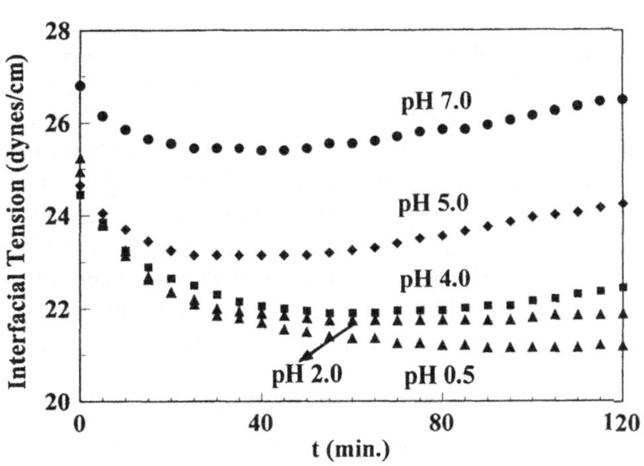

Figure 29. Interfacial tension $\gamma(t)$ versus t for 0.1 wt% asphaltene/toluene solution equilibrated with HCl solutions of various pH values.

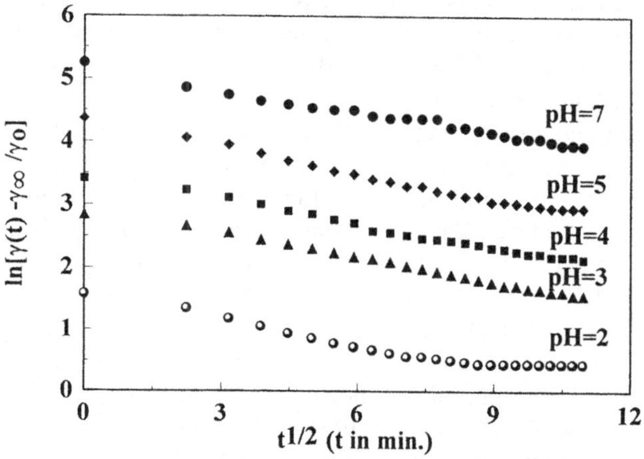

Figure 30. $\ln[\gamma(t) - \gamma_\infty]$ versus $t^{1/2}$ for 0.025 wt% asphaltene.

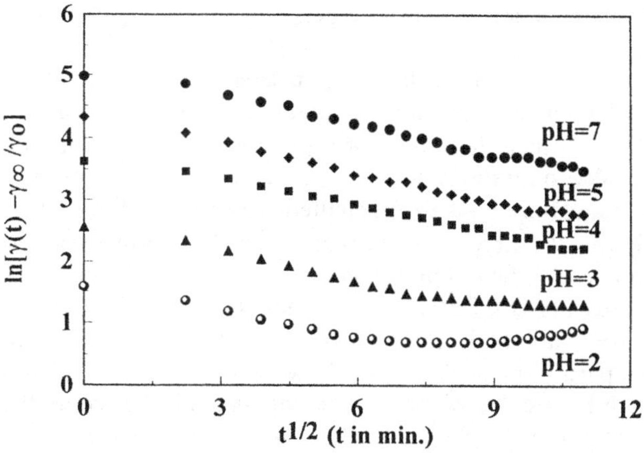

Figure 31. Same plot as Figure 4 for 0.05 wt% asphaltene.

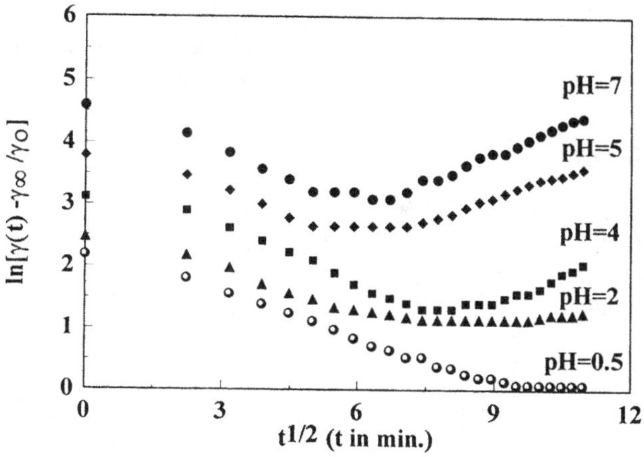

Figure 32. Same plot as Figure 4 for 0.1 wt% asphaltene.

the reduction of the DIFT is truly due to the asphaltene adsorption to the interface, then the kinetic process may be diffusion controlled, at least for the first ~ 40 minutes. As time goes beyond this time period, the DIFT behaves differently depending on the asphaltene concentration and the pH of the aqueous phase.

For asphaltene concentration below the critical micelle concentration (CMC, here we assume the CMC to be similar to the CMC in pyridine, i.e., 0.05 wt%), the DIFT tends to level off and reach the new equilibrium state. In this concentration region, the lower the pH of the aqueous phase, the faster the equilibrium can be reached. This is true for both 0.025 and 0.05 wt % examined here, except for the extreme case (i.e., pH = 2 and asphaltene = 0.05 wt%) where DIFT increased again when t > ~ 65 minutes.

When asphaltene concentration was above CMC at 0.1 wt % (see Figure 32), different features were observed. First, the DIFT decreased linearly and reached a minimum. It then stayed at the minimum for a period of time and increased thereafter. This behavior is true for all pH, except for the pH = 0.5 case where the DIFT leveled off and did not rise again at the end of the measurement. Secondly, we found that the DIFT reached the minimum faster when the pH of the aqueous phase was higher. This was opposite to the cases with asphaltene concentration below or near CMC. Finally, the increase of DIFT after the minimum also exhibited a linear behavior when plotted according to Eq.(48) and (49) (see Figure 30-32). These three major features delivers some underlying physical and/or chemical processes to be analyzed in the following.

In the analysis of the DIFT data we again separated the data into two parts. The first part was for the asphaltene concentration below or near CMC cases and the second part for the cases above CMC. Since the DIFT for asphaltene concentration below or near CMC decayed monotonically, only the β parameter (see Eq.(48)) is of interest. β represents the DIFT reduction rate, and thus the rate of asphaltene adsorption. When concentration is higher than CMC (see Figure 29), three parameters are needed to characterize the processes. They are: (1) the β parameter for the initial period (this is the period when DIFT is decaying), (2) the slope of the increasing DIFT after the minimum (based on Figure 30 to 32) and (3) the time duration the DIFT stays at the minimum (referred to as the lifetime from what follows).

For the first part of the analysis, we show the β values (extracted from the slopes of Figure 27 and 28) in Figure 33. With the error bars taken into account the β value remains fairly constant (~ 0.02 min^{-1}) for all pH values. In the second part (> CMC, the 0.1 wt % case), we show in Figure 34 the α and the β values extracted from the decaying and the

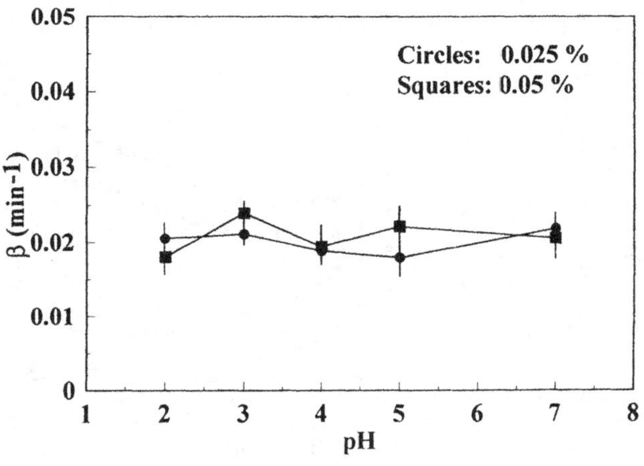

Figure 33. β extracted from Figure 4 and 5 as a function of pH.

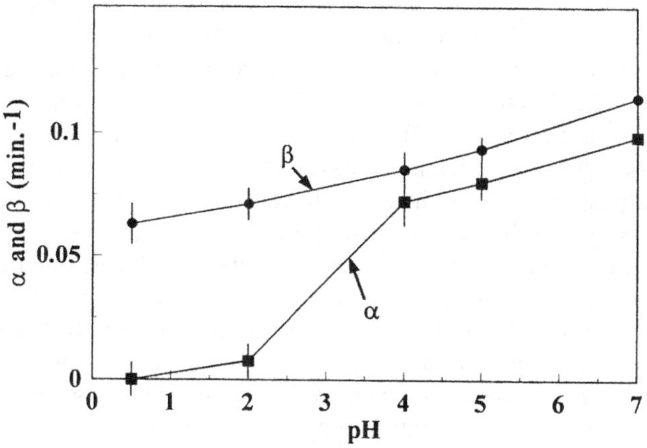

Figure 34. α and β values extracted from the decaying and increasing slopes of Figure 6.

increasing slopes (see Figure 32) respectively. One notes that the α and β values are approximately parallel to each other for pH ≥ 4, while for pH < 4, β is considerable larger. This may indicate an onset of some process. We will discuss this point in the following discuss section.

In Figure 35 we show the lifetime as a function of pH. As one can see, it increases drastically for pH < 4. This correlates to the sharp decrease of α in Figure 34, again, indicative of the occurrence of some process. In the following, we shall try to interpret these observed phenomena together with the α and β plot (Figure 34) using a combination of two processes.

For asphaltene concentration below CMC, the analyzed data (Figure 30-32) show that the process is similar to a diffusion controlled process with very slow rates. Physically, there are two scenarios that may result in these DIFT behaviors. The first scenario is that the reduction of the DIFT is a pure diffusion process with asphaltene monomers diffuse slowly to the interface [48]. This kinetic slowness is likely due to little amount of energy the system can gain by diffusing asphaltene monomers to the interface. In other words, the asphaltene

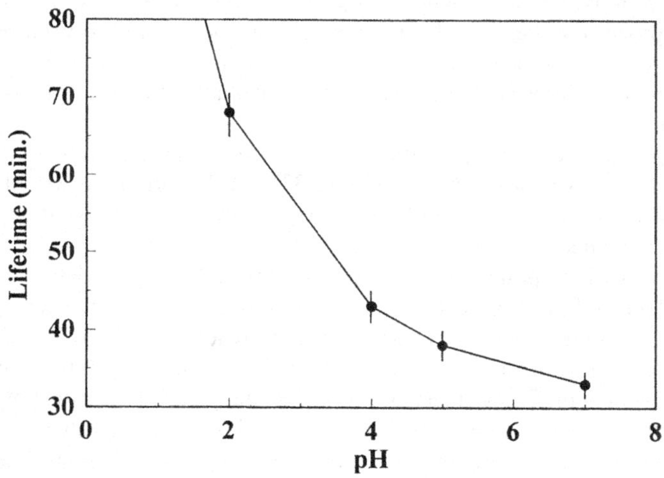

Figure 35. DIFT lifetime at the minimum as a function pH.

monomers, although consisting of two portions of dual nature to form reversed micelles in toluene, do not have enough hydrophobicity contrast between these two portions to compensate the high interfacial tension energy. As a result, the potential difference created between being in the bulk and at the interface is not significant, as far as the asphaltene is concerned. This leads to a slow diffusion of asphaltenes toward the interface. Thus, the adsorption is slow. We believe this is the case, since the major free energy contribution to the asphaltenes micellization is not hydrophobic energy as we have discussed [11,29].

The other scenario that may explain the data is that the kinetic process may not be solely diffusion controlled. Instead, it may involve two processes, one being the diffusion of the monomers to the interface, and the other being the arrangement of the structurally different asphaltene molecules at the interface. If the arrangement of the asphaltenes at the interface is a process slower than the diffusion process, then it becomes the bottle neck of the kinetic process. This molecular arrangement process may also be describable by an exponential form (see Eq.(48)) similar to the diffusion limited process. If this is the case, DIFT will also show linearity in the plots like Figure 30 and 31.

As for the possible occurrence of chemical reactions, we can not judge from the DIFT data observed, at least for the cases with asphaltene concentration below or near CMC. If there are reactions occurring at the interface, the meniscus line should not remain well defined at the end of each DIFT measurement as observed. Experimentally, the DIFT curve cannot show the occurrence of the reactions unless (1) the reactions are slow and (2) the reaction products change the interfacial tension to a measurable level. If both conditions are met, one may observe two distinctive DIFT kinetic processes with two slopes shown in Figure 29 or Figure 30.

The second part of the data analysis is for the case of 0.1 wt% (above CMC), where DIFT shows both decreasing and increasing features as a function of time. To explain the DIFT dependence on pH, Figure 34 is an important plot. α and β in Figure 34 represent the increasing and the decreasing rates respectively. The first point noticed in Figure 34 is that α and β are parallel to each other when pH ≥ 4. Quantitatively, they differ slightly. What this indicates is that the decrease and the increase of DIFT may represent two opposite processes. In this case, β represents the rate of the forward process and α the reversed one. The reversed process is likely a desorption process. If this is the case, the forward process should be an adsorption process. It also indicates that, there is either none or negligible chemical reactions occurring in this pH range. We thus suggest that it is a simple adsorption and desorption process. Additionally, the asphaltene adsorption is the dominate kinetic process instead of the diffusion process. Otherwise, α and β would not be parallel to each other. The slight difference between α and β is because some asphaltene molecules (likely the more polar ones) may stay at the interface which reduces the asphaltene concentration at the interface for desorption.

One also notices that β is much greater than those β values obtained for the concentration lower or near CMC (see Figure 33). At this point we speculate it may have something to do with the existence of the asphaltene micelles, but we do not know how to interpret this difference.

The most striking point that needs to be discussed here is the data for pH < 4 where α and β start to deviate. In fact, α decreases to nearly zero at pH $= 0.5$. We believe such a drastic change of α value signifies the onset of some chemical reactions [49,50]. The asphaltenes, after such reaction, become more active at the interface, thus, would stay at the interface instead of going back to the bulk phase. We therefore conclude that pH $= 4$ is a critical pH value for some chemical reactions to occur, above which only physical processes involved.

The scenario of the entire equilibrium process for the 0.1 % case may be described as follows. The asphaltene monomers and/or micelles diffuse to the interface. If the potential created by the interface is lower than the activation energy (the pH ≥ 4 cases), the asphaltene molecules adsorb to the interface, and saturate the interface (the adsorption rate controls the DIFT reduction rate). The residual molecules/micelles then diffuse back to the oil phase, making the DIFT to increase again. On the other hand, if the potential created is higher than the activation energy (the pH < 4 cases) the reactions are initiated. When the reactions are completed, the reaction products exhibit more hydrophilic-hydrophobic dual nature thus, remaining at the interface. This makes the DIFT stay at the minimum for a longer period of time (the lifetime, see Figure 35) until some un-reacted asphaltenes desorb from the interface and diffuse back to the oil phase (see the pH = 2 case in Figure 34). At this point the DIFT rises again. The asphaltene amount that diffuse back to the oil phase should be smaller than the forward process (many of the asphaltenes adsorbed to the interface have involved in reactions or simply have higher polarity). Thus, the desorption rate is lower (α is smaller than β). When pH further decreases to 0.5, nearly all the asphaltene molecules (or micelles) that diffuse to the interface are involved in the reactions. Therefore, the lifetime (see Figure 35) increases drastically and the α obtained is nearly zero.

The above qualitative interpretation is what we can achieve at this point. More accurate dynamic experiments may be needed in order to provide a more concrete conclusion.

VII. RHEOLOGY OF THE BASE TREATED ASPHALTENE MICELLAR SOLUTIONS

As we have mentioned earlier, the self-association of asphaltenes likely arises from the interaction between the charges carried by the heteroatoms. The DIFT experiments further indicate that the surface activity of asphaltenes may be modified via adsorbing asphaltenes to an aqueous phase of various pH values. Shiffert et al. [11] took another route to investigate the flocculation of heavy oil (certainly contains asphaltenes) and charges. These authors applied the electrophoretic technique to study the surface electrical charges and flocculation of heavy oil from the distillation residues in various solvents. The solvents selected serve as either electron donors or electron acceptors. The surface charge density predominates the stability of the asphaltene colloids, which stems from the transformation between the asphaltene colloids and the solvents. They found that the surface charges vary upon addition of electron donors or acceptors. This brings up an idea that if one treats asphaltenes solution by caustic solutions or the acidic solutions, the asphaltene charges may be varied. As a result, the rheological properties of the asphaltene micellar solution may be altered [51].

In the previous section, we discussed the asphaltene surface activity at the oil/acid interface. The main issue for the oil/acid interface is the reactions. It appears to occur when pH is sufficiently low. In the case of oil/caustic solution case, it is different. The two phases remain well separated, but the interfacial tension is lowered upon asphaltene adsorption. Thus, the issue here is how the asphaltene solution is affected by this two-phase equilibrium treatment. In this section we would present our rheological study for the based (NaOH) treated Ratawi asphaltene/toluene solution.

In this experiment, NaOH was used to provide negative charges. The NaOH treatment was done by preparing a caustic solution of known normality and an asphaltene solution (toluene as solvent). These two immiscible solutions were then mixed using high shear vortex mixer for one minute, followed by an overnight equilibrium. The resulting sample is a two-phase solution with well defined meniscus line. A this point, we took the oil phase and measured the low shear viscosity.

The viscosity was measured using the Brookfield viscometer with a 0-1000 x 10^{-3} Pas working range and a 0.1% accuracy of the full scale. Details of this viscometer has been

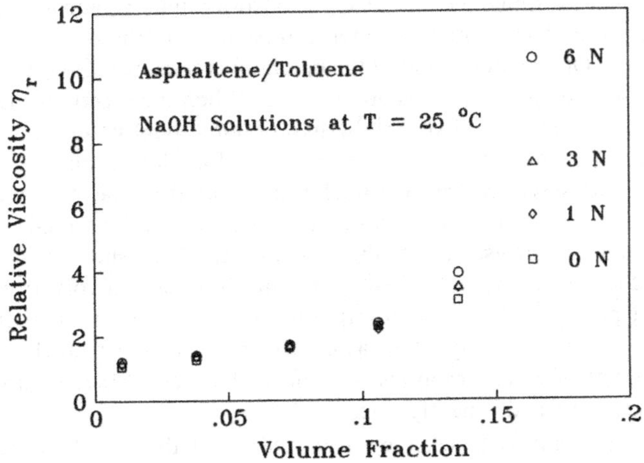

Figure 36. Relative viscosity for non-treated and treated samples as a function of volume fraction. Caustic solution apparently introduces a great impact on the viscosity through modification of the asphaltene micellar charges.

described in IV-A. Six asphaltene volume fractions were measured for each NaOH treated series and four caustic solutions were used (0 N, 1 N, 3 N, and 6 N). The measurements were performed at room temperature. Both NaOH treated and non-treated asphaltene/toluene solutions were measured for comparison.

As a general trend, the viscosity was found to increase after treatment (see Figure 36). This is an indication that the treated asphaltene colloids may exhibit stronger (and possibly longer range) inter-colloid interactions. In the following, we shall apply a series of available theories to interpret the data. Among them, the two fluid model [20] developed for an interacting system appear to be the best. With this model, we were able to quantify the inter-micellar interactions. The dependence of the inter-colloid potential as a function of NaOH normality (N) was also determined.

Since the asphaltene solutions behave similarly to a surfactant based system, their viscosity data may be analyzed using the theories developed for analyzing the viscosity data of colloidal dispersions, micellar solutions or microemulsion systems. As one can see from Figure 36, the Einstein equation is not applicable, due to the strong interactions. Clearly, the key issue here is to understand the impact of the charges carried by the asphaltene colloids after NaOH treatment. Thus, the effect of interaction on the rheological behavior will be the focal point in the analysis. In the following we will briefly describe three equations we used for this analysis.

The first equation used was the Dougherty and Krieger [52] equation which was proposed for a high volume fraction uncharged system,

$$\frac{\eta}{\eta_s} = \left[1-\left(\frac{\phi}{\phi_m}\right)\right]^{-[\eta]\phi_m} \tag{50}$$

where $[\eta]$ is the intrinsic viscosity,

$$[\eta] = \lim_{\phi \to 0} \frac{\frac{\eta}{\eta_s}-1}{\phi} \tag{51}$$

ϕ_m is again the maximum attainable volume fraction. For hexagonal or face center cubic (fcc) packing ϕ_m is 0.74, whilst for body center cubic (bcc) it is 0.68, and 0.52 for simple cubic packing. This equation, although given empirically, was derived later by Ball and Richmond [53] within the mean field argument and lately by Brady [54]. The ϕ/ϕ_m factor was created in the derivation as the scaled volume. This equation has been successfully applied to many dispersed systems. However, experimental works often find that the best $[\eta]$ to fit a hard sphere system is 2.0 to 2.1 instead of 2.5. This dilemma makes Eq.(2) an empirical equation more than a derived equation.

In addition to the Dougherty-Krieger equation, the Mooney equation was used. This was because the polydispersity was not explicitly included in the Dougherty-Krieger equation. Since our systems exhibit polydispersity [43], its effect needs to be taken into account explicitly. The Mooney equation reads [55],

$$\frac{\eta}{\eta_s} = \exp\left[\frac{[\eta]\phi}{1-\lambda(\frac{\phi}{\phi_m})}\right] \tag{52}$$

where λ is a measure of the polydispersity ($\lambda = 1$ for monodisperse systems). This equation not only takes into account the polydispersity, it is applicable for high volume fraction. Unfortunately, it is limited to uncharged systems, where no long range interactions are involved.

In order to take into account the interaction effect, we again adopted the Grimson and Barker equation (see Eq.(8)) and used Λ, ϕ_m, and n as the adjustable parameters. However, one has to note that the hydrodynamic interaction was not accounted for in Eq.(8). Thus, the accuracy of this analysis depends on the relative contributions of the inter-particle interactions and the hydrodynamic interaction. The other implicit assumption was that the particles are more or less spherical with the viscosity of the disperse phase much higher than that of the solvent (i.e, the particles are close to hard spheres).

Before we used these equation to analyze the viscosity data, we plotted the viscosity data according to the form Bedeaux suggested [56] so as to magnify the effect of inter-particle interaction which we believed to be the main contribution,

$$\frac{\eta_r - 1}{\eta_r + \frac{3}{2}} = \phi\ [1 + S(\phi)] \tag{53}$$

where $\eta_r = \eta/\eta_s$ is the relative viscosity, and $S(\phi)$ is the term account for the interactions. This representation is sensitive to the inter-particle interactions. Since $S(\phi)$ represents the interactions effect, one can plot $S(\phi)$ as a function of ϕ to see how the interaction varies. This representation will be used for characterization of the interactions using these three equations.

To analyze the viscosity data, we first applied the Pal and Rhodes equation as shown in Figure 37. For the non-treated case, a straight line was obtained indicating that the asphaltenes are hydrodynamically spherical-like. However, after the NaOH treatment, the deviation starts to show at both low and high volume fraction regimes. This may be attributed to the re-distribution of the asphaltene aggregates or to the change of the charges carried by the asphaltene aggregates. In order to understand what really happens to the asphaltene aggregates after NaOH treatment, we analyzed the data with the Dougherty-Krieger equation (Eq.(50)) and the Mooney equation (Eq.(52)). First, we assumed the particles remain spherical-like (assume the deviation from a straight line in the Pal and Rhodes plot (Figure 37) is due to inter-particle interactions). Then, we assumed the local structural arrangement

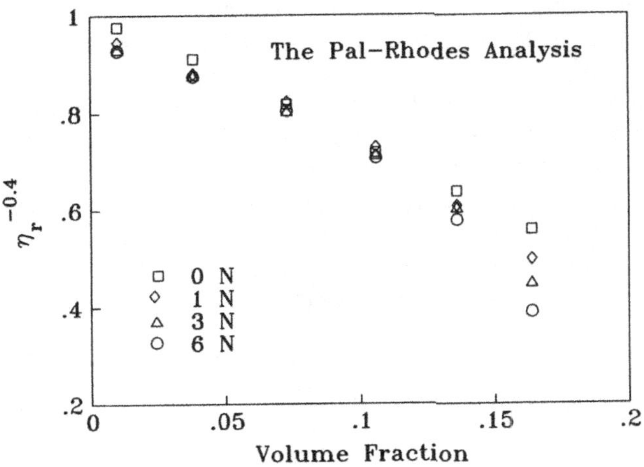

Figure 37. The Pal and Rhode analysis of Figure 36.

is according to the random packing (i.e, the maximum packing volume fraction is equal to 0.68). With these two assumptions, we fitted the Dougherty-Krieger analysis ([η] was used as an adjustable parameter) to the experimental data. [η] can not be obtained experimentally, because of the long range interaction, as we have shown in Pal and Rhodes analysis (see Figure 37).

To perform the Mooney analysis, we used the polydispersity parameter λ and [η] as adjustable parameters to fit the data. Figure 38a to 38d show both the Dougherty-Krieger and Mooney fittings. Because asphaltene aggregates are known to be polydisperse, the Dougherty-Krieger equation was not expected to give a good fit, as one can see from Figure 38. This is particularly true for base treated samples (see Figure 38b to 38d). At this point, one may question for keeping $\phi_m = 0.68$ in the Dougherty-Krieger analysis. We argue that ϕ_m should not be adjusted for different concentrations, because this equation does not take into account the inter-particle interactions, which may have significant effect on varying ϕ_m when concentration varies. We therefore selected 0.68 for ϕ_m. Later in the Grimson-Barker analysis, we will discuss the inter-particle interactions and how they change ϕ_m .

The quality of the fits for the Mooney analysis was better than the Dougherty-Krieger analysis, mainly because the polydispersity effect was taken into account. However, the Mooney equation does not account for the interactions. Thus, the fittings were not satisfactory, when inter-particle interactions are appreciable. In the following, we shall discuss the effect of interactions.

We evaluated the effect of interactions on both Dougherty-Krieger and Mooney analyses according to the representation suggested by Bedeaux [56]. Figure 39a to 39d show the results for all the samples investigated. The Dougherty-Krieger analysis, as expected, differs from the experiment substantially. The Mooney representation gives reasonable agreement with the experimental data for non-treated samples, as well as the high concentration cases of the treated sample. This is an interesting phenomenon, but is explainable. The scenario we like to propose is as follow: for non-treated samples, the interactions are similar to the hard sphere interaction. This is true, as one can see from the Pal and Rhodes analysis. In this case, the Mooney equation should be satisfactory, since it does take into account the polydispersity. Therefore, the agreement should also hold in the Bedeaux representation, even though it was originally for interactions purpose.

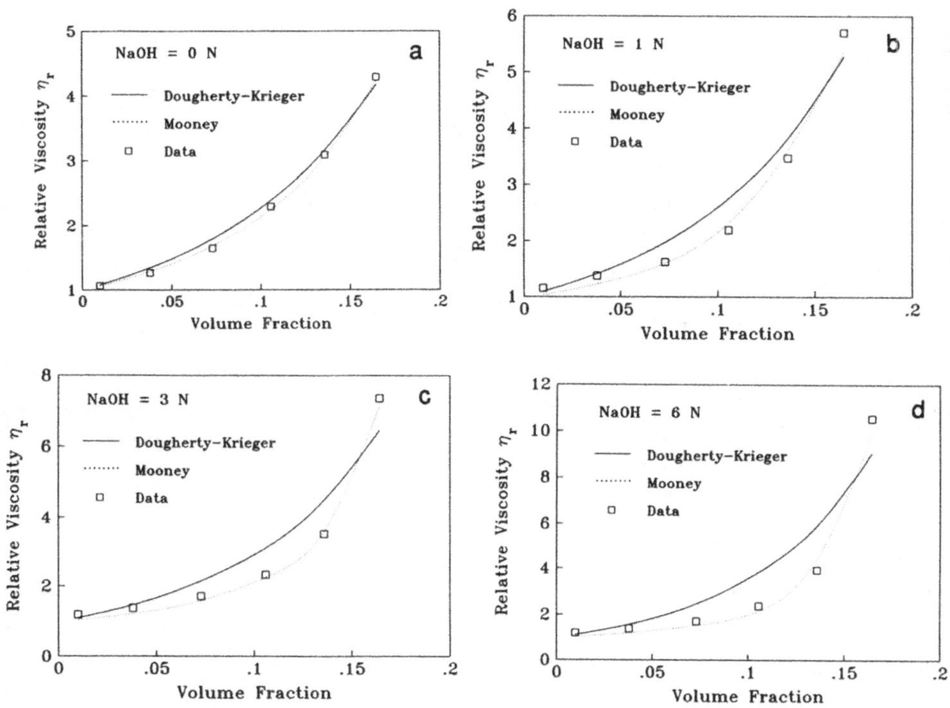

Figure 38. Dougherty-Krieger and Mooney analyses of the relative viscosity (Figure 36). Figure 38a to 38d correspond to the non-treated, 1N, 3N, and 6N treated series, respectively.

However, as the inter-particle interactions increase, the deviation increases (particularly for long range interactions). This interaction effect is suppressed when concentration is high, because of the screening effect. This is commonly observed in simple electrolytes or ionic micellar solutions. Thus, Mooney equation is expected to be reasonable again. Similar to the Dougherty-Krieger equation, Mooney equation does not account for inter-particle interactions, thus, one should not use ϕ_m as an adjustable parameter for data fitting.

In the Grimson-Barker's two fluid model analysis, the situation is different. It explicitly expresses the viscosity as a function of inter-particle interactions, for a given fairly generalized potential. Because the long range interaction potential is explicitly included in this equation, the particle packing is restricted to an "interaction boundary" which naturally reduces the maximum packing volume fraction ϕ_m. Thus, in using this equation for analysis, ϕ_m should be considered as a variable. Figure 40a to Figure 40d show the results of the analysis. The extracted parameter values are tabulated in Table V, together with the Dougherty-Krieger and the Mooney analyses.

The extracted parameter values strongly indicate that the interactions becomes longer and longer ranged, as NaOH normality increases (n decreases as normality increases). As the interaction becomes longer ranged, the "interaction boundary" is expected to become larger than the particle size, which in turn results in a smaller ϕ_m. The extracted ϕ_m's did show this trend. As for the amplitude factor Λ, it supposes to somewhat represent the effect of interaction on viscosity, but it is difficult to give a physical explanation here. This is because n and ϕ_m vary simultaneously. Thus, Λ

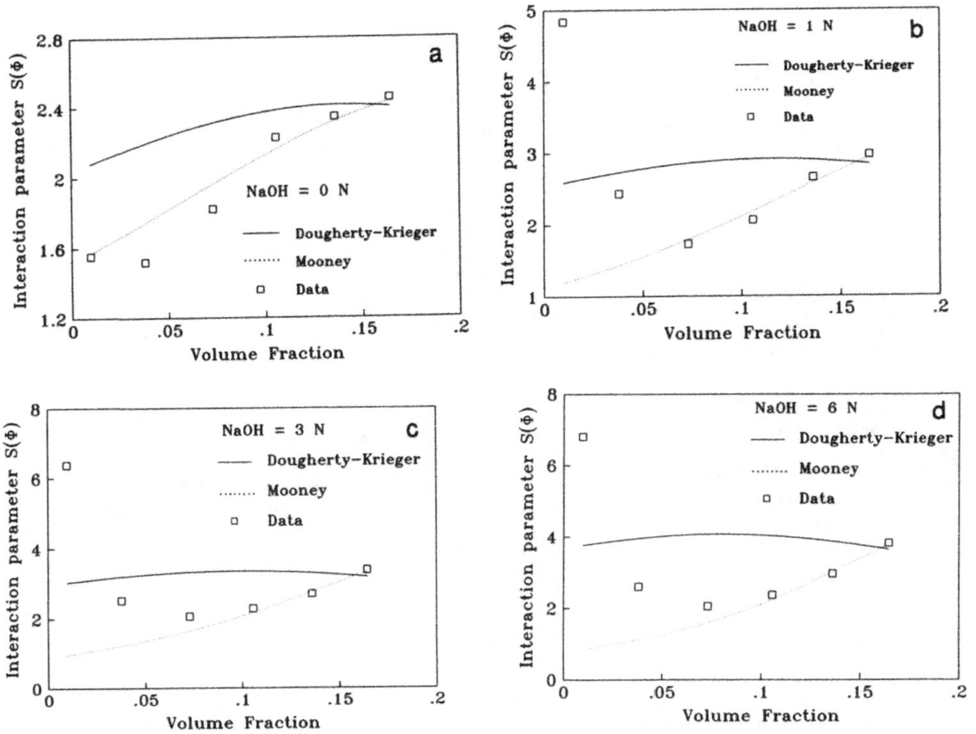

Figure 39. Analysis of $S(\phi)$. Figure 39a to 39d correspond to the same samples as in Figure 38 (see section VII for details).

should be taken as an accommodation factor for data fittings. This is different from the case discussed in IV-A where n remained nearly constant (~ 2.0). Figure 41a to Figure 41d show the $S(\phi)$ representation with both measured and calculated η_r (using the Grimson-Barker equation, Eq.(53). The agreement is generally good, except at low concentration regime, which is likely due to the polydispersity effect.

Table V. The extracted parameter values from three analyses

NaOH (N)	$[\eta]_{D\text{-}K}$	λ_{MNY}	$[\eta]_{MNY}$	$\Lambda_{G\text{-}B}$	$\phi_{m\text{-}B\text{-}K}$	$n_{B\text{-}K}$
0	7.6	1.2	6.3	0.61	0.46	2.1
1	8.8	2.0	5.3	0.29	0.36	1.8
3	9.9	2.5	4.7	0.17	0.20	1.3
6	11.7	2.8	4.4	0.13	0.19	1.2

$[\eta]_{D\text{-}K}$ = intrinsic viscosity extracted from the Dougherty-Krieger analysis.
λ_{MNY} = Mooney's polydispersity parameter (see Eq.(52)).
$[\eta]_{MNY}$ = intrinsic viscosity extracted from the Mooney analysis.
$\Lambda_{G\text{-}BB}$ = Coupling constant of the Grimson-Barker equation (see Eq.(8)).
$\phi_{m\text{-}B\text{-}K}$ = Maximum packing volume fraction extracted from the Grimson-Barker analysis.
$n_{B\text{-}K}$ = the exponent values of the Grimson-Barker potential extracted from data fittings.

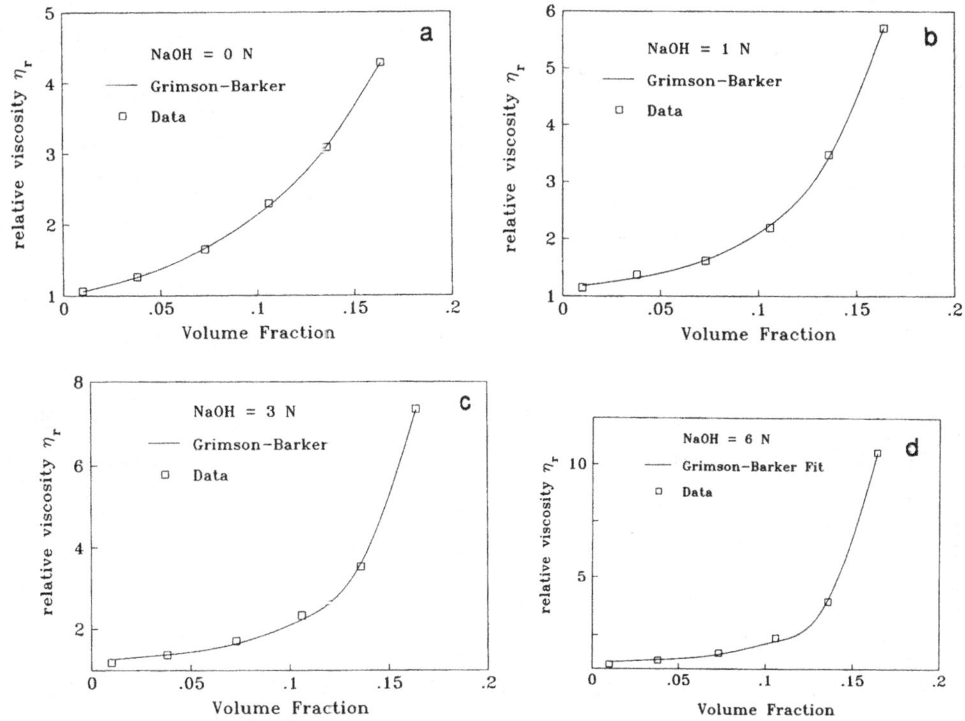

Figure 40. The Grimson-Barker analysis. Figure 40a to 40d correspond to the same samples described in Figure 38.

Although Grimson-Barker equation satisfactorily explains the data. One question remains unanswered. For most of the hard sphere interaction systems, the $S(\phi)$ shows a concave down curve (see ref. 55, Figure 8), while in our case, the curves are concave up for ϕ up to ~ 0.17. Is it a typical behavior for a long range interacting system ?

VIII. DISCUSSION AND FUTURE PERSPECTIVES

Asphaltenes have drawn attention to petroleum scientists for nearly a century. From the works that have been done, we do thank the wisdom of the researchers from the early 20th century who hypothesized such a minute microscopic model for such a complex material. In particular, the colloidal concept of Nellensteyn from which Pfeiffer and Saal proposed the peptized asphaltene micellar model. Through the state of the art equipment we have today, we have gained much more knowledge about this mysterious material. Most importantly, we understand that asphaltenes exhibit similar physical properties to the surfactant systems, which lays a good ground for its potential applications.

The future perspective lies in the control of the asphaltene properties to suit its industrial needs. For example, if the asphaltene surface activity can be manipulated and controlled, one may be able to use it as an *in situ* natural surfactant to replace the added surfactants in many petroleum processing. A foreseeable obstacle is to control the asphaltene surface activity, because its self-association propensity is not originated from the hydropho-

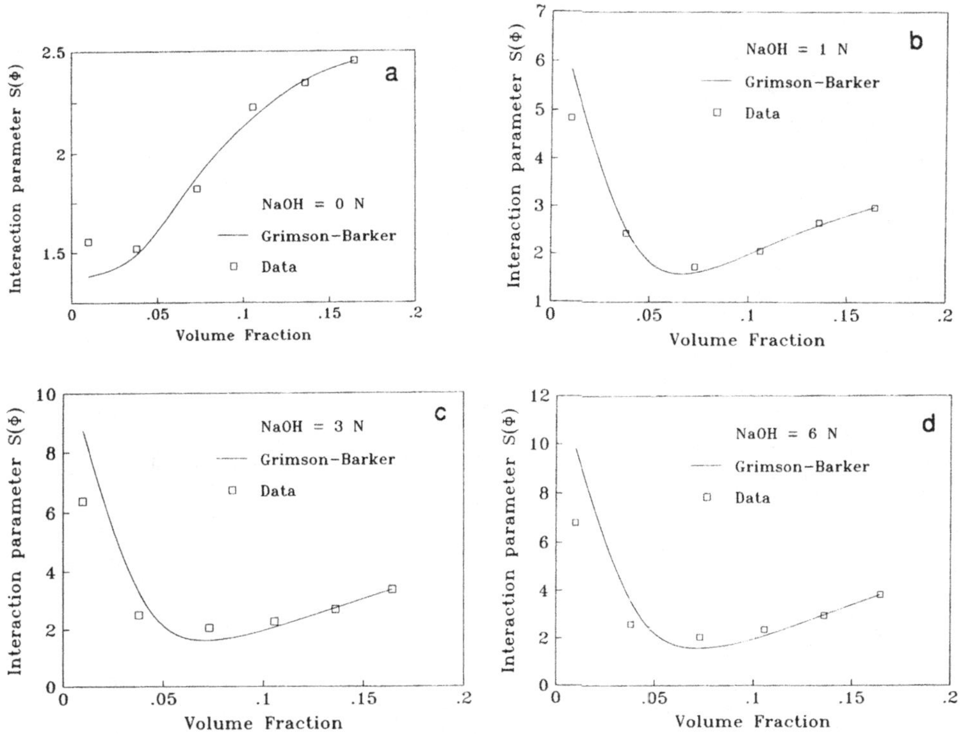

Figure 41. Same plot as Figure 39 using Grimson-Barker equation for analysis.

bicity-hydrophilicity imbalance. This fundamental difference makes further research necessary, in order to better manipulate the asphaltene properties for industrial applications.

IX. ACKNOWLEDGMENT

The authors like to thank M. M. De Tar and M. B. Shields for their assistance in measuring the data presented here. The SANS works have benefited from the use of the Intense Pulsed Neutron Source (IPNS) at Argonne National Laboratory (ANL) and at National Institute of Standards and Technology (NIST). The IPNS is operated under the auspices of the US Department of Energy, BES-Materials Science, under Contract W-31-109-Eng-38, to whom thanks are extended for the use of their facilities. We also acknowledge the support of the National Institute of Standards and Technology, U.S Department of Commerce, in providing the facilities used in this experiment.

X. REFERENCES

1. J. P. Pfeiffer, *The Properties of Asphaltic Bitumen*. Elsevier, Amsterdam, 285 pp. (1950); J. G. Speight, *The chemistry and Technology of Petroleum*. Marcel Dekker, New York, (1980); J. G. Speight, *Fuel Science and Technology Handbook*. Marcel Dekker, New York 1193 pp (1990).
2. T. F. Yen, *Am. Chem. Soc., Div. Petrol. Chem. Preprint*, ***17(1)***: 102-104 (1972); T.F. Yen, *Energy Source*, **1(4)**: 447-463 (1974); T.F. Yen, The role of asphaltene in heavy crude and tar sands. In: R.F. Meyer and

C.T. Steele (Editors), *The Future of Heavy Crude and Tar Sands*, McGraw-Hill, New York, pp 174-179 (1980); J.W. Bunger and N.C. Li, *Chemistry of Asphaltenes*. Advances in Chemistry Series 195. American Chemical Society, Washington D.C. (1981); T.F. Yen and G.V. Chilingarian (editor) *Asphaltenes and Asphalts, 1*, Elservier, Amstercam (1994); M.K. Sharma and T.F. Yen (editors) *Asphaltene Particles in Fossil Fuel Exploration, Recovery, Refining, and Production Processes*, Plenum Press, New York (1994).
3. F.J. Nellensteyn, *Chem. Weekblad, 28*, 313 (1931); F. J. Nellensteyn and N. M. Roodenburg, *Chem.-Zeiyung*, **545**, 819 (1930).
4. C. Mack, *Phys. Chem.*, **36**, 2901 (1932).
5. J. P. Pfeiffer and R. N. J. Saal, *J. Phys. Chem.*, **44**, 139 (1940).
6. M. J. Rosen, *Surfactants and Interfacial Phenomena*, 2nd ed., John Wiley & Sons, New York (1988).
7. E. Y. Sheu, M. M. De Tar, D. A. Storm, and S. J. DeCanio, *Fuel*, **71**, 299 (1992).
8. S. I. Anderson and K. S. Birdi, *J. Coll. Int. Sci.*, **142**, 497 (1991).
9. J.P. Dickie and T.F. Yen, *Anal. Chem.*, **39**(14) 1487-1852 (1972); T.F. Yen, *Adv. Chem. Ser.*, **195:** 39-51 (1981).
10. Marusk H. P., and Rao, B. M. L., *Fuel Sci. & Tech. Int.*, **5(2)** 119 (1987).
11. B. Shiffert, J. Kuczinski, and E J. Papirer, *J. Coll. Int. Sci.*, **135**, 107 (1990).
12. E.Y. Sheu, M.M. De Tar and D.A. Storm, Surface activity and dynamics of asphaltenes, In: M.K. Sharma and T.F. Yen (editors) *Asphaltene Particles in Fossil Fuel Exploration, Recovery, Refining, and Production Processes*, 115 pp, Plenum Press, New York (1994).
13. S.E. Taylor, *Fuel*, **71**, 1338 (1992).
14. E.Y. Sheu and D.A. Storm, *Fuel*, **73**, 1368 (1994).
15. E.Y. Sheu, M.M. De Tar and D.A. Storm, *Fuel 70*, 1151 (1991); E.Y. Sheu, M.M. De Tar, and D.A. Storm, *Fuel Sci. Technolo. Int.*, **10(4-6)**, 607-647 (1992).
16. J. Hunt, unpublished data (private communication).
17. R. Pal and E. Rhodes, *J. Rheolcgy, 33*, 1021 (1989).
18. H. Eiler, *Kolloid-Z Z. Polym.*, **97**, 313 (1941).
19. G.A. Campbell and G. Forgacs, *Phys. Rev. A*, **41**, 8 (1990).
20. M .J. Grimson and G. C. Barker, *Europhys. Lett.*, **3**, 511 (1987).
21. S.P. Das and G.F. Masenko, *Phys. Rev. Lett.*, **54**, 118 (1985).
22. E. Y. Sheu, *Phys. Rev. A.*, **45**, 2428 (1992).
23. L. A. Feigin and D. I. Svergun, *Structure Analysis By Small Angle X-ray and Neutron Scattering*, Plenum Press, New York (1987).
24. L. S. Ornstein and F. Zernike, *Proc. Akad. Sci.*, **17**, 793 (1914).
25. J. K. Percus and G. J. Yeciv, *Phys. Rev.*, **110**, 1 (1958).
26. J. B. Hayter and J. Penfold, *J. Chem. Soc. Faraday Trans. I*, **77**, 1851 (1981).
27. R. J. Baxter, *J. Chem. Phys.*, **52**, 4559 (1970).
28. L. Blum and J. S. Høye, *J. Phys. Chem.*, **81**, 1311 (1977).
29. E.Y. Sheu, K.S. Liang, S.K. Sinha, and R.E. Overfield, *J. Coll. Int. Sci.*, **153**, 399 (1992).
30. S. H. Chen and J. Teixeira, *Phys. Rev. Lett.*, **57**, 2583 (1985).
31. D. Stauffer, in *On Growth and Form*, edited by H. E. Stanley and N. Ostrowsky, Martinus Nijhoff Publisher, New York (1986).
32. H. E. Stanley and N. Ostrowsky eds, *On Growth and Form*, Martinus Nijhoff Publisher, New York (1986).
33. D. P. Landau and F. Family, Eds., *Kinetics of Aggregation and Gelation*, North Holland, Amsterdam, (1984).
34. R. Pynn and A. Skeltorp, Eds. *Scaling Phenomena in Disordered System*, Plenum New York (1986).
35. J. Feder, *Fractal*, Plenum, New York (1988).
36. S. H. Chen, J. Rouch and P. Tartaglia, *Croatica Chemica Acta*, **65(2)** 353 (1992).
37. P. Raghunathan, *Fractal dimension in polymeric amorphous materials determined by electron spin relaxation measurements: A study of asphaltene polymers*. In. C. L. Khetrapal and G. Govil (Editors), *Magnetic Resonance*, page 324-332, Narosa Publishing House, New Delhi, India (1991);P. Raghunathan, *Chem. Phys. Lett.*, **182,** 331 (1991).
38. P. Ekwall, In. *Advances in Liquid Crystal;* G. H. Brown, Ed., Academic Press, New York 1, (1975).
39. M. Borkovec, *J. Chem. Phys.*, **91(19)** 6268 (1989).
40. A. S. Janardhan and G. Mansoori, *J. Petrol. Sci. Eng.*, **9,** 17 (1993).
41. E.Y. Sheu, M.M. De Tar, and D.A. Storm, *Fuel*, **73**, 45 (1994).
42. Cole, K.S., Cole, R.H., *J. Chem Phys.*, **9** 341 (1941)
43. E.Y. Sheu, D.A. Storm and M.M. De Tar, *J. Non-crystal. Sloids*, 131-133, **347** (1991).
44. Battacharya, S., Stokes, J.P., Kim, M.W., and Huang, J.S., *Phys. Rev. Let.*, **55** (1985) 1884.
45. Ponton, A., Bose, T. K., and Delbos, G., J. Chem. Phys., 94 (1991) 6879

46. Storm, D. A., "Temperature Dependent Rheological Study of Vacuum Residue" presented at the Peterson Asphalt Research Conference, the 13th Annual Meeting, Laramie, Wyoming, July 12-14, 1993
47. Vogler, E.A., *J. Coll. Int. Sci.*, **133**, 228 (1989).
48. E.M. Trujillo, *Soc. Petro. Eng. AIME*, **645**, Aug. (1983).
49. H.A. Nasr-El-Din, B.F. Hawkins and K.A. Green, *Preproint, Int. Symp. on Oilfield and Geothermal Chemistry*, Feb. 20-22, Anaheim, CA., paper SPE 21028 (1991).
50. H.A. Nasr-El-Din and K.C. Taylor, *Coll. Surfaces*, **66**, 23 (1990).
51. E.Y. Sheu, M.B. Shields, and D.A. Storm, *Fuel*, **73**, 1766 (1994).
52. I.M. Krieger and T.J. Dougherty, *Trans. Soc. Rheology* **3**, 137-152 (1959).
53. R. C. Ball and P. Richmon, *Phys. Chem. Liq.*, **9**, 99 (1980).
54. J. Brady, *J. Chem Phys.*, **99**, 569 (1993).
55. M. Mooney, *J. Coll. Int. Sci.*, **6**, 162 (1951).
56. D. Bedeaux, *J. Coll. Int. Sci.*, **118,** 80 (1987).

Chapter II

SULFUR AND NITROGEN MOLECULAR STRUCTURES IN ASPHALTENES AND RELATED MATERIALS QUANTIFIED BY XANES SPECTROSCOPY

Oliver C. Mullins

Schlumberger-Doll Research
Ridgefield, Connecticut 06877

INTRODUCTION

Asphaltenes are an important class of natural materials with many distinctive properties.[1-4] Asphaltenes strongly affect the chemical and physical properties of petroleum, and, as such, have a significant impact on the production, transportation and refining of petroleum. The viscosity and adhesion of petroleum markedly increase with greater asphaltene content, causing significant difficulties in the processing of petroleum. The asphaltene fraction is the least valuable component of crude oil, and its relatively high resistance to cracking further decreases the yield of valuable distillate fractions such as gasoline. The degradation of catalysts is accelerated by the presence of asphaltenes due to both their heavy metal content and their tarry nature when combined with resins. Also, asphalts represent a hazard when released (unintentionally) into the environment due to their resistance to biodegradation and their tenacious adhesive powers.

Many of the properties which make asphaltene an undesired component of crude oil, add considerable value for particular purposes. Asphalt is the material of choice for highway surfaces; obviously, asphaltenes are a critical component, albeit somewhat chemically altered from their nascent form. Here, bioinertness, rheological toughness, and strong adhesion are mandatory. These properties make asphalts ideally suited to a variety of protective coating applications where exposure to weather extremes is the norm. Petroleum asphaltenes also find use in oil-well drilling fluids for fluid-loss control and wellbore-stabilization, paving the way for the production of more crude oil. Thus, although asphalt and asphaltenes are the least desired components of crude oil, they remain economically important and are widely used.

Asphaltenes often elude strict attempts at definition and classification.[1-3] The term asphaltene is generally used to convey chemical identity; yet, asphaltenes are actually defined in terms of their solubility. Some variation exists in the exact solubility definition leading to somewhat disparate reported properties of asphaltenes. Generally, asphaltenes are

Asphaltenes: Fundamentals and Applications, Edited by E. Y. Sheu and O. C. Mullins
Plenum Press, New York, 1995

defined to be insoluble in a particular alkane (e.g. n-heptane) and soluble in a particular nonpolar, polarizable solvent (e.g. carbon tetrachloride). In addition, different source materials, such as petroleum and coal tars, are used to generate asphaltenes, thus producing understandably significant variation in the chemical properties in the family of asphaltenes. In this chapter, the term asphaltenes refers to petroleum asphaltenes unless otherwise specifically stated.

In spite of the classification of asphaltenes in terms of their solubility, their chemical properties are generally fairly uniform, especially if only one source material is considered. Petroleum asphaltenes are predominantly composed of hydrogen and carbon, typically greater than 90 atomic-%. ^{13}C nuclear magnetic resonance (NMR) and x-ray studies show that the carbon distribution is typically about 40% aromatic and the remainder, saturated.[2] However, the hydrogen is found primarily (>90%) on saturated carbons, predominantly methylene groups, as determined by infrared (IR) and NMR spectroscopy.[2] X-ray diffraction studies indicate that asphaltenes possess a few stacked layers of condensed aromatic ring systems;[1] ^{13}C NMR and IR results indicate that these ring systems contain alicyclic and alkyl substituents.[2] These spectroscopic results can be analyzed further yielding a detailed picture of the distribution of carbon and hydrogen in asphaltenes. For instance, the fraction of aromatic carbons on the periphery and the interior of the fused ring systems can be estimated.[2,3] By combining results from a variety of methods, chemical structures are suggested which are expected to be close to actual asphaltene molecular structures for the distribution of carbon and hydrogen.[2,3] However, the quantification of moieties containing heteroatoms is more problematic.

Asphaltenes generally contain small quantities of heteroatoms, amounting to a few atomic percent, although the exact quantities can vary appreciably.[1-4] The significance of the heteroatom content of asphaltenes greatly outweighs their mass fractions. The heteroatoms help determine the identity of asphaltenes by strongly affecting their solubility characteristics, hence their classification. The heteroatomic chemical functions are often polar and sometimes charged yielding strong intra- and intermolecular interactions. The acid-base chemistry of asphaltenes is mediated through heteroatomic sites, and the small but significant metallic fraction of asphaltenes is complexed by heteroatoms.[5] Some of the more reactive sites of asphaltenes are associated with the heteroatoms. Thus, understanding the heteroatom chemistry is an essential component of asphaltene characterization.

The identification, but not quantification, of the chemical moieties containing oxygen has been somewhat accessible through the use of IR. In spite of the small oxygen concentrations, the relatively strong and isolated vibrational bands of some oxygen-containing functional groups permit their identification. The carbonyl group is readily identified in the IR spectra of asphaltenes. Asphaltenes show IR absorption where OH and NH produce characteristic signatures but are not resolvable from each other.[1,4] Some evidence of ether groups is also found. As will be discussed, x-ray methods have been used to identify sulfoxide groups in some asphaltenes. A substantial difficulty with the analysis of oxygen is atmospheric oxidation. In order to determine the original oxygen-containing moieties, the asphaltene and its corresponding crude oil would have to be kept strictly anoxic. This difficulty coupled with the small quantities of oxygen in asphaltenes introduce uncertainty and impeded quantification of oxygen functional groups in asphaltenes.

The investigation of the sulfur and especially nitrogen chemistry of asphaltenes has been problematic.[2] Difficulties are associated with applying spectroscopic techniques which are not element specific to the investigation of elements present in small percentages. With these methods, identification of some of the functional groups may be accomplished, but quantification of all functional groups for a trace element is prohibitive. The application of separation techniques such as chromatographic methods to asphaltenes is often hindered by their large molecular weights (~1000 amu) and many, diverse types of molecules. High resolution separation methods generate a huge number of components which cannot be individually analyzed, or quantified. Even gross separations of asphaltenes (e.g. basic and nonbasic nitrogen) yield ambiguous results. Chemically reactive methods are fraught with uncertainties for complex, high molecular weight mixtures such as asphaltenes and the analysis of their heteroatom content is only exacerbated by their small mass fractions.

The quantitative determinations of sulfur and nitrogen chemical moieties in complex materials have recently been addressed through the use of x-ray methods. In the past ten years, the development of x-ray methods, particularly X-ray Absorption Near Edge Structure (XANES) spectroscopy, has allowed for detailed chemical analyses of complex materials such as asphaltenes. XANES is element specific; by using x-rays over a particular energy range, only the element of interest provides significant signal. Moderately low elemental concentrations (10ppt - 100ppm, depending on the specific case) are compatible with XANES measurements. The technique is largely nondestructive, and no separation is required, the whole sample is studied and all chemical forms of the element of interest contribute to the signal, unlike some types of spectroscopy. XANES chemical analyses rely on different chemical groups of an element yielding different spectroscopic resonances. To a large degree, both sulfur and nitrogen compounds of interest to fossil fuel materials produce resolvable spectroscopic signatures. Thus, XANES spectroscopy provides a powerful method of heteroatom analysis of asphaltenes and related organic materials.

In this chapter, we describe the application of XANES to the study of sulfur and nitrogen molecular structures in asphaltenes and related materials. First, processes associated with x-ray absorption are summarized. The broad utility of x-ray spectroscopy becomes evident, especially when combined with synchrotron facilities which continue to evolve towards higher performance. The sulfur XANES spectra of several simple compounds are shown to illustrate systematic behavior. Utilization of these systematics has been very fruitful in the study of asphaltenes as well as coals. More recently, high resolution nitrogen XANES studies have been performed on various compounds. These results are briefly reviewed with emphasis on the different types of nitrogen-containing aromatic structures. Again, systematics are found in nitrogen XANES spectroscopy which have been applied to asphaltenes as well as kerogens, bitumens and coals. These XANES results are compared to those from a related technique, x-ray photoelectron spectroscopy (XPS); general agreement is found but specific differences are noted. Finally, a summary of the XANES results in the study of sulfur and nitrogen molecular structures in asphaltenes illustrates the power of this method.

X-RAY ABSORPTION AND XANES SPECTROSCOPY

X-ray absorption corresponds to the excitation of inner shell electrons of atoms to unoccupied or partially filled states.[6-8] When the x-ray energy exceeds the binding energy of an inner shell electron, then direct photoionization can result. The energy required to ionize the K-shell (1s) electrons of atoms varies with atomic number Z as Z^2 in accordance with the Rydberg equation (with shielding corrections). X-rays are, of course, sufficiently energetic to ionize valence shell electrons; however, the cross section for this process is small. The overlap integral between the initial state and the continuum electron final state is small if the two states have grossly different kinetic energies. For energies much above the ionization threshold, the photoionization cross section decreases rapidly with increasing photon energy. Substantial oscillator strength for either resonant transitions or direct ionization involving core electrons occurs near the ionization threshold; with the bound and continuum resonances generally being within ~30 eV of the ionization threshold. XANES corresponds to the study of x-ray spectra near the ionization threshold.

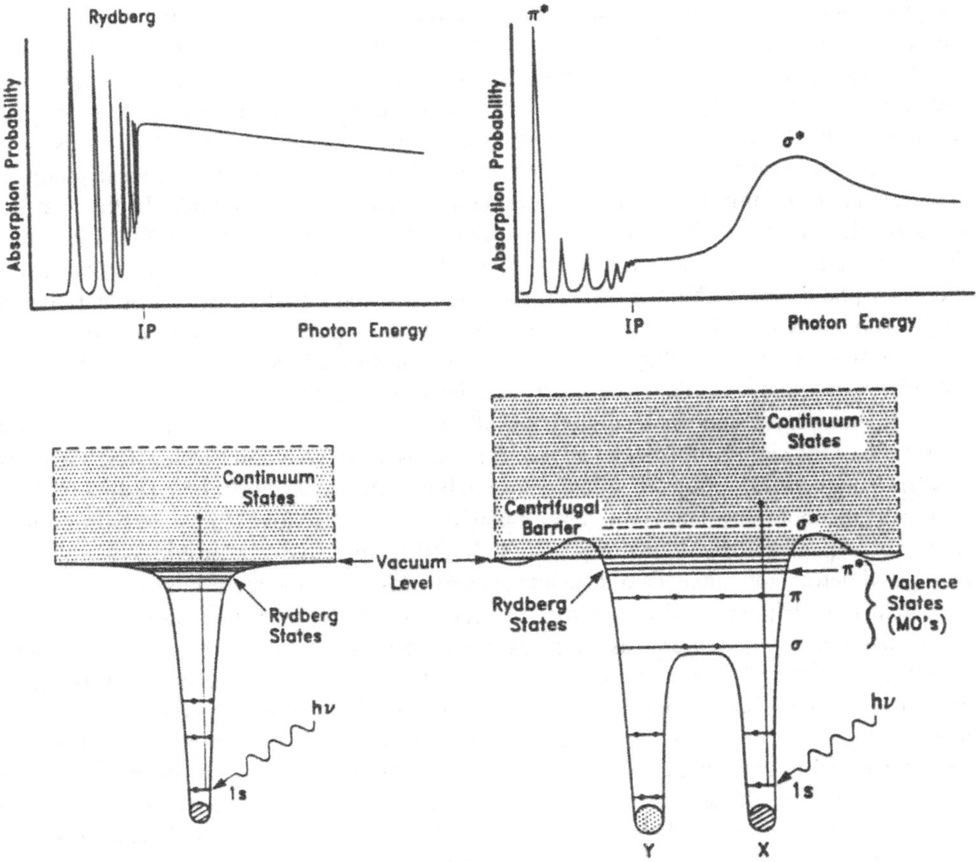

Figure 1. Schematic potentials (bottom) and K-shell x-ray absorption spectra (top) of an atom (left) and a diatomic molecule (right) (reproduced from reference 8). The atomic spectrum consists of the excitation of a Rydberg series below the ionization potential (IP) and the excitation of the structureless continuum above the IP. Molecules exhibit these excitations and others as well. The excitation of an unoccupied valence orbital, the π* state, is often predominant as shown. The excitation of a σ* shape resonance in the continuum is also shown. This resonance is a consequence of the modification of the continuum wavefunctions by the centrifugal barrier.

Figure 1 shows a schematic of the K-edge x-ray absorption spectrum for an individual atom and for a diatomic molecule with both σ and π bonding. In the x-ray spectrum of the atom, excitation of a Rydberg series is depicted; the cross section for higher Rydberg states decreases corresponding to their smaller overlap with the core 1s state. At sufficiently high energy, the 1s electron is ejected directly into the continuum and the absorption probability shows a step behavior, of course, at the ionization potential (IP). For the diatomic molecule, excitation of a Rydberg series and the continuum occurs as in the atomic case. Molecules also show additional transitions. The excitation of a π* state is often very prominent as shown. Excitation of a σ* shape resonance in the continuum is also shown. This resonance is a consequence of the modification of the continuum wavefunctions by the centrifugal barrier.

The number, energy and magnitude of XANES resonances depend explicitly on the chemical identity of the molecule. Consequently, x-ray spectroscopy can be used to investigate chemical bonding and structure by analysis of resonant energies and intensities. Furthermore, chemical groups, for example, pyridines, often show similar and distinctive

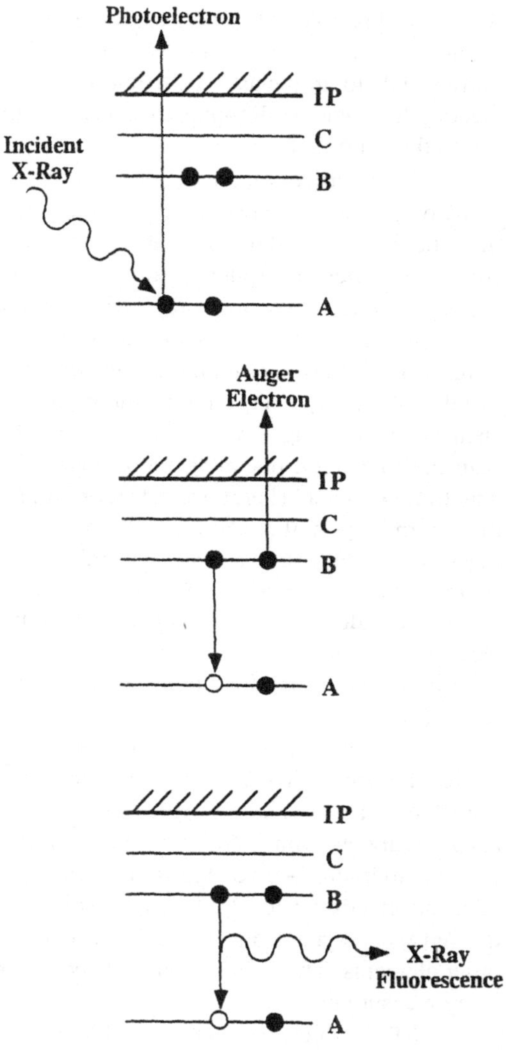

Figure 2. Schematic representation of x-ray absorption and relaxation processes. The incident x-ray creates an electron vacancy in state A, shown here by direct ionization (as opposed to the excitation of the autoionizing state C). The vacancy is filled by an electron dropping into state A from state B. The released energy is carried away either by emission of an Auger electron or by an x-ray fluorescence photon.

XANES spectra. Thus, samples containing unknown chemical groups can be analyzed using XANES in terms of an appropriate set of model compounds, much in the way IR spectroscopy is used for chemical group identification in samples. A significant strength of x-ray methods is the ability to study individual elements. Inner shell x-ray ionization edges of elements within a sample are often well separated. In this case, all resonances near an inner shell edge of a particular element correspond to that element. Often, the only spectral interference encountered from other elements of the sample is a nearly flat or linear background ionization signal. Because of this specificity, x-ray measurements have high sensitivity, and can be applied to dilute systems.

Fig. 2 depicts the processes which occur with absorption of an incident x-ray and subsequent relaxation of the atom. In fig. 2, direct photoionization of orbital A is shown where the incident x-ray energy is greater than the ionization potential (IP) for state A. For smaller incident x-ray energies, the transition from A to the empty orbital C can take place yielding a resonance below the IP (not depicted). The x-ray absorption event produces a hole in level A depicted by an open circle in fig. 2. An electron from a higher energy level, B, can drop into this hole. The energy released in this process is carried off either by an Auger electron or by a x-ray photon, which is x-ray fluorescence. The Auger process essentially corresponds to an inelastic collision of two electrons, both in level B in fig. 2, where one electron loses energy dropping to level A and the other electron acquires this energy, exciting it into the continuum.

These processes are the basis of the detection of x-ray absorption. In analogy with many optical spectrometers, x-ray spectra can be determined by measuring the reduction of the x-ray beam flux through a sample as a function of x-ray energy. This reduction in beam flux corresponds to the photoabsorption process in fig. 2. However, in some cases, the sample has a stopping power for x-rays which is too high to allow useful measurement of the transmitted x-ray intensity. X-ray absorption can also be monitored by measuring the induced ionization. The primary ionizing event depicted in fig. 2 yields a photoelectron with energy equal to the x-ray energy minus the binding energy. For excitation of autoionizing states such as state C in fig. 2, there is no primary photoelectron. For x-ray energies slightly larger than the ionization potential, the primary photoelectron has very little energy, thus little penetrating power. Therefore, detection of only the primary photoelectron is not useful for the determination of XANES spectra of bulk samples. In addition to the primary ionization event, x-ray absorption spawns the emission of electrons from the Auger process, depicted in fig. 2. Auger electrons from inner shell transitions are generally fairly energetic and cause substantial additional ionization in the sample. These secondary electrons dominate the electron emission from the sample resulting from x-ray absorption. Detection of x-ray absorption by measuring total electron emission or "electron yield" is a widely used technique.

X-ray absorption can also be monitored by detecting the resulting x-ray fluorescence, depicted in fig. 2. X-ray fluorescence competes with Auger emission in the deexcitation of the atom with a core hole. Auger emission is favored over x-ray fluorescence for low Z and higher-lying electron shells. The energy of the fluorescence x-ray photon is somewhat less than the ionizing x-ray as the fluorescence process generally involves transitions between adjacent electronic levels. Energy-resolved x-ray detection allows differentiation of the desired x-ray signal from elastically scattered x-rays as well as from characteristic x-rays of other elements. Thus, fluorescence detection increases the specificity and sensitivity of the x-ray measurement.

XPS, which is related to XANES, utilizes information carried by the primary photoelectron. X-rays with energy exceeding the ionization potential are used producing primary photoelectrons of sufficient energy to escape the sample. The binding energy of the electrons is dependent on the chemical environment of the element; by monitoring the energy of the emitted primary photoelectrons, chemical information is acquired. Primary photoelectrons which have suffered no inelastic collisions must be detected. Photoelectrons have a very short scattering length in condensed phase; thus, XPS is essentially a surface investigative technique.

Experimental constraints can preclude the use of certain methods in the determination of the absorbance, A. The energy dependent absorption coefficient ε for a particular element of concentration c can be determined by measuring I_0 and I, the incident and transmitted intensities of light, for a sample of pathlength l.

$$-\mathrm{Log}(I/I_0) = \varepsilon\, c\, l = A \tag{1}$$

The absorbance A is often plotted vs energy and is linear in the absorption coefficient. For samples of sufficiently low absorption, Eq. 1 can be expanded and the absorbed intensity, $I_0 - I = \Delta I$, becomes linear with absorbance.

$$\Delta I/I_o \propto A \tag{2}$$

A difficulty in measuring the transmitted x-ray intensity to obtain A in Eq. 1 is that the sample is presumed to be uniform in cross sectional area such that different x-rays penetrating the sample intersect the same thickness and composition of materials; otherwise a complicated sum is required to relate A with ε. For solid samples, such uniformity can be difficult to achieve, especially considering that the penetration length particularly for low energy x-rays is very short. Also, for thick samples, the transmitted flux falls to very low values, thereby hindering the accurate determination of the absorbance from the transmitted intensity.

The determination of the absorbance by fluorescence detection or electron yield detection can eliminate difficulties associated with measuring the transmitted intensity, I. In x-ray spectroscopy, one must account for the absorption by the element in question as well as (structureless) absorption of the other elements of the sample. We consider the detection of x-ray fluorescence from the selected element in the sample of thickness t. The assumed geometry is where the fluorescence x-ray travels the same path, but opposite direction, as the incident excitation x-ray. One obtains the intensity of fluorescence, I_F, in terms of the energy-dependent absorption cross section of the selected element σ_s, and the absorption cross section of the background material, σ_B.[9]

$$I_F \propto \frac{\phi I_o \sigma_s(E_i)}{\sigma_s(E_i) + \sigma_B(E_i) + \sigma_B(E_F)} \times \left(1 - \exp\left\{-[\sigma_s(E_i) + \sigma_B(E_i) + \sigma_B(E_F)]t\right\}\right) \tag{3}$$

ϕ is the quantum yield for x-ray fluorescence. The energy of the incident x-rays is E_i, and the fluorescence x-rays, E_F. For very thin samples, this expression reduces to the linear approximation shown in Eq. 2 (where $\varepsilon\, c = \sigma \log(e)$).

$$I_F \propto \phi\, I_o\, \sigma_s(E_i) \tag{4}$$

In addition, for very thick samples, which are dilute in the selected element, $\sigma_B(E_i) + \sigma_B(E_F) >> \sigma_s(E_i)$, thus, Eq. 4 also applies.[9] This important result allows $\sigma_s(E_i)$ and the absorbance to be extracted directly and linearly from spectra collected from thick samples which are dilute in the selected element. Sample uniformity is no longer a requirement. This approximation follows because the background absorption is so strong that the incident x-ray beam only penetrates a thin section of the sample, thereby yielding the thin sample approximation. A similar analysis applies to the detection of the ionization signal resulting from Auger electron emission. In fact, for the emitted electron, the absorption cross section, σ_B, is much larger than for the emitted x-ray, so the linear approximation works for higher concentrations of the selected element.

XANES spectroscopy is widely used to relate spectra with electronic, physical or chemical properties of pure compounds.[7,8] High resolution x-ray spectra can be compared with theoretical calculations providing a detailed probe of atomic and molecular structure.[10] Individual resonances can be assigned by the comparison of theory and experiment. We consider a sulfur example because of the central role of sulfur XANES in this chapter. The complex spectrum of SF_6, fig. 3, has been the object of extensive theoretical effort using a variety of theoretical formalisms.[11] The large resonance, B in fig. 3, is always assigned to the 1s-3p(t_{1u}) transition of sulfur, based both on the position and magnitude of the resonance. Resonances marked A and C have been assigned to forbidden transitions 1s-3s(a_{1g}) and 1s-3d(t_{2g}) respectively, and occur through a vibronic assist.[12,13] There is disagreement as to whether the feature D is also a forbidden transition, 1s-3d(e_g), and feature E, the direct ionization above the potential barrier,[11-13] or whether features D and E are doubly excited

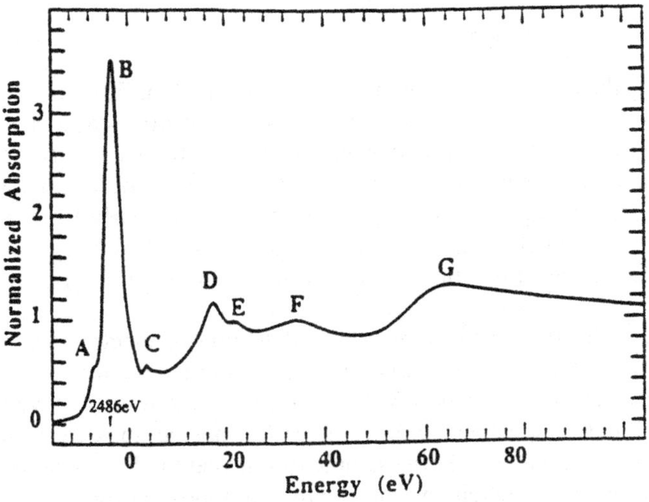

Figure 3. The sulfur K-edge XANES spectrum of SF_6.[11] The large resonance B is assigned to the allowed 1s-3p(t_{1u}) transition. The assignment of the smaller resonances has been the subject of extensive theoretical effort as discussed in the text.

autoionizing resonances of the form $1s^{-1}v^{-1}3p(t_{1u})v^*$ where v and v* are occupied and unoccupied valence orbitals.[11,14] Features F and G have recently been assigned to a scattering resonance and an octahedral shape resonance.[11] The example of SF_6 illustrates the power of combining theory and experiment. Nevertheless, the challenge of predicting XANES spectra from theory is evident.

In addition, XANES spectroscopy is frequently used to determine the chemical bonding of trace elements in complex chemical systems. For these studies, model compounds of known structure are used to generate a data base of XANES spectra. The spectral effect of various chemical properties such as oxidation state and bonding identity, type and geometry are thereby determined. The XANES spectra of the unknown sample is then analyzed in terms of this set of model compounds. Such methodology has been employed in biological systems to analyze the metal bonding structures in a variety of enzymes.[15] In these biological cases, the geometry and even the identity of nearest neighbors can be in doubt. The application of XANES methods to molecular structures of sulfur and nitrogen in asphaltenes is somewhat different. For this case, the uncertainties largely correspond to the identities and quantities, not the geometries, of particular molecular structures which are present in these samples. For both types of applications, structural determination or speciation, the selection of adequate model compounds is imperative. The following sections in this chapter will describe in detail, XANES analyses of chemical composition of complex, natural materials.

The magnitude of the resonances in a XANES spectrum can be normalized by the direct photoionization signal, allowing the comparison of resonances in different spectra. Resonances in XANES spectra are associated with a variety of processes. The magnitude of these various resonances depends critically on the wave functions of the final state and thus the chemical identity of the particular compound under study. For x-ray transitions, the radial part of the excited state wavefunctions which overlaps with the core is only slowly varying with energy, thus, is similar for different states of the same symmetry. The normalized cross sections for excitation of these states are close in value.[16] It is the large difference in the amplitude (or density of states) of the excited state wavefunctions which yield large differences in cross section.[16] Continuum states of the same energy all have the same normalization; thus excitation of the core to the continuum yields a cross section independent of chemical identity.

In addition to XANES, there has been some application of extended x-ray absorption fine structure (EXAFS) to fossil fuel samples for the determination of bond lengths, thereby helping to identify constituents. Here we briefly mention the basis of the technique; many detailed reviews of EXAFS are available in the literature.[8,17] EXAFS is frequently used as an aid in determining the distance, identity, and geometry of neighboring atoms.[6,8,10] EXAFS becomes a useful measurement at energies significantly above the ionization threshold where valence resonances are no longer observed. Wavefunctions of the outgoing photoelectrons incorporate single and multiple scattering of the photoelectron off of nearest neighbor atoms. The interference produced by this scattering results in a modulation of the photoionization cross section $\mu(k)$ as a function of electron wave vector k. Multiple scattering can contribute significantly to the oscillations at energies closer to the threshold where the photoelectron energy is small; at higher photoelectron energies, single scattering tends to dominate.[8] Generally, EXAFS data are treated as shown in Eq. 5 by obtaining the normalized oscillatory part of the cross section $\chi(k)$.

$$\chi(k) = \frac{\mu_o - \mu(k)}{\mu_o} \tag{5}$$

where μ_o is an energy-averaged cross section in the smoothly varying portion of the spectrum. $\chi(k)$ is then plotted vs the k vector of the photoelectron. With several approximations, Eq. 6 is obtained for $\chi(k)$, where f(k) is the backscattering amplitude, R is the atomic spacing, σ^2 is the mean square fluctuation of R, and $\alpha(k)$ is the electron phase shift.[7,8]

$$\chi(k) = \frac{|f(k)|}{kR^2} \exp(-2\sigma^2 k^2) \sin[2kR + \alpha(k)] \tag{6}$$

Figure 4 shows the EXAFS oscillations obtained in the x-ray absorption spectrum of Br_2 along with the results of a theoretical calculation.[10] The Fourier transform of this plot

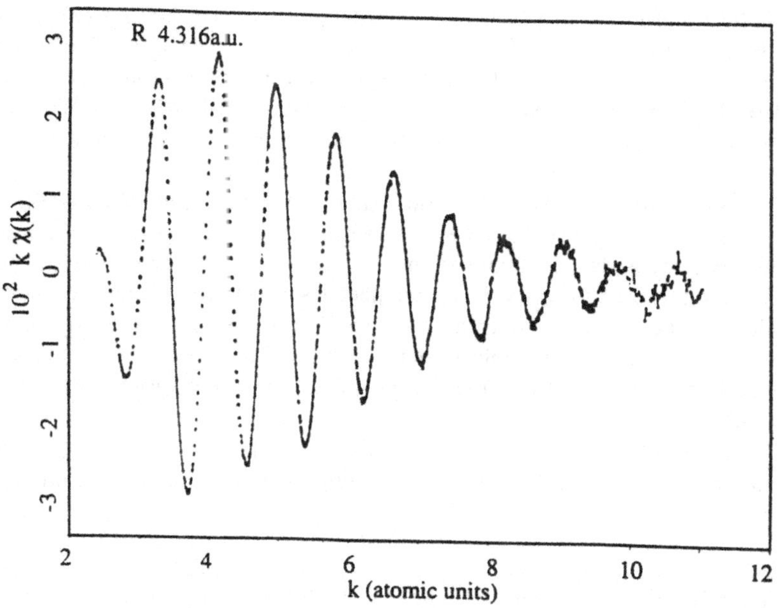

Figure 4. The EXAFS spectrum of Br_2, the plot of the oscillation of the ionization signal χ (multiplied by k) vs the wavevector k of the photoelectron.[10] Theory and experiment are both shown and are in excellent agreement. The oscillations result from interferences in the scattering of the photoelectron off the adjacent atom.

yields the bond length. An approximate bond length is obtained from the phase term in Eq. 6 and can be obtained directly from the data in fig. 4 using Eq. 7.

$$R\Delta k = \pi \qquad (7)$$

where $k(\text{Å}^{-1}) = 0.54(E(eV))^{1/2}$; E is the photoelectron energy. This equation illustrates that successive scattering resonances roughly correspond to satisfying the same boundary conditions with additional half cycles of the electron wavefunction.

STORAGE RINGS AND BEAMLINES

Synchrotron radiation is the electromagnetic radiation emitted when relativistic charged particles are accelerated perpendicular to their direction of motion. Synchrotron radiation from electron or positron storage rings provides an extremely bright and versatile source of radiation over a very broad energy range and is particularly well suited for soft and hard x-ray regions.[6-8] Relativistic electrons with energies in the GeV range are maintained in a repeating orbit in the storage ring. As the electrons circulate, bending magnets in the curved arcs and the insertion devices, wigglers and undulators, in the straight sections produce prodigious quantities of synchrotron radiation. This radiation is sharply peaked in the forward direction and linearly polarized, for the bending magnets, in the plane of the ring. The radiation is directed along individual beamlines which are optimized for specific objectives (spectroscopy over a particular spectral range, imaging, polarization studies, x-ray diffraction, small angle scattering, etc.). The x-ray brightness, flux and bandwidth provided by storage rings are orders of magnitude greater than from conventional sources. Furthermore, synchrotron radiation pulses are short in time allowing dynamic studies to be performed. Synchrotron facilities are designed to provide a bright source for both hard (1.5 keV-100 keV) and soft (40 eV-1.5 keV) x-ray regions. Synchrotron technology continues to advance at a rapid pace promising an even brighter future. Figure 5 shows the floor plan of the synchrotron radiation facility at the National Synchrotron Light Source (NSLS) at Brookhaven National Laboratory. The larger, hard x-ray ring and the smaller, soft x-ray ring each possess many tangential beamlines which support a wide variety of experiments.

A summary of synchrotron facilities has recently been published in the Journal of Synchrotron Radiation, a journal devoted to new aspects of synchrotron source development, instrumentation, methods and applications.[18] Fig. 6 lists the evolution of storage-ring synchrotron radiation sources over several decades illustrating their increasing number world-wide and the range of machine energies.[18] The large increase in the number of synchrotron sources at considerable cost reflects both the successful achievements and anticipated advances of synchrotron radiation applied to a wide variety of problems across the sciences.

The electron energy E(GeV), listed in fig. 6, is related to the critical photon energy E_c(eV) and the field strength B(tesla) of the bending magnets as shown in Eq. 8.[6]

$$E_c = 670\ BE^2 \qquad (8)$$

Field strengths are typically about one tesla; for B = 1 tesla and E = 2 GeV, Ec = 2.68 keV. One half of the total energy radiated is at photon energies above E_c. The photon emission energy with the maximum flux is at half the critical energy. In addition, the decrease of flux at high photon energy is expressed in terms of the critical energy. A heuristic rule is that there is good flux up to ~$4E_c$.

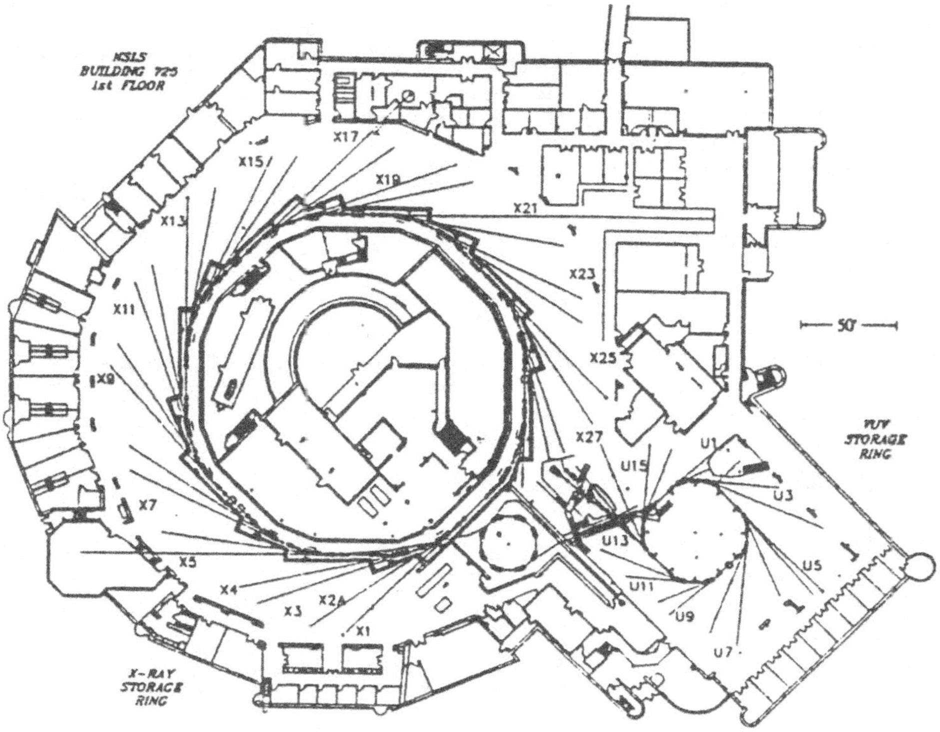

Figure 5. Overview of the synchrotron radiation facility at the National Synchrotron Light Source (NSLS) at Brookhaven National Laboratory. The larger, hard x-ray storage ring (2.5 GeV) and the smaller, soft x-ray storage ring (750 MeV) are surrounded by numerous beamlines which support a wide variety of experiments. (Courtesy NSLS.)

Synchrotrons emit intense radiation from the x-ray (or even γ-ray) region to the far infrared. Thus, dispersion is critical for the utilization of synchrotron radiation. An interesting and helpful consideration in x-ray optics is that the index of refraction of materials is very slightly less than one for x-rays; this results from anomalous dispersion. Consequently, in optimal circumstances, grazing incidence, where the x-ray path is nearly parallel to the reflecting surface, corresponds to total external reflection for x-rays yielding high reflectivities. This is analogous to total internal reflection for the standard case where $n>1$ for dielectrics. Generally, the dispersion of hard x-rays is achieved using Bragg reflections in crystal monochromators.[7] The penetration length of hard x-rays is reasonably large and the wavelengths of hard x-rays are on the order of lattice constants in crystals yielding efficient reflection when the Bragg condition is satisfied. Figure 7 schematically depicts the layout and description of the optical components for X-19A at the NSLS. This beamline has been used to collect much of the sulfur XANES data discussed in this chapter.

Soft x-ray dispersion is generally accomplished using grazing incidence on gratings. Crystal monochromators are generally not as effective because soft x-ray wavelengths are much longer than the crystal spacings. Also, absorption of soft x-rays by the crystals would result in significant attenuation. The dispersion of x-rays with grating monochromators follows the principles used for optical radiation. A variety of schemes are employed to disperse soft x-rays efficiently.[7] Figure 8 shows a schematic

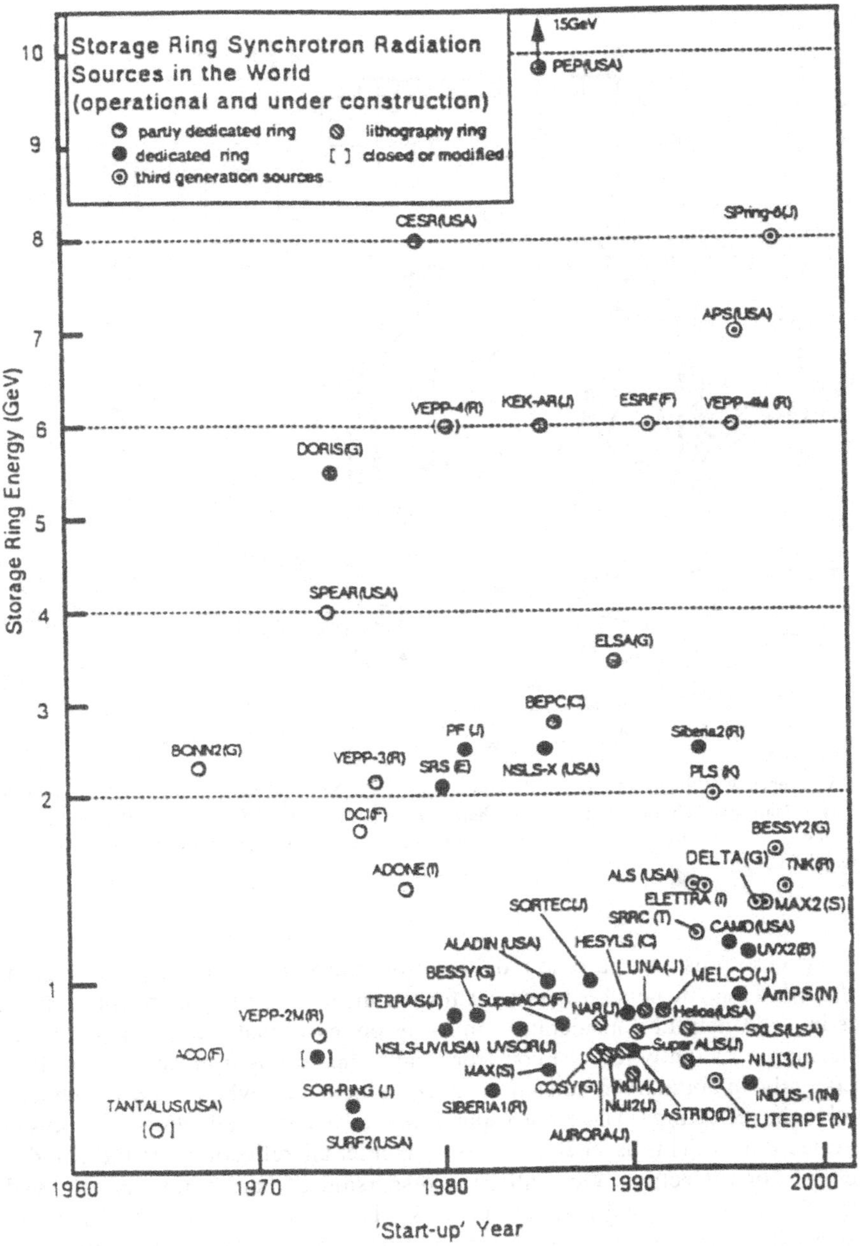

Figure 6. The evolution of storage-ring synchrotron radiation sources over the years showing their increasing number and range of machine energies.[18]

of beamline U4B, the NSLS "Dragon" beamline[19-21] of Bell labs which was used to collect all of the nitrogen XANES data discussed in this chapter. A critical component of this beamline is the movable exit slit of the monochromator which allows the wavelength-dependent exit focal length to be tracked. The design of this beamline allows for high flux at high resolution.

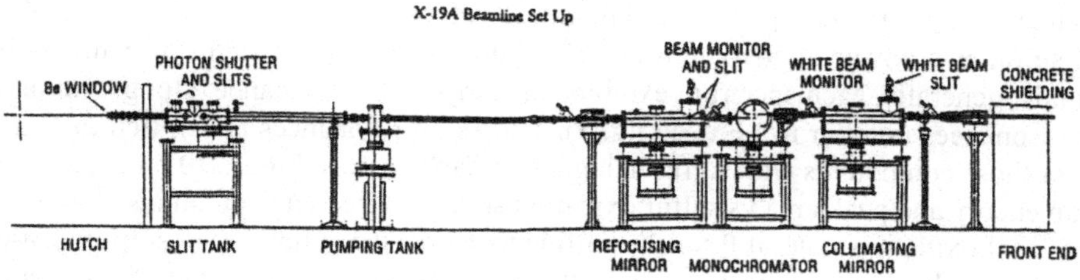

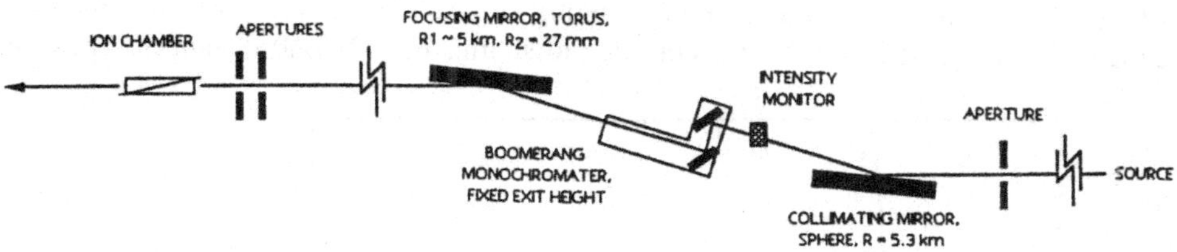

Figure 7. Schematic representation of the hard x-ray beamline X-19A of the NSLS. The synchrotron radiation is collimated by a grazing incidence spherical mirror. Energy dispersion of the broad band synchrotron beam is accomplished by Bragg reflections in the double crystal monochromator. Different frequencies emerge at different heights; beam direction is maintained by synchronously adjusting the crystal positions and angles. Detuning the crystals reduces throughput more for higher harmonics yielding the predominance of the first order of diffraction for the transmitted beam. (Courtesy of Dr. L. Furenlid, NSLS.)

SULFUR XANES SPECTRA OF SIMPLE COMPOUNDS

The first sulfur XANES studies of fossil fuel materials were performed on coals;[22,23] the spectra indicate the presence of both pyrite and organic sulfur in coal. With improved energy resolution, the ability to differentiate different organic forms of sulfur is achieved.

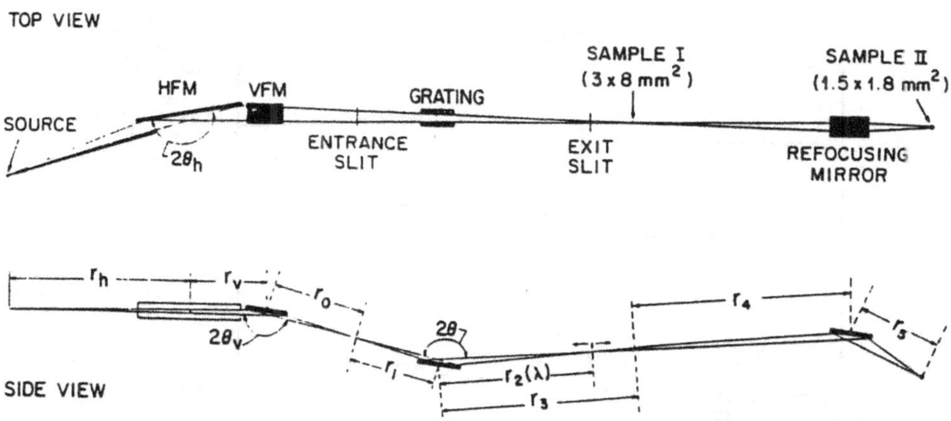

Figure 8. Schematic representation of the soft x-ray Bell Labs beamline U4B at the NSLS.[19-21] The horizontal focusing mirror (HFM) images the source inside the storage ring onto the sample position. The vertical focusing mirror (VFM) focuses the x-ray beam into the entrance slit of the monochromator. The spherical grating diffracts and focuses the beam onto the exit slit; grating rotation selects different wavelengths and the movable exit slit accounts for the angle, thus, wavelength dependence of the output focal length. This beamline achieves a resolving power of >10,000. The refocusing mirror allows for a second position of a sample.

In particular, the systematic dependence of the energy of the 1s-3p transition on the oxidation state of sulfur is a powerful analytic tool.[24-26] Figure 9 shows the spectra of many sulfur compounds; generally, each spectrum exhibits one large 1s-3p resonance, along with other smaller resonances (similar to SF_6, see fig. 3). The 1s-3p resonances are shifted to higher energy for those compounds with sulfur in higher oxidation states. Figure 10 plots the 1s-3p transition energy and peak area vs sulfur oxidation state for selected compounds.[27] In going from a formal oxidation state of 0 for the sulfide to +6 for the sulfate, the 1s-3p transition energy increases by 10 eV, a sizable and easily measured shift. With increasing oxidation state, thus, reduced valence orbital population, the 1s orbital becomes more tightly bound while the 3p orbital is subject to the counteracting effects of decreased screening due to 3p vacancies but increased ligand repulsion, due to shorter ligand bond distances. Thus, the

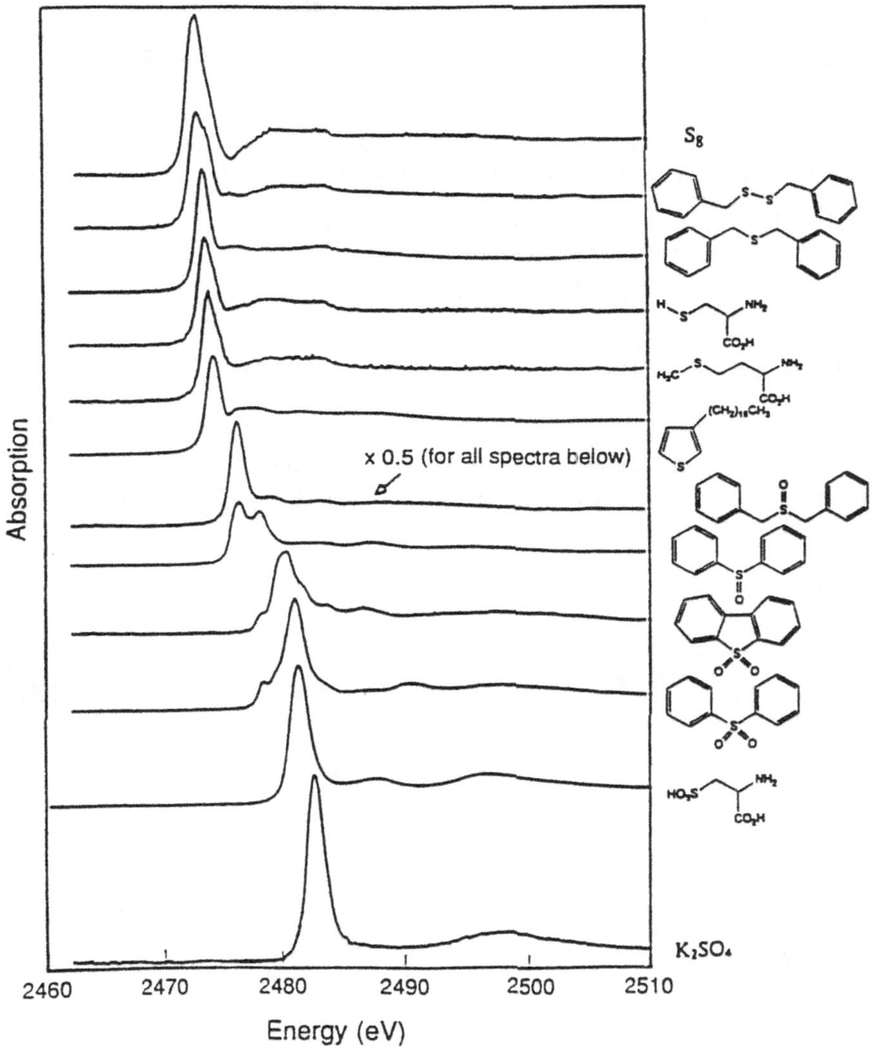

Figure 9. The sulfur K-edge XANES spectra for a series of sulfur compounds.[30] The 1s-3p transition shifts to higher energy and grows in magnitude with larger sulfur oxidation states. In addition, a small but measurable energy difference is found between sulfidic and thiophenic sulfur. These systematic findings are very useful in the analysis of sulfur molecular structures in complex materials such as asphaltenes.

prediction of the change in the 1s-3p transition energy with changing oxidation state is not obvious, a priori. Apparently, for higher oxidation states, the more tightly bound orbital (1s) undergoes a greater reduction in energy than the valence orbital yielding increased 1s-3p transition energies. Such a shift in x-ray resonances with oxidation state have been observed in other systems.[28]

In addition to the change in resonance energy with oxidation state, the resonance amplitude also changes in a systematic way. The peak areas in figure 9 are seen to increase with increasing oxidation state.[27] It is the ratio of the peak area (or height if the widths are comparable) to the edge jump which must be considered; the peak area vs sulfur oxidation state is also plotted in fig. 10. The increase in peak amplitude with oxidation state is readily understood using a simple model. The 1s-3p transition can occur as long as there is some vacancy in the 3p orbital. With increasing sulfur oxidation state, the 3p electron population is reduced and the vacancy increased, thereby increasing the amplitude of the 1s-3p transition. This simple model does not indicate how covalency should be considered and is not meant to be quantitative. Nevertheless, this simple description provides insight for the x-ray spectral systematics for a series of compounds.

Figure 9 shows that the different oxides with their different oxidation states are readily resolvable by virtue of the large energy changes in peak positions. However, two chemically distinct classes of sulfur-containing compounds, thiophenes and sulfides, have the same formal oxidation state of zero. The thiophenes are aromatic, while the sulfides are saturated. These two classes of compounds are individually quite important in materials such as asphaltenes, and their discrimination is essential. As expected, the 1s-3p resonances for thiophenes and sulfides are quite similar in energy. Nevertheless, the sulfides have a slightly lower resonance energy than the thiophenes. Perhaps the thiophene sulfur loses electron density due to delocalization of the $2p_z$ electrons in the aromatic ring, thereby yielding a slightly higher resonance energy for the thiophenes. (We will explore a similar effect with aromatic rings containing nitrogen.) In any event, the small energy separations are sufficient to allow for discrimination between the aromatic and saturated fractions of sulfur in known mixtures of compounds[25,27,29,30] and, by inference, in natural substances.

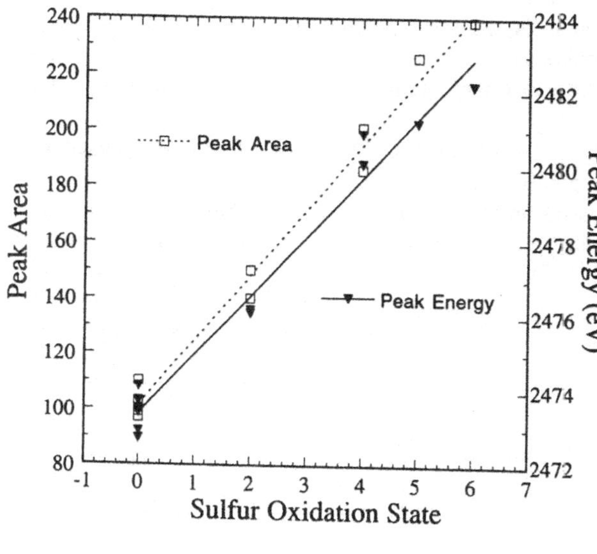

Figure 10. The sulfur 1s-3p peak areas and energies vs sulfur oxidation states for a series of compounds.[27] The 1s-3p transition shifts to higher energies and larger magnitudes for larger oxidation states of sulfur, making sulfur XANES useful for analysis of complex materials such as asphaltenes.

SULFUR XANES SPECTRA OF ASPHALTENES AND RELATED MATERIALS

Significant effort has been devoted to resolving the organic sulfur components of fossil fuel samples. Figure 11a shows the sulfur XANES spectra of three petroleum asphaltenes and also shows the third derivative of these spectra.[25] Fig. 11b shows the spectra, along with their third derivatives, of 3 different mixtures of, from top to bottom, pure dibenzothiophene, a 1:1 molar mixture of dibenzothiophene and dibenzyl sulfide, and pure dibenzyl sulfide.[25] Generally, the absolute energy scale is uncertain in XANES spectra; the energy scale is normally referenced against a standard. The spectra in fig. 11 are shifted to lower energy by 2.2 eV compared to the spectra in fig. 9 and 12. The asphaltene third derivative spectra were analyzed using the third derivative spectra of dibenzothiophene and dibenzyl sulfide only giving the results shown in Table 1.

Significant variation is found for the asphaltenes, and, generally, thiophene is present in greater quantity than sulfide.[25] The sulfur EXAFS spectrum of asphaltene 1 gave a predominant bond length, 1.74 Å, which is equal to that of dibenzothiophene, thus supporting the XANES analysis.[25]

In this asphaltene study, a third derivative technique was used on the spectra to help resolve sulfides and thiophenes.[25] However, with such a technique, normalization problems can arise as is evident with sample 2 where the sum does not add to 100%. Normalization is problematic for the third derivative techniques especially for three or more contributing chemical species, because the magnitude of the edge jump is not evident. In addition, third derivative methods are sensitive to the shape and sharpness of spectral peaks, which vary for model compounds within a chemical class. Comparison of third derivative spectra of a particular model compound, such as dibenzothiophene, with spectra of asphaltenes, which presumably contain a mixture of thiophenes, is somewhat uncertain.

A XANES study on petroleum asphaltenes was performed which included a search for oxidized as well as reduced forms of sulfur.[30] Figure 12 shows the sulfur XANES spectra for three asphaltenes with differing quantities of oxidized sulfur.[30] The deconvolution of the asphaltene spectra principally in terms of thiophenes, sulfides and sulfoxides is also shown in fig. 12. The sulfoxide fraction in these samples is evident as the peak at 2476 eV. Other sulfur oxides were also included. Table 2 provides the results of this study.

Analysis of model compounds indicates that errors are a few percent.[30] This work confirms that in petroleum asphaltenes, thiophenes and, to a lesser extent, sulfides are generally the most prominent forms of sulfur. It is reassuring that with different spectral analysis techniques and even with somewhat different model compounds, two sulfur XANES studies of asphaltenes yield similar conclusions regarding the thiophene and sulfide fractions.

In addition, sulfoxides are found to be present in small quantities in many asphaltenes. The sulfoxide resonance is well separated from those of other forms sulfur (with different oxidation states). One of the thiophene model compounds, dibenzothiophene, does have a small peak which overlaps with the sulfoxide peak, but this model also has another, larger

Table 1. Thiophene and sulfide fractions in three asphaltenes

Asphaltene	Thiophene%	Sulfide%
1	~100	~0
2	43	37
3	50	50

peak at higher energy by 3.3 eV where the asphaltenes show no peak. Thus, the presence of sulfoxides in the asphaltenes is evident directly in the data (see fig. 12). One asphaltene from a very heavy California crude oil contains 44% sulfoxide. No other oxides of sulfur were evident in significant quantities. As expected no significant inorganic sulfur is found in the asphaltenes. The asphaltenes were remeasured after one year of storage of the asphaltene

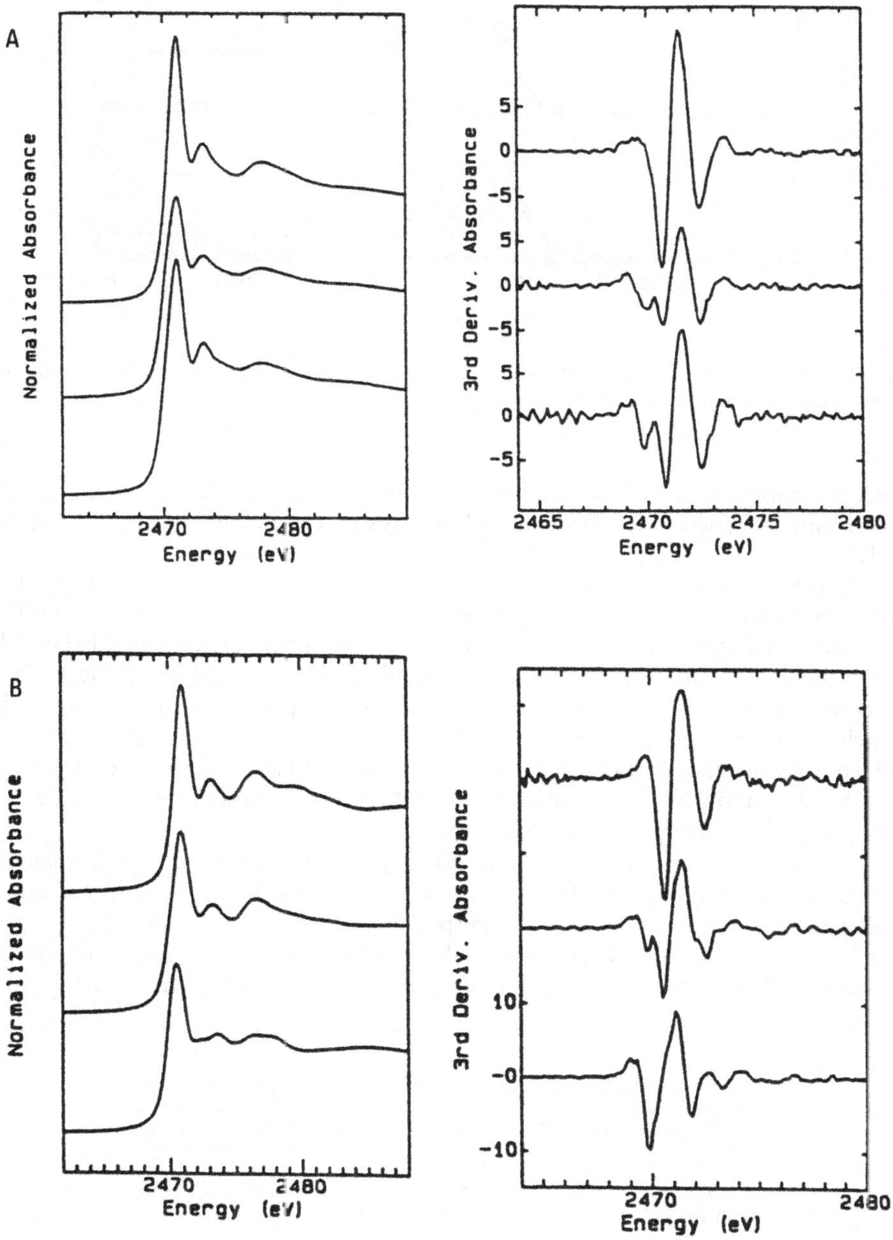

Figure 11. Sulfur XANES spectra of A) three asphaltenes along with the third derivative of the spectra.[25] The sulfur XANES spectra and their third derivatives are shown in B) for top to bottom, pure dibenzothiophene, 1:1 dibenzothiophene and dibenzyl sulfide and pure dibenzyl sulfide.[25] The asphaltene spectra can be analyzed in terms of a thiophene and a sulfide component.

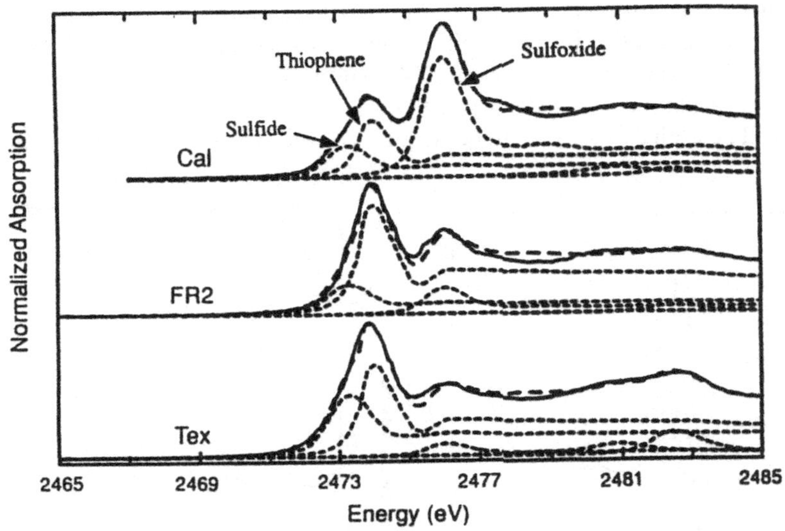

Figure 12. Sulfur XANES spectra of three asphaltenes.[30] Large peaks resulting from sulfide, thiophene and sulfoxide groups are identified and can be quantified.

without precautions to remove atmospheric oxygen.[30] No change in sulfur components was found. If atmospheric oxidation of the sulfur takes place, perhaps the labile sites react quickly.

Generally, in asphaltenes, large sulfoxide fractions correlate with small sulfide fractions as shown in fig. 13a.[30] The sulfoxide mole fraction, normalized by the sulfur and oxygen elemental mole fractions measures the extent to which oxygen is bonded to sulfur. Fig. 13b shows this sulfur-oxygen correlation increases with decreasing sulfide.[30] Thus, it appears that oxidation of asphaltenes is associated with the consumption of sulfide and the production of sulfoxides. Sulfides, particularly alkyl sulfides, are known to be susceptible to oxidation producing sulfoxides and to a lesser extent, sulfones.[31] For instance, when coals are subjected to mildly oxidizing conditions, the sulfides are converted to sulfur oxides while the thiophenes remain largely untouched.[32]

Asphaltenes are isolated from crude oils, but, for these cases, much later than when the crude oils were produced. It is unclear whether the sulfoxides were present in the asphaltenes before production of the corresponding crude oil or whether the sulfoxides resulted from air oxidation. It is unlikely that the large sulfoxide fraction of the CAL asphaltene resulted from air oxidation subsequent to crude oil production. All crude oil and

Table 2. Sulfur speciation (%) in eight asphaltenes

Asphaltene	Thiophene	Sulfide	Sulfoxide	Sulfone	Sulfate
Fr2 ST3	67	20	11	1	<1
Fr1 ST1	64	25	9	1	2
Kw1 BG5	61	35	3	<1	<1
Can Sales	60	32	8	<1	<1
Kw3 W40	56	34	7	2	<1
Kw2 UG8	52	43	4	2	<1
Tex Frio	51	38	4	3	4
Cal	36	16	44	3	<1

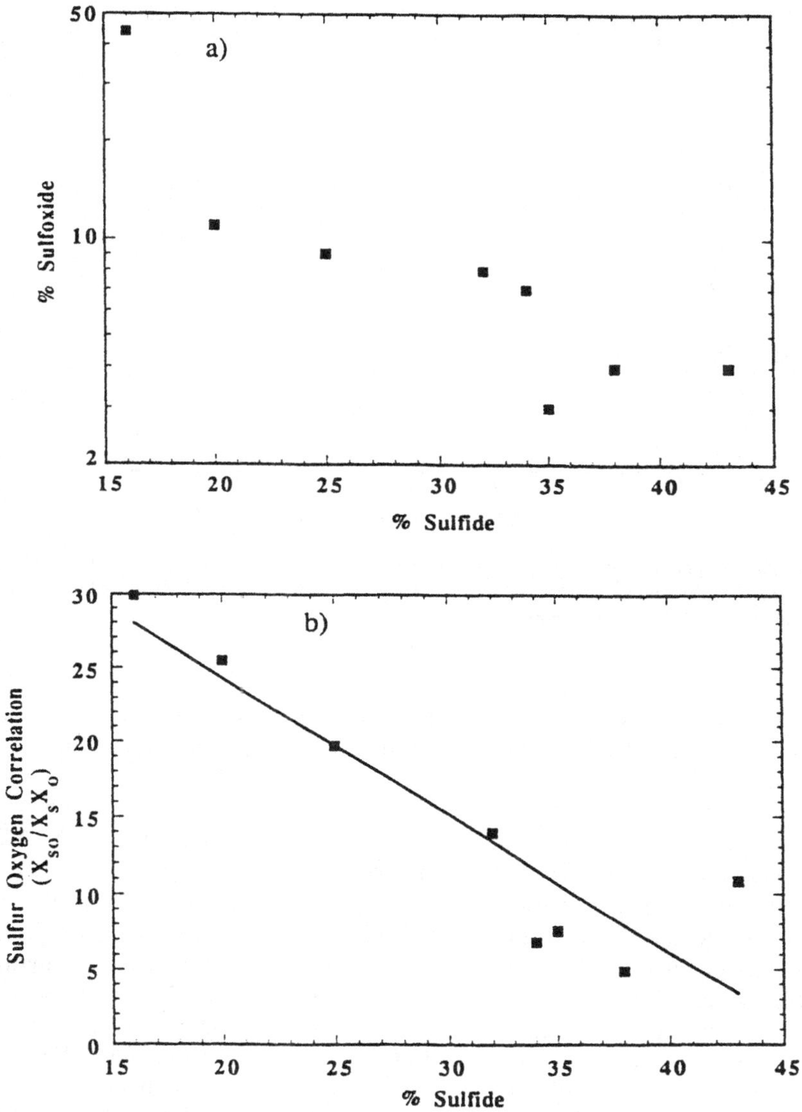

Figure 13. Increasing sulfoxide in asphaltenes (a) is associated with decreasing sulfide.[30] The sulfoxide fraction normalized by the sulfur and oxygen fractions (b) is found to increase systematically with decreasing sulfide, implying that oxidation occurs preferentially at the sulfide site in asphaltenes.

asphaltene samples were handled similarly, yet only CAL has a very large sulfoxide fraction. The very large fraction of sulfoxide in this heavy crude oil sample is consistent with some degree of oxidation of the crude oil in the reservoir, perhaps by contact with oxygen-containing waters or by microbial degradation.[4,33]

Crude oil generally becomes lighter with greater thermal maturation, thereby yielding less asphaltene.[4] Other factors also affect the quantity of asphaltene such as source material of the crude oil; hence, the quantity of asphaltene is an indicator, but not a direct measure of thermal maturation. Fig. 14 shows a plot of the thiophene fraction of the asphaltene vs the asphaltene fraction of the corresponding crude oils. Greater thermal maturation (smaller

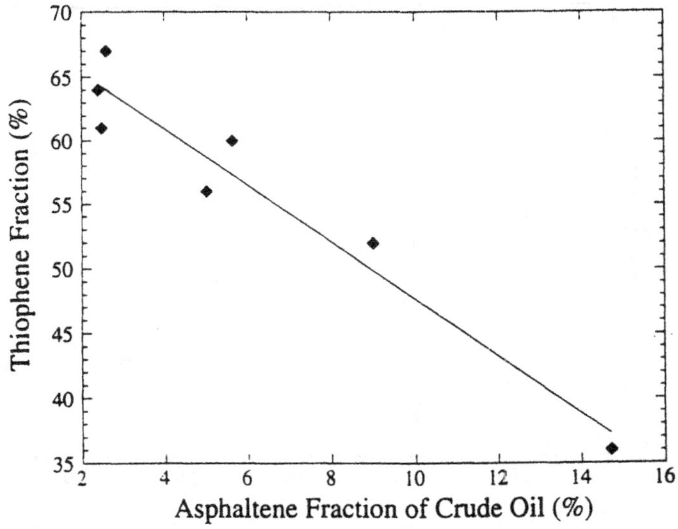

Figure 14. The thiophene fraction is generally larger for asphaltenes extracted from crude oils with small asphaltene fraction. Smaller asphaltene fractions are an indicator of thermal maturation. Thus, larger thiophene fractions correlate with increased thermal maturation.

asphaltene fraction) is associated with significantly higher thiophene fractions of the sulfur. Apparently, the more labile sulfides are lost or converted to thiophenes with thermal maturation.

Crude oils contain asphaltenes as well as many smaller, sulfur-containing molecules. It is instructive to compare sulfur analyses of crude oils and asphaltenes. A XANES sulfur analysis of various crude oils reveals that thiophenes dominate; also, certain crude oils have substantial sulfides and, to a lesser extent sulfoxides.[27] The results from this work are expressed in a ternary diagram in fig. 15. The larger thiophene fractions were found to correlate with greater thermal maturation as determined by the increased aromatization of steroid biomarkers. This result is similar to that obtained for asphaltenes. In only one group of crude oils, atmospheric oxidation over the period of a year was found to increase the sulfoxide content at the expense of the sulfide content. Other crude oils did not show this behavior. Not surprisingly, the sulfur moieties of crude oils closely match those of asphaltenes. Reduced sulfur species dominate and aromatics are found in greater abundance than saturated species, especially with greater thermal maturation. For both asphaltenes and crude oils, sulfoxides are the only significant oxidized sulfur species. The ambiguity of the origin of the sulfoxides before or after production of the crude oil is found both for the asphaltenes and crude oils.

Extensive sulfur XANES work has been performed on coal. Coal is a sedimentary rock, with a complex and inhomogeneous structure, consisting primarily of organic matter as well as mineral matter and water.[34-36] The organic part of coal is composed of fossilized plant remains called macerals, which metamorphose over geological time under the influence of elevated temperature and pressure. Mineral grains and dissolved ions also find their way into the coal beds, forming the inorganic component. The macerals include vitrinite, derived primarily from lignin from the woody plant material, exinite from lipids and waxy material, and inertinite possibly from material which was charred at the time of deposition.

The extent of metamorphism is described by the rank of a coal; a higher rank coal is more extensively metamorphosed than a lower rank coal. Lower rank coals have a lower

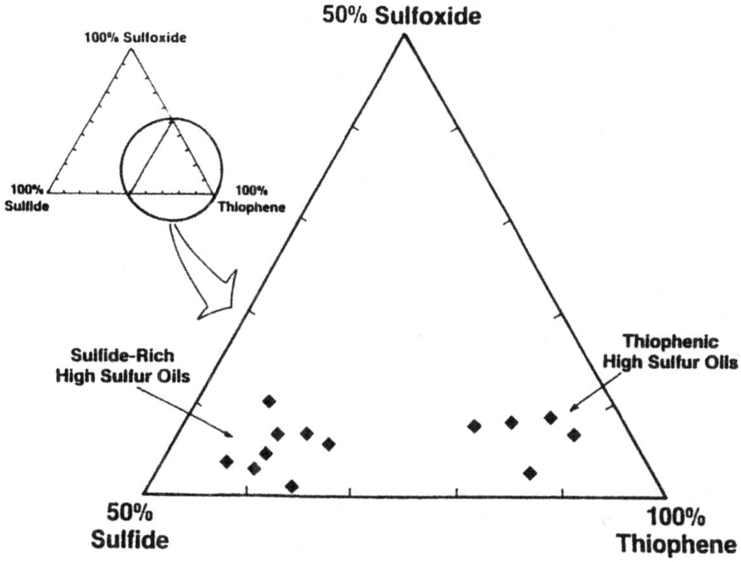

Figure 15. Ternary diagram showing the sulfur composition of several high sulfur crude oils.[27] Crude oils and asphaltenes have the same sulfur-containing components in roughly comparable quantities.

carbon and higher oxygen content than higher rank coals. ^{13}C NMR, IR, as well as various chemical studies indicate that the carbon aromaticity in coal increases with an increase of carbon content and coal rank.[34-36] The aromatic carbon content of a low rank coal can be as low as 40% of total carbon, and the same figure for a high rank coal can be as high as 90%.[35,36] Hydrogen occurs in coal in both aromatic and aliphatic fractions; the aromatic hydrogen content also increases with increasing rank.[35,36] As the rank of coal increases up to the anthracite level, the porosity and the moisture content of coal decreases.[34] Oxygen is the principal heteroatom present in coal; the organic structures include furan analogues, ethers which are often attached to aromatic systems, and phenols.[35,36]

In spite of the significant differences between asphaltenes and coals, it is instructive to consider the principal findings regarding sulfur molecular structure in coals. Here, we review a few of the most important findings of this informative body of work. The capability of probing fossil fuels using synchrotron methods was first shown with sulfur analyses of coals.[22,23] XANES studies have been particularly fruitful in quantifying the molecular structures of sulfur in coals. Third derivative methods were used to determine the thiophene and sulfide fractions in two coals; both components are present in large quantities.[37] XPS results generally agree with XANES results, but specific quantitative differences were found particularly for a coal sample with a large pyrite fraction.[38] Perhaps the surface which is probed by XPS differs from the bulk which is probed by XANES. Coal macerals and other related samples were investigated allowing the difficulties associated with pyrite to be removed.[26] An EXAFS analysis of sulfur bond lengths in coals indicates the predominance of thiophenic structures.[26] An extensive multicomponent study of many coals has been performed.[29] Eight coals from the Argonne Premium Coal Sample Bank at Argonne National Laboratory[39] were studied. These coal samples have been carefully obtained and preserved and allow different investigators to study the same samples. These samples vary in rank from lignite to low volatile bituminous. Specifically, the samples are a low volatile bituminous coal Pocahontas No. 3, VA (POC), a medium volatile bituminous coal Upper Freeport, PA

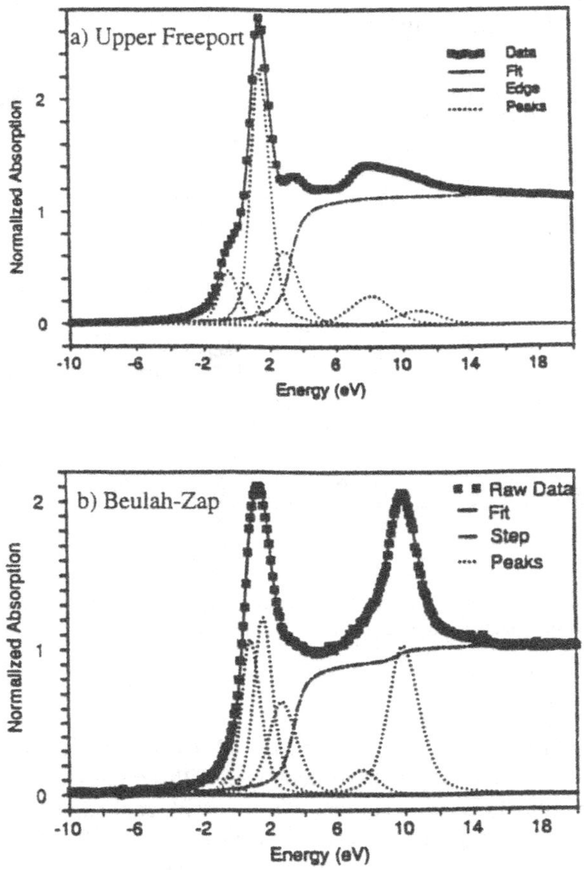

Figure 16. The sulfur XANES spectra of two coals,[29] the energy scale is referenced to elemental sulfur. Figure 16a) shows coal Upper Freeport (UF) with a relatively large fraction of pyrite as evidenced by the lowest energy peak, and 16b) shows coal Beulah-Zap (ND) with a relatively large fraction of sulfate as evidenced by the large peak at 10 eV. The peak amplitude of sulfate is much larger than the peak amplitude of pyrite.

(UF), four high volatile bituminous coals, Pittsburgh #8, PA (PITT), Lewiston-Stockton, WV (WV), Blind Canyon, Utah (UT), and Illinois #6, IL (IL), a subbituminous coal Wyodak-Anderson, WY (WY), and a lignite Beulah-Zap, ND (ND).

Fig. 16 shows the sulfur XANES spectra of two coals with the energy scale referenced to the peak of elemental sulfur. The coal in fig. 16a shows a large thiophene peak. Also, a substantial pyrite peak (the lowest energy peak) implies a significant pyrite fraction because, in mixtures of compounds, large quantities of pyrite produce only modest pyrite peaks.[29] The second coal spectrum, fig. 16b shows a large sulfate peak at 10 eV. Table 3 lists the results of this coal analysis in order of decreasing rank.[29]

Coals possess a large inorganic fraction of sulfur, primarily pyrite, and, in the lowest rank coal, sulfate. The inorganic components of sulfur have been characterized by other methods such as Mossbauer spectroscopy, but the analysis of the organic components of sulfur in coal have been most effective using XANES.[40] Large and variable quantities of pyrite are present in coals.[29] Pyrite is found in abundance in source rocks for petroleum, but the processes of catagenisis and crude oil migration, along with their solubility classification precludes any significant inorganic sulfur as a component of asphaltenes.

The most striking trend with coal rank is that higher rank coals have lower sulfide fractions and, as a percentage of organic sulfur, higher thiophene fractions. Increased maturation yields a greater predominance of the aromatic fraction of the organic sulfur as it does for carbon. A similar correlation of increased maturity with increased aromatic sulfur was also found with crude oils,[27] and asphaltenes although the effect is not as dramatic as

Table 3. Sulfur speciation (%) in the Argonne Premium Coals

Coal	Pyrite	Sulfide	Thiophene (% organic S)	Sulfoxide	Sulfone	Sulfate
POC	24	0	75 (100)	0	1	0
Upp. Freeport	62	6	32 (84)	0	1	0
Pittsburgh #8	52	13	35 (73)	0	1	0
Lewis-Stock. WV	26	16	56 (77)	1	1	0
Blind Canyon UT	40	15	46 (75)	0	0	0
Illinois #6	48	19	33 (63)	0	0	0
Wyodak-Anders WY	24	29	46 (61)	0	0	0
Beulah Zap ND	29	28	30 (50)	2	0	12

with coals. This increase in the aromatic fraction can result in part from the conversion of sulfide or inorganic sulfur into aromatic sulfur and in part due to the loss of sulfide perhaps as H_2S.[41] In addition to the examination of coals of varying rank, individual coal macerals have also been examined.[29] Although the differences are not large, exinite which has a higher hydrogen content than vitrinite and inertinite, correspondingly, is found to have a somewhat higher sulfide fraction of its organic sulfur. However, inertinite also has a larger sulfide fraction than vitrinite; no obvious explanation accounts for this observation.

Coals are seen to have almost no oxidized sulfur component with the exception of the lowest rank coal. With the low rank coal, the oxidized sulfur is a sulfate, thus inorganic. These coal samples obtained from the Argonne Premium Coal Sample Bank have been collected and stored in the absence of oxygen. As the oxidized sulfur is only found for the lowest rank coal, and the sample exposure to the atmosphere is minimal, this sulfate is probably not due to oxidation in sample handling. The maturation of coals is associated with the loss of oxygen and is consistent with the lack of oxidized sulfur in higher rank coals.

The primary difference in the distribution of sulfur between coals and asphaltenes is the presence of inorganic sulfur in coals mostly in the form of pyrite. The organic sulfur of both coals and asphaltenes is found primarily in thiophenes and, to a lesser extent, sulfides. Significant variation in these fractions is found. For asphaltenes, crude oils, and coals, the thiophene fraction increases with thermal maturation. Asphaltenes also have varying quantities of sulfoxides which are not found in coals.

At this time, almost all XANES and EXAFS work on fossil fuel samples has utilized the K-edge. The L-edge of sulfur can be a useful analytic tool; however, the very low energy of the x-rays, 163 eV for the $L_{2,3}$ edges, makes this technique challenging. The sulfur $L_{2,3}$ edges of model compounds and several coals have been compared.[42] A quantitative analysis of the coals was not attempted; nevertheless, contributions to the coal spectra from organic and inorganic components are evident.

NITROGEN XANES SPECTRA OF SIMPLE COMPOUNDS

The ability to perform soft x-ray XANES on nitrogen has recently been dramatically improved. The Bell labs beamline U4B at the NSLS provides high flux at very high resolution[19-21] - a powerful combination. The development of windowless, multielement Ge detectors allows for detection of soft x-ray fluorescence, enhancing elemental specificity of the measurement.[43] Figure 17 shows the XANES spectrum of N_2.[44] Because of the high resolution of the beamline coupled with the lack of spectral broadening in the gas phase sample, fine structure is resolved in this spectrum. The large resonance at 400 eV corresponds to the 1s-π^* transition. This resonance is a general feature of nitrogen XANES spectra for

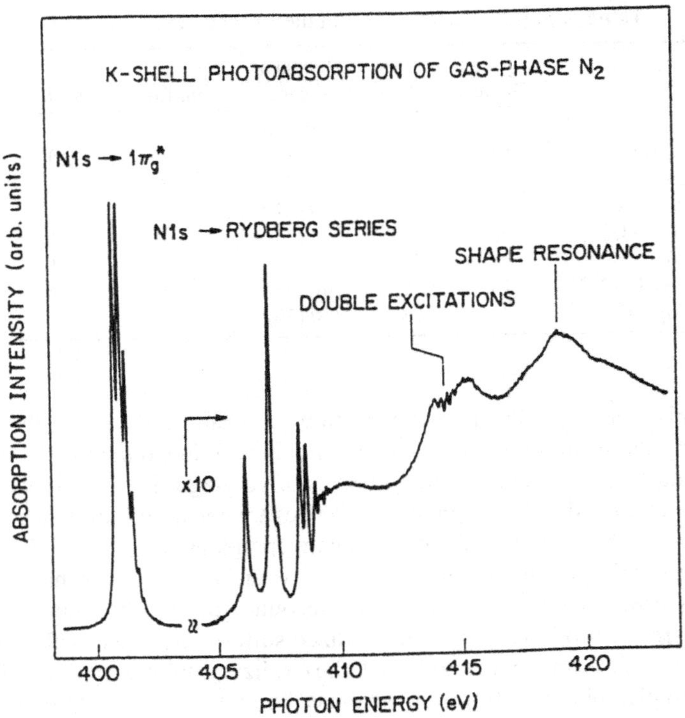

Figure 17. Nitrogen XANES spectrum of N_2.[44] The 1s-π* transition at 401 eV is resolvable into individual vibronic components. The large π* resonance is characteristic for molecules containing π-bonded nitrogen. Other types of transitions are assigned in the figure.

nitrogen atoms involved in π bonding,[45] although generally without the vibrational fine structure in condensed phase. Individual vibrational lines are resolved in this gas phase spectrum giving the vibrational splitting in the excited state with the core hole. For N_2, the value of the vibrational constant ω_e of the excited state, 1897 cm^{-1}, is less than for the ground state, 2363 cm^{-1}, because the bond order in the ground state is 3 while in the excited state, 2.5.[44] Resonances assigned as Rydberg series are evident in fig. 17 in the energy range 406 eV to 410 eV. These resonances will only be observed in gas phase samples, not condensed phase where the large Rydberg orbits would collide with nearest neighbors.

A large shape resonance is observed at 420 eV and is a general feature of nitrogen XANES spectra. Shape resonances can result when centrifugal barriers modify the molecular potential as shown in fig. 1. The σ shape resonance in N_2 has been identified as an f-wave shape resonance.[16,46] The energy of the shape resonance with respect to the ionization potential is a strong function of the bond length;[45,47] for the very short bond of N_2, the shape resonance is at high energy. Other resonances observed in Fig. 17 have been identified as excitation of doubly excited states,[44] which occurs primarily through electron correlation effects.

Nitrogen XANES spectra can be used to identify and quantify different nitrogen chemical classes in complex samples. Figure 18 shows the nitrogen XANES spectra of analogues of a pyridine (acridine), a pyrrole (cabazole) and a saturated amine (tribenzyl hexahydrotriazine) along with the spectrum of an asphaltene.[48,49] The asphaltene shows absorption features which can be identified in the model compounds. It is important to understand the spectra of these model compounds in order to use them to analyze asphaltene

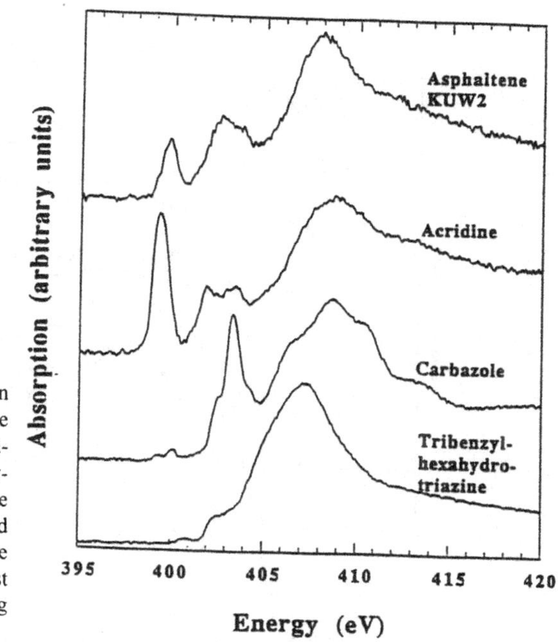

Figure 18. Nitrogen XANES spectra of an asphaltene along with the spectra of a pyridine analogue, acridine; a pyrrole analogue, carbazole and a saturated amine, tribenzyl hexahydrotriazine.[49] The aromatic compounds have both π^* and σ resonances while the saturated amine has only a σ resonance. Analysis of the spectra show that asphaltene nitrogen is almost entirely aromatic with pyrrole analogues being dominant.

spectra. The spectra of all compounds appear to contain a shape resonance at ~408 eV. In addition, the spectra of compounds containing aromatic nitrogen have one or more π^* resonances. For instance, the pyridines have a large π^* resonance at ~400 eV. However, as expected, the saturated amines show no π^* resonance, just the shape resonance, thereby supporting the assignment of these resonances.[16,45-47]

Both pyridine and pyrrole analogues appear to be important components of the nitrogen in asphaltenes. Figure 19 shows the XANES spectra of several pyridine and pyrrole analogues.[49] All pyridines show a strong π^* resonance at ~400 eV; while all pyrroles show strong peaks at ~403 eV. The class of compound, not subtleties such as different substitution, is predominant in determining the location of the π^* resonance. Nevertheless, substituents do have important effects as seen in the comparison of the spectrum of 2-phenylindole with those of other pyrrole analogues.

With sulfur, we have seen that higher oxidation states of sulfur correspond to larger energies of the 1s-3p transition. Similar reasoning appears to govern the relative location of the π^* resonances of pyridine and pyrrole.[49] Fig. 19 shows that the π^* resonance of the pyridines is about 3 eV lower than that of the pyrroles. Although the nitrogen in pyridine and in pyrrole have the same formal oxidation state, a closer look reveals that the relative locations of the π^* resonances in nitrogen conform to the trends found with sulfur. The lone pair of electrons on the nitrogen in pyridine is in an sp^2 orbital and not involved in the aromatic bonding; the pyridine nitrogen possesses a formal double bond. The localization of the lone pair on the nitrogen coupled with the high electronegativity of nitrogen yields a slightly negative charge on the nitrogen. Thus, the nitrogen is basic and, correspondingly, has a low energy π^* resonance.

For pyrrole, the nitrogen is also sp^2 hybridized, but here, all three sp^2 orbitals are involved in σ bonding; the pyrrole nitrogen does not possess a formal double bond. In pyrrole, the lone pair of electrons of nitrogen is in the p_z orbital, and thus is involved in the π system of the 5-membered aromatic ring. Of the six π electrons, two come from the nitrogen atom and four from each of the carbons. In the sharing of π electrons, the nitrogen atom loses

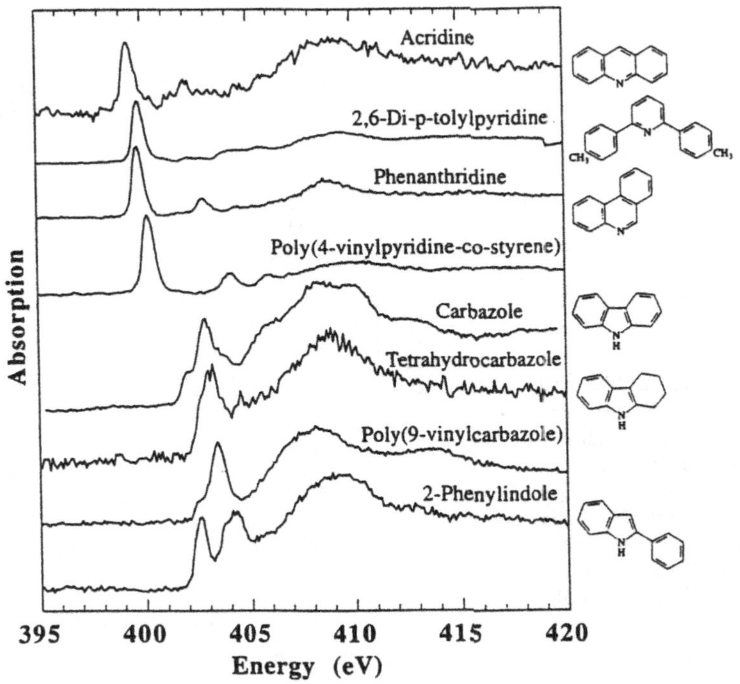

Figure 19. Nitrogen XANES spectra of several pyridine and pyrrole analogues.[49] The large π^* resonance of all pyridine compounds is at ~399.5 eV; while the large π^* resonance of all pyrrole compounds is at ~403.5 eV. The more basic nitrogen of pyridine correlates with a lower energy π^* resonance relative to pyrrole. This large energy difference allows facile identification and quantification of pyridinic and pyrrolic nitrogen in complex materials such as asphaltenes.

electron density; the electron density rearranges towards the five-fold symmetry of the cyclopentadienyl anion. This aromatic effect outweighs the electronegativity effects; the pyrrole nitrogen is not basic. The reduced electron density on the nitrogen atom in pyrrole yields a large blue shift in the π^* resonance compared to pyridine.

Porphyrins provide interesting examples where the number of atoms in the ring does not correlate with the nitrogen type. In addition, porphyrins are an important component of some asphaltenes. The difference in the π^* peak position of nitrogen in pyridine vs pyrrole is determined to first order by the orbital description of the lone pair of electrons, not by the number of atoms in the aromatic ring. If the lone pair is in the sp^2 orbital, then the nitrogen is basic and the π^* resonance is red shifted (~400 eV); if the lone pair is in the p_z orbital and thus shared in the aromatic π system, then the nitrogen is not basic and the resonance appears closer to 403 eV. Fig. 20 shows the nitrogen XANES spectra of a free-base and a metallated porphyrin.[50] The free-base porphyrin spectrum exhibits both the pyrrolic and pyridinic π^* resonances corresponding to the two types of nitrogen seen in the free-base porphyrin molecule (cf. fig. 20a). The metalloporphyrins are know to have four-fold rotational symmetry.[51] Nevertheless, fig. 20b shows that the metalloporphyrin XANES spectrum retains the two distinct nitrogen types, pyrrolic and pyridinic, as do the individual resonance structures, perhaps due to symmetry breaking by the core hole.[8]

The comparison of x-ray and optical spectroscopy of porphyrins offers an interesting view of the different excitation processes. In the optical spectrum, the resonances are π - π^*, while for the x-ray spectrum the resonances 1s-π^*. As shown in fig. 21A, the free-base

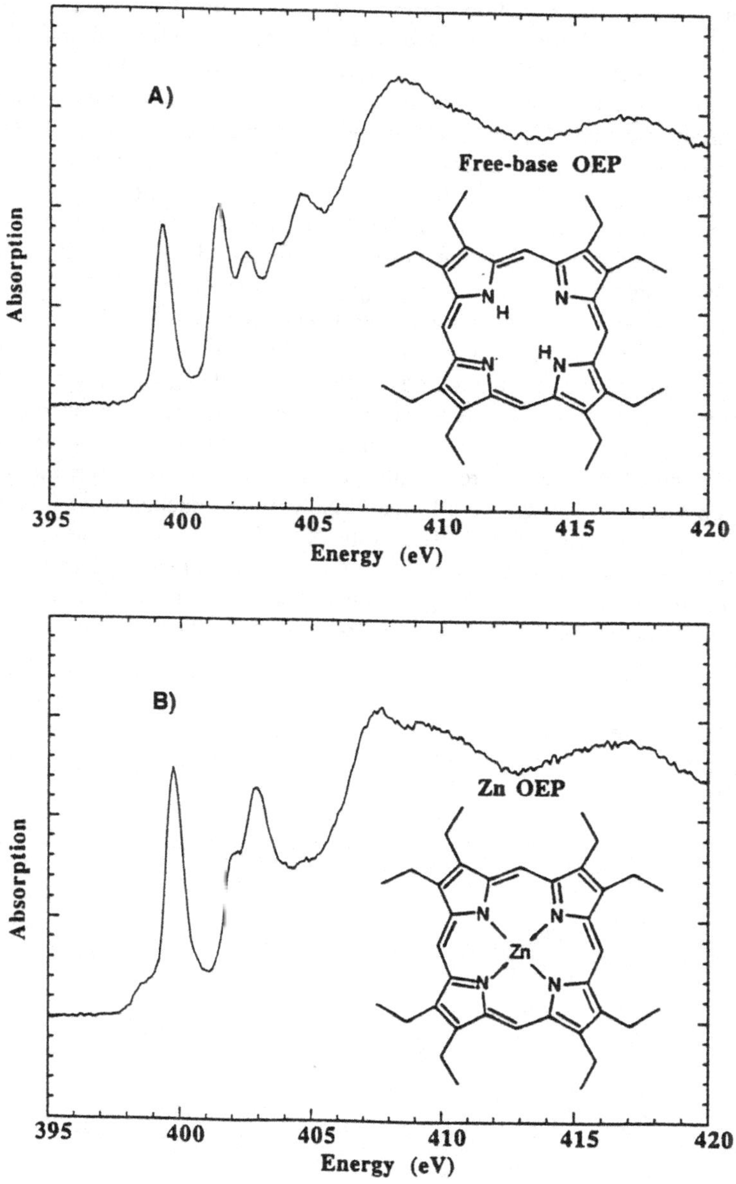

Figure 20. Nitrogen XANES spectra of (A) octaethylporphyrin (OEP), a free-base porphyrin and (B) zinc OEP. The free-base porphyrin formally has the two types of nitrogen, pyrrolic and pyridinic, and its XANES spectrum exhibits peaks for each of the two nitrogen types. Metalloporphyrins are known to possess four-fold rotational symmetry. Nevertheless, the XANES spectrum of the metalloporphyrin also exhibits both types of nitrogen, perhaps due to symmetry breaking by the core hole.

porphyrin has effectively only a two fold rotational axis of symmetry; this lower symmetry results in the splitting of the Q electronic band into a Q_x and Q_y band; each band of the two electronic transitions has its own (00) and (10) vibrational transitions yielding four bands in the visible for the free-base porphyrin. On the other hand, the optical spectrum of the metalloporphyrin (fig. 21B) shows only one Q band with its two vibronic components. The theoretical analyses of the metalloporphyrins (with no core hole) show that the four 5-membered rings are essentially equivalent, giving the porphyrin a four-fold rotational symmetry axis.[51] However, the XANES spectra of both free-base porphyrin and metalloporphyin show two distinct types of nitrogen, implying two-fold rotation symmetry. The different apparent symmetries of the optical and x-ray spectra of metalloporphyrins may relate to the core hole. In optical spectroscopy, an electron from a π molecular orbital is excited; this orbital reflects the four-fold symmetry of the metalloporphyrin. The state excited in the x-ray measurement is a 1s orbital of one of the four nitrogen atoms. Possibly, the core hole on the excited nitrogen atom destroys the four-fold symmetry of the porphyrin.[8] Thus, the XANES spectrum of the metalloporphyrins is characterized by lower symmetry, thereby yielding resonances associated with the two types of nitrogen.

The DNA bases also provide interesting examples where the location of the lone pair of electrons, not number of atoms in the ring, correlates with the nitrogen type.[52] These bases are quite complicated with many functional groups; nevertheless, their nitrogen XANES spectra exhibit underlying systematics. We consider a very simple model of the sp^2 hybridized nitrogen of the DNA bases where the nitrogen is considered to be pyrrolic if bonded to hydrogen and pyridinic if possessing a formal double bond, independent of the number of members in the aromatic ring. Fig. 22 shows the XANES spectra for the nucleotides and base structures of the five nucleic acid bases.[52] The tautomers listed are know to predominate.[53] Adenine has four aromatic nitrogen atoms, three are pyridinic and one is pyrrolic. (We ignore the amine nitrogen which does not produce a resonance in the relevant energy

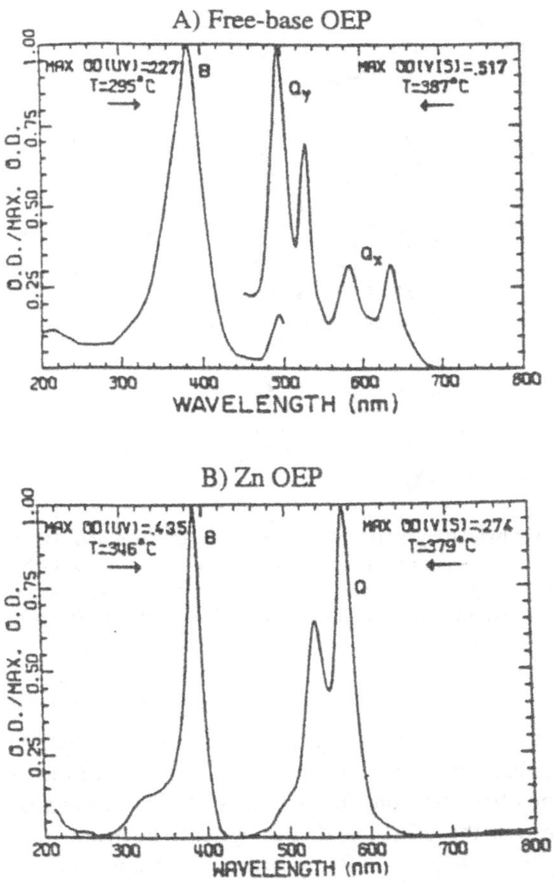

Figure 21. Optical absorption spectra of (A) free-base octaethylporphyrin (OEP) and (B) zinc OEP.[51] The lower symmetry of the free-base porphyrin is reflected in the splitting of the Q band into Q_x and Q_y components, while the four-fold rotational symmetry of the metalloporphyrin is indicated by the Q band degeneracy.[51]

range.) Correspondingly, the XANES spectrum shows a large low-energy peak, the pyridine peak, and a small peak at 403 eV, the pyrrole peak. The pyridine peak of adenine and the other bases is shifted somewhat with respect to the pyridine models shown in fig. 19. This shift is probably associated with the inductive and resonant effects of the many heteroatoms in these ring systems; nevertheless, the general understanding of peak locations is evident. Guanine and cytosine have equal numbers of pyridinic and pyrrolic nitrogens, and each has two roughly equal π^* peaks, although those of the cytosine nucleotide overlap. Thymine and uracil only have pyrrolic nitrogens (in 6-membered rings) and the XANES spectra of the

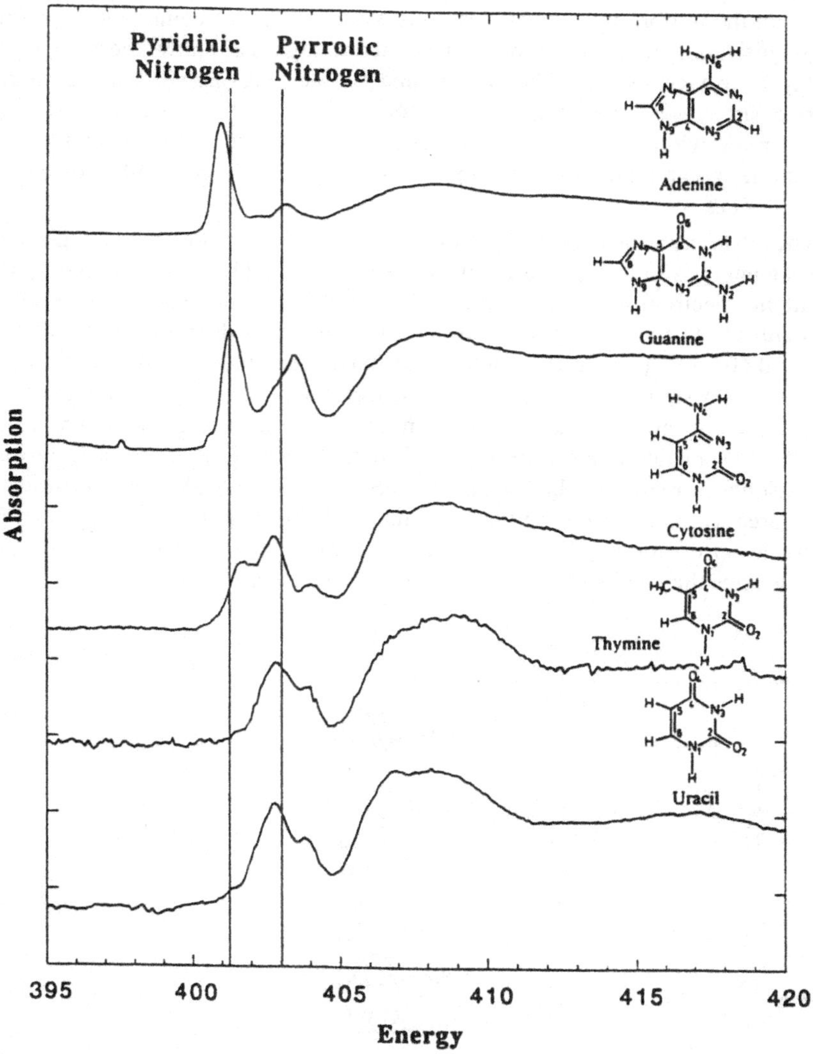

Figure 22. Nitrogen XANES spectra of the nucleotides of the five nucleic acid bases.[52] Assignment of the π^* resonances into pyrrolic and pyridinic nitrogen can be accomplished even with these complicated molecules; however, there is some energy shift particularly of the pyridinic peak relative to simpler compounds. With respect to the aromatic ring nitrogen, adenine contains 3:1 pyridine to pyrrole; guanine and cytosine, 1:1 and thymine and uracil, only pyrrolic nitrogen. The sizes of the pyridine and pyrrole π^* peaks are in accord with these ratios. The orbital description of the lone pair of electrons, not the number of atoms in the ring determines whether the nitrogen is pyrrolic or pyridinic.

nucleotides of these bases lack the pyridine peak and show a large pyrrolic peak. It is encouraging that heteroatomic ring systems as complicated as the DNA bases and porphyrins are amenable to such a simple model regarding the classification of aromatic nitrogen.

NITROGEN XANES SPECTRA OF ASPHALTENES AND RELATED MATERIALS

The large separation of resonances for different classes of nitrogen compounds coupled with the uniformity of peak positions within a class of compounds is quite useful for the quantification of nitrogen types in asphaltenes. Figure 23 shows the nitrogen XANES spectra for seven asphaltenes.[49] The pyridine and pyrrole peaks are noted in the figure. These peaks are clearly resolvable, in spite of the complexities of the asphaltenes. However, the asphaltene peak widths particularly for pyrrolic nitrogen are larger than for individual pyrrolic model compounds; this probably results from the large number of contributing pyrrole analogues.

The quantitative analysis of the nitrogen in asphaltenes[49] proceeds along similar lines used for the analysis of sulfur in asphaltenes.[27,29,30] Figure 24 shows the decomposition of an asphaltene spectrum into constituent peaks.[49] The asphaltene resonance at 399.5 eV is assigned entirely to pyridine. The small contribution of pyridine analogues to asphaltene absorption at 403.5 eV (see fig 19) is subtracted and the remainder and bulk of the absorption at this energy is assigned to pyrrole. The presence of alkyl amines in asphaltenes could be detected by the appearance of large σ resonances without the corresponding π^* resonances (see fig. 18). After subtracting contributions from the spectra of pyridine, and pyrrole to the σ-shape resonance region of the asphaltene spectrum (E ~408 eV), the remainder of this resonance area was attributed to saturated amine.[49] The percentages of the pyrrole, pyridine and saturated nitrogen components present in the asphaltene samples are then calculated. Table 4 tabulates the results of the nitrogen analysis of the asphaltenes.[49]

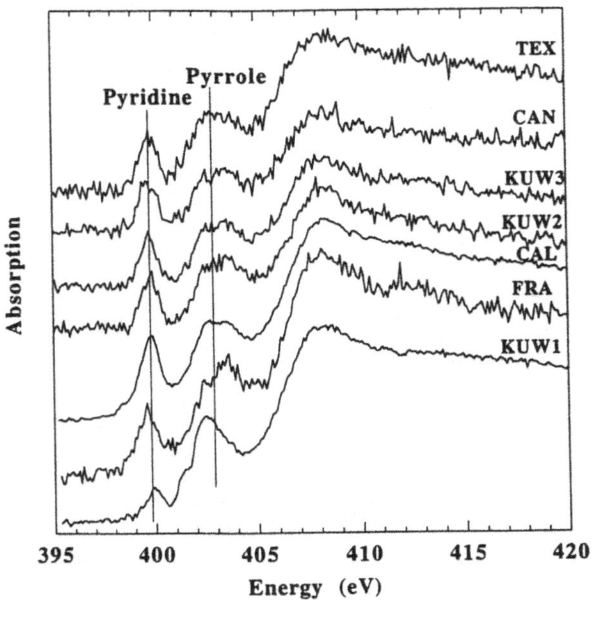

Figure 23. Nitrogen XANES spectra of seven petroleum asphaltenes.[49] Pyridinic and pyrrolic nitrogen peaks are identified directly in the spectra. The pyrrolic peak area is generally larger for the asphaltenes; thus, pyrrole is seen to dominate.

Table 4. Nitrogen speciation (%) in asphaltenes[a]

Asphaltene samples	Pyrrole	Pyridine	Saturated amine
FRA ST1	76	22	3
Kw1 BG5 7% porphyrin[b]	87	2	2
Can Sales	74	26	0
Kw3 W40	77	22	2
Kw2 UG8	79	22	3
Kw2 UG8[c]	71	26	3
Tex Frio	80	18	0
CAL 3% porphyrin[b]	65	30	2

[a]Obtained from fluorescence yield detection unless otherwise stated.
[b]Estimated from the metal content for two asphaltenes.
[c]Obtained from electron yield detection.

Almost all of the nitrogen found in the four asphaltenes studied is aromatic. There is little evidence of saturated amines. Furthermore, pyrrolic nitrogen in these samples is uniformly predominant. The smaller pyridine fraction shows significant relative variability. These asphaltenes are the same set for which the sulfur analysis was performed (see Table 2) and are listed in order of decreasing thiophene fraction. No trends are apparent, although the asphaltene with the greatest oxidized sulfur fraction by far, CAL, has the smallest pyrrole and largest pyridine fractions. The nitrogen fractions do not correlate with the asphaltene fraction of the crude oil; thus they do not appear to be strongly altered by thermal maturation. This is not too surprising; unlike sulfides, neither pyrroles nor pyridines are very labile and would not be expected to degrade with greater maturation.

For two asphaltenes in Table 4, a porphyrin fraction is listed. To obtain this quantity, all vanadium and nickel was assumed to be complexed with porphyrin or porphyrin-like macrocycles; and all porphyrin was assumed to be metallated with V and Ni.[49] Using the vanadium, nickel and nitrogen fractions for these two asphaltenes obtained from elemental analyses, the corresponding porphyrin nitrogen fraction was estimated.[49] When a metalloporphyrin spectrum is analyzed in terms of its pyrrolic and pyridinic nitrogen using nitrogen

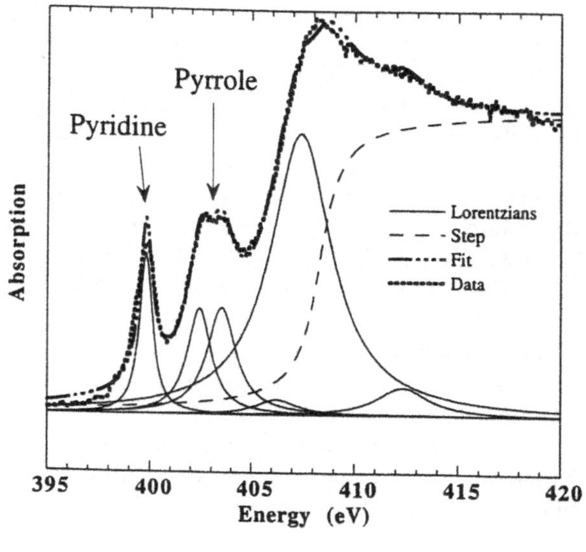

Figure 24. Decomposition of a nitrogen XANES spectrum of an asphaltene into pyrrolic and pyridinic constituents.[49] In analyzing the data, the small contribution of pyridine analogues to absorption in the pyrrole peak region must be properly taken into account. Also, the different π^* peak areas of pyrrole and pyridine compounds must be considered.

model compounds, the nitrogen is found to be evenly split, as expected.[54] The spectral contribution of the porphyrin could then be subtracted to determine the remaining nitrogen speciation. The fraction of porphyrin nitrogen is found to be small; nevertheless, for Kuw1, the porphyrin fraction is not much smaller than the pyridine fraction. That is, about one half of the total pyridine nitrogen in this asphaltene is in metalloporphyrin. The porphyrin analysis is a reminder that it is the disposition of the lone pair of electrons which determines whether a nitrogen is pyrrolic or pyridinic, not the number of atoms in the aromatic ring.

It is instructive to compare electron yield results with fluorescence yield results. Fluorescence detection allows a much greater depth of investigation in the sample than electron yield due to the short penetration and escape length of electrons. For the analysis of fossil fuel samples, an issue of importance is whether surface effects may distort the results for electron yield spectra. Surface effects could result if surface oxidation or some other surface reaction is prevalent. In addition, the samples themselves may inherently have some surface to bulk variations.

In addition to surface effects, there may be other differences between electron yield and fluorescence results. Fluorescence emission and detection are anisotropic, as is evident from orientation studies of molecules on surfaces,[45,55] whereas electron yield measurements are generally isotropic. Furthermore, there is a possibility that the ratio between Auger electron emission and fluorescence emission is affected by excitation of quasibound or bound resonances, which would result in variations between electron yield and fluorescence spectra. Fluorescence branching ratios among the different fluorescent decay channels are known to depend on the particular resonance excited.[56] This could give rise to differential self-absorption with fluorescence detection; one, but not another resonance could produce a fluorescence x-ray which is resonantly absorbed in the sample. Thickness effects are more severe for fluorescence measurements, particularly for model compounds. To mitigate some of these effects, separate analyses of electron yield and fluorescence data sets were performed. Thus, fluorescence results of the model compounds were used to analyze the fluorescence excitation spectra of asphaltenes, and likewise, the electron yield spectra of the model compounds were used to analyze the electron yield spectra of asphaltenes.

Table 4 lists the results for asphaltene Kw2 obtained from analysis of electron yield and fluorescence yield data. The results for the two different detection methods are close, indicating that surface effects are not very large. General agreement in the two separate analyses lends credence to the results. Nevertheless, the electron yield results do produce a larger estimate than the fluorescence results for the pyridine fraction by 8%. This difference may correspond to the error or uncertainty in the estimation. If the asphaltene surface differs from the bulk, then the difference between the electron yield and fluorescence results could be due to the smaller depth of investigation of the electron yield measurement.

Kerogens and bitumens, which are the source materials for crude oils and asphaltenes, have also been studied by nitrogen XANES methods. Kerogen is the solid, insoluble organic fraction of petroleum source rocks. Kerogen from Green River oil shales is one of the most studied kerogens, and is classified as a Type I kerogen. The H/C ratio in an immature Green River kerogen is typically 1.5, and the O/C ratio is less than 0.1.[57] There are a large number of aliphatic chains in Type I kerogens, and polycyclic aromatic hydrocarbons and heteroatom-containing compounds are present in relatively small amounts. Bitumen is the dark brown, viscous, solvent-extractable organic part of shales and petroleum source rocks. The bitumen content of an oil shale is typically 5% of the total organic matter.[1] The H/C ratio is roughly 1.5 in Type I bitumens as in kerogen, and the number-average molecular weights are typically around 1,200.[1]

Fig. 25 shows the nitrogen XANES spectra of three oil shale kerogens and three bitumens.[54] In the isolation of the kerogen from the source rock, all kerogen samples were first debitumenized by solvent extraction. One kerogen sample (K-3) was then demineralized

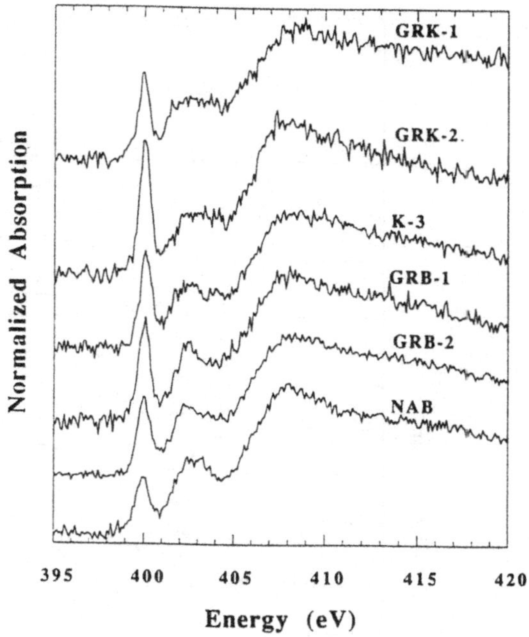

Figure 25. Nitrogen XANES spectra of three kerogens and three bitumens.[54] In addition to pyridinic and pyrrolic nitrogen, these samples also have lesser quantities of pyridone and aromatic amines.

with HF which removes the silicate minerals as well as the carbonates, while the other two kerogens were demineralized only with HCl, thus are still mixed with silicate minerals. These two kerogens are from Green River Shales which are properly called marls; their mineral content is largely carbonate.

Generally, the XANES features in these spectra are similar to the asphaltenes. The large resonances at 399.5 eV are readily identified as pyridine contributions and the large absorption features at 403.5 eV are assigned to pyrrole. In addition, unlike most of the asphaltenes, the kerogens and bitumens show significant absorption slightly below 402 eV. Pyridone (2-hydroxypyridine) can account for this absorption.[54] Pyridones are known to be components in crude oils[58] and asphalts[59,60] and, thus, are plausible components of kerogens and bitumens. The pyridone and pyrrole resonances are quite close to each other. It is important to establish if overlap exists in peak locations for these two nitrogen structures. Figure 26 shows that the lowest energy (π^*) resonances of the four pyridones are always below 402.2 eV, spanning about 0.5 eV, while for pyrroles the corresponding resonances are always above 402.2 eV, spanning about 1 eV. The spectra ranges of pyridones and pyrroles meet at 402.2 eV, but do not overlap much. This result is not surprising because the two structures are chemically quite distinct and distinct structures often produce resonances in different regions.

The pyridone compounds can exist in two tautomeric forms, the keto (or amide) form and the enol (or hydroxypyridine) form. The enol is clearly aromatic and the keto form has delocalized π electrons due to the lone pair of electrons of the nitrogen in the p_z orbital. IR spectroscopic studies have shown that pyridones exist predominantly in the keto form.[61] The XANES data support this finding; the keto form would be expected to yield a π^* resonance close to pyrrole whereas the enol form would yield a resonance close to pyridine. For the pyridones, the resonance is close to pyrrole. In any event, all of the pyridone analogues which have been investigated by XANES spectroscopy produce a large π^* resonance at nearly the same energy. In the determination of the pyridone analogue content in the kerogens and bitumens, the assumption is made that the pyridone model compounds, whatever their

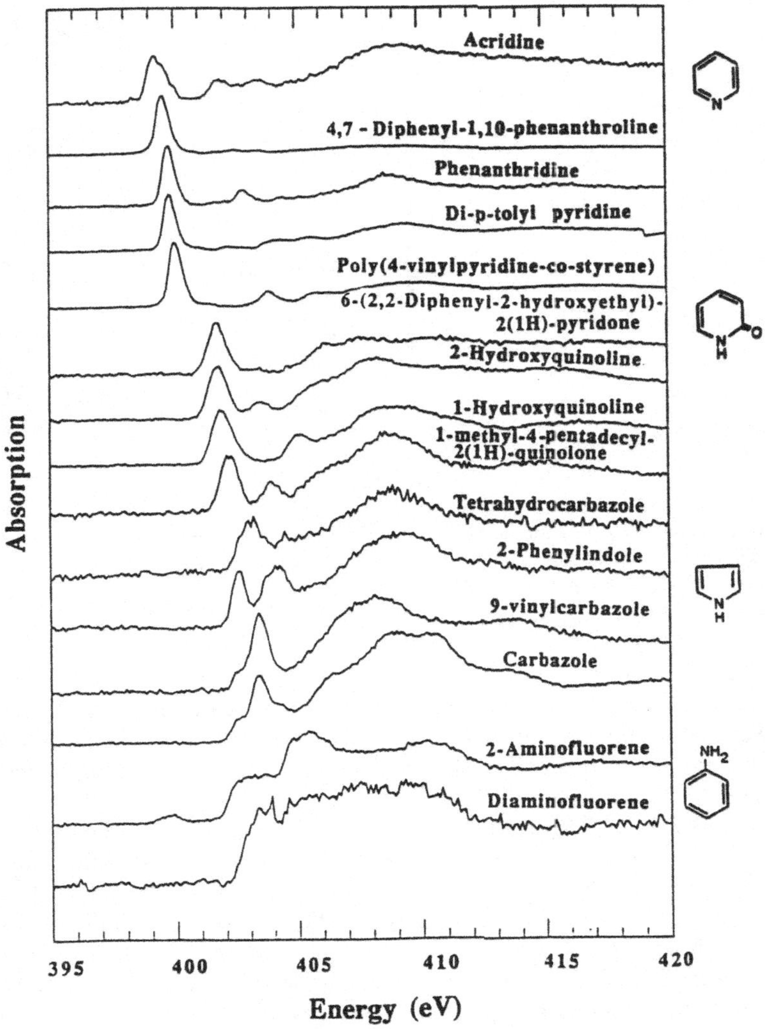

Figure 26. Nitrogen XANES spectra of pyridines, pyridones, pyrroles and aromatic amines.[68] These nitrogen structures are important constituents of kerogens, bitumens and coals. The pyridine π^* resonance is isolated from all others, thus pyridines are readily identifiable. The π^* resonance of pyridones are slightly lower in energy than those of pyrroles allowing differentiation of these nitrogen types in materials such as kerogens. The pyridine derivative quaternary pyridinium has a π^* absorption peak which is nearly indistinguishable from pyridone.[50] Aromatic amines exhibit absorption at 405 eV where other nitrogen model compounds exhibit a spectral valley.

tautomeric form and hydrogen bonding, represent good models for the same chemical function in the fossil fuel samples.

Figure 26 also shows the spectra of aromatic amines. These aromatic amines show strong absorption at approximately 405 eV. For the kerogen and bitumen spectra, the assumption is made that absorption at 405 eV which is not resultant from pyridines, pyridones and pyrroles is due to aromatic amines. Such an analysis presupposes that the aromatic amine fraction is not large. Table 5 lists the results of the analysis of kerogens and bitumens.[54]

Table 5. Nitrogen speciation (%) for kerogens and bitumens

Samples	Pyridine	Pyridone[b]	Pyrrole	Ar. amine	Sat. amine	% N
KerogenGRK-1	20	23	42	15	0	1.09
BitumenGRB-1	27	11	49	11	2	0.83
KerogenGRK-2	34	11	42	2	11	1.03
BitumenGRB-2	20	19	49	12	0	1.47
KerogenK-3	24	20	45	10	0	1.81
Bitumen NAB	14	11	67	8	0	2.60
Bitumen NAB 21% porphyrin[a]	6	11	55	9	0	2.60

[a]Porphyrin fraction estimated from V, Ni content.
[b]Includes pyridinium[50].

Most of the nitrogen is present in aromatic forms; pyrrole is the most dominant nitrogen structure in the kerogens and bitumens, with pyridine being the second most dominant structure. The pyridone fractions are not too large but are always present. The aromatic amine fractions are generally small and the saturated amine is essentially negligible except for one sample. Perhaps, the 11% saturated amine is an overestimation for this sample. Asphaltenes and kerogens are similar with regard to the two dominant fractions; pyrrole and pyridine. Kerogens and bitumens additionally have small quantities of pyridones and aromatic amines. The kerogen and bitumen samples are very similar with respect to their nitrogen molecular structures. Bitumens are much easier to isolate than kerogens; thus, bitumens can be used as models for kerogens with regard to the study of their nitrogen chemistry.

The NAB sample is estimated by the vanadium and nickel content to have 21% of its nitrogen in porphyrin.[54] The optical spectrum of this sample shows a significant Soret band confirming the presence of porphyrins. Close examination of the XANES spectrum of NAB indicates a feature which is probably caused by the porphyrin fraction.[54] The NAB bitumen was analyzed both with and without explicitly accounting for the porphyrin. The results of the two fitting procedures are in good agreement; by accounting for the estimated 21% metalloporphyrin, the pyrrole and pyridine fractions are reduced equally while the other nitrogen fractions are unaffected. In this bitumen, more than half of the pyridine fraction resides in the porphyrin rings.

Asphaltene nitrogen has also been explored using XPS.[62] Figure 27 shows the XPS spectrum of an asphaltene with the assigned contributing peaks. Two separate analyses were performed on this same data by the same investigators. In one analysis all nitrogen was assigned to pyrrole and pyridine, and in the other analysis, four types of nitrogen were examined, pyrrole, pyridine, saturated amine, and quaternary pyridinium nitrogen. Two analyses were performed because of the uncertainties associated with fitting a single peak with four composite peaks.[62] Both analyses indicate that pyrrole dominates. In the second analysis pyrrole spans a range of 42% to 66% for different asphaltenes. Pyridine is the second most important component spanning a range of 16 to 35%. In addition, for some of these asphaltenes, saturated amine nitrogen was found to be present in large quantity, up to 25%.[62] Furthermore, the XPS study indicates that small quantities (<11%) of pyridinium nitrogen are present.[62] The asphaltene XANES data in fig. 23 was reanalyzed to determine the sum of pyridone plus pyridinium; the XANES π^* peak of pyridone overlaps with that of quaternary pyridinium compounds.[50] Aromatic amines were also included in the data fitting. These asphaltenes show low concentrations of pyridone plus pyridinium (<6%) and of aromatic amines (<4%).[54] Thus, the XANES analysis shows that the contribution of both pyridone and pyridinium nitrogen in these asphaltenes is quite small. Note that inorganic quaternary nitrogen does not exhibit a π^* resonance so is not confused with pyridone in XANES analyses.[50]

The XANES and XPS results on different sets of asphaltenes agree with regard to the dominance of pyrrole analogues followed by pyridine analogues. However, there are discrepancies. The XPS study tentatively indicates the presence of large quantities of saturated amine nitrogen whereas the XANES study finds virtually no saturated amines in the asphaltenes. Perhaps, the difference in the results is just due to the differences in the asphaltenes. Nevertheless, the XPS study was aggressive in fitting a single composite XPS peak in terms of four overlapping components. In particular, Fig. 27 shows that the saturated amine XPS peak lies between the large pyrrole and pyridine peaks, and the peak separations are much less than the widths. Thus, the error in the XPS determination of the saturated amine component is probably large.

In addition, the XPS results imply the presence of small amounts of pyridinium nitrogen, generally around 7% in asphaltenes. The XANES results do not indicate any large quantity of pyridone (or pyridinium) nitrogen;[54] these two pyridine analogues have very similar XANES spectra.[50] Small quantities (<6%) of pyridone and pyridinium nitrogen are consistent with the XANES asphaltene results.[54] Thus, the XANES and XPS results are consistent regarding these two pyridine derivatives, although the uncertainties for both methods are probably comparable to the quantities of these pyridine derivatives found.

An important difference in the determination of pyridine and pyrrole fractions between XANES and XPS is related to porphyrin structures. As we have discussed, in XANES, metalloporphyrins and, to a large extent, free-base porphyrins are found to have roughly equal quantities of pyrrolic and pyridinic nitrogen. For samples with significant porphyrin, a large quantity of the pyridinic nitrogen can actually be contained in porphyrin rings. XPS data on free-base porphyrins also show roughly equal quantities of pyrrolic and

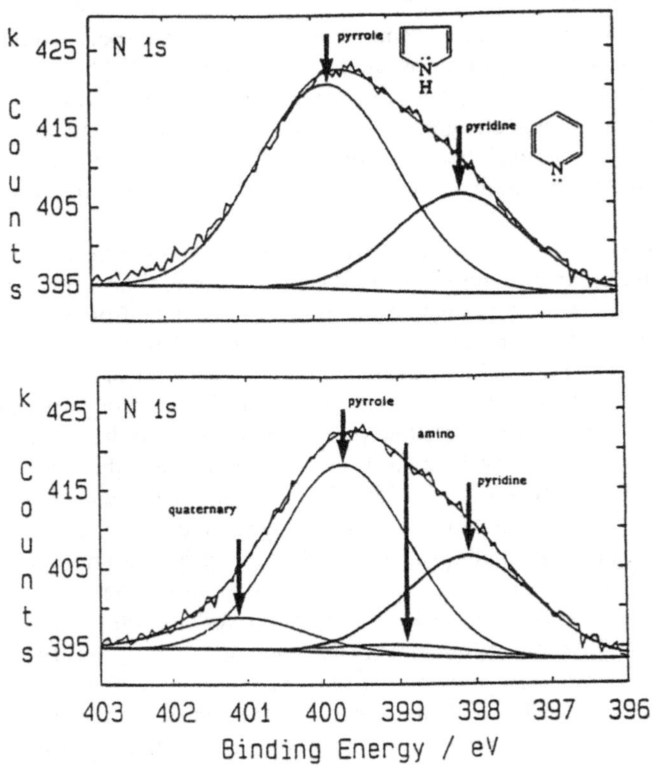

Figure 27. Nitrogen XPS spectrum of an asphaltene.[62] This spectrum is analyzed with a conservative approach where only pyrrole and pyridine are assumed to be present. For the same spectrum, a more aggressive analysis is also shown where the additional components saturated amine and pyridinium nitrogen are also assumed to be present. Conclusions relating to pyrrole and pyridine are most robust while conclusions relating to the saturated amine, with its peak between the peaks of the two dominant nitrogen species, are least robust.

pyridinic nitrogen.[63,64] However, the XPS spectra of metalloporphyrins shows predominantly or solely a pyridine peak, with generally a very small peak in the pyrrole region; this secondary peak is assigned to electron shake-up satellites.[65] Thus, for the same compound, XPS and XANES give a much different interpretation in terms of pyrrolic and pyridinic nitrogen. This may be an advantage of XANES for nitrogen classification; it is more consistent to classify the nitrogen in both free-base and metalloporphyrins in the same manner as in XANES analyses, but unlike XPS analyses. Metalloporphyrins and other compounds originating from chlorophyll contribute to the nitrogen molecular structures found in asphaltenes. Consequently, some differences between XANES and XPS analyses of asphaltenes can be expected. Nevertheless, both XANES[52,54] and XPS[63,64,66,67] appear to accurately reflect the pyrrolic and pyridinic nitrogen fractions in other complex heterocycles.

As with sulfur, it is instructive to compare nitrogen structures found in asphaltenes with those found in coals.[68] Figure 28 shows the XANES spectra of the eight Argonne Premium Coals in order of the coal rank, the highest rank being at the top.[68] The spectra of the different coals were analyzed using the model compounds shown in fig. 26. A preliminary analysis showed that the coals have very small quantities of saturated amines so this group was eliminated from the fitting procedure.[68] Thus, four different chemical groups, pyridines, pyridones, pyrroles, and aromatic amines were used to fit the coal spectra. The results of the analysis are listed in Table 6.[68]

As with asphaltenes, kerogens and bitumens, the pyrroles are predominant. Generally, the quantity of pyridines are quite large in all of the fossil fuel samples. In contrast to asphaltenes, but as in kerogens and bitumens, coals contain small quantities of pyridones and aromatic amines. However, for low rank coals, a dramatic decrease in pyridine is found. The decrease in pyridine in low rank coals correlates with a large increase in the pyridone fraction as well as a large increase in oxygen for low rank coals. Fig. 29 graphically illustrates this correlation. The loss of oxygen is known to occur with coal maturation. It appears that the oxygen in pyridone is driven off with maturation resulting in the production of pyridine. The small pyridine fraction and large pyridone fraction for low rank coals, in contrast to high rank coals is evident directly in the raw data. Fig. 30 shows a low rank and high rank coal.

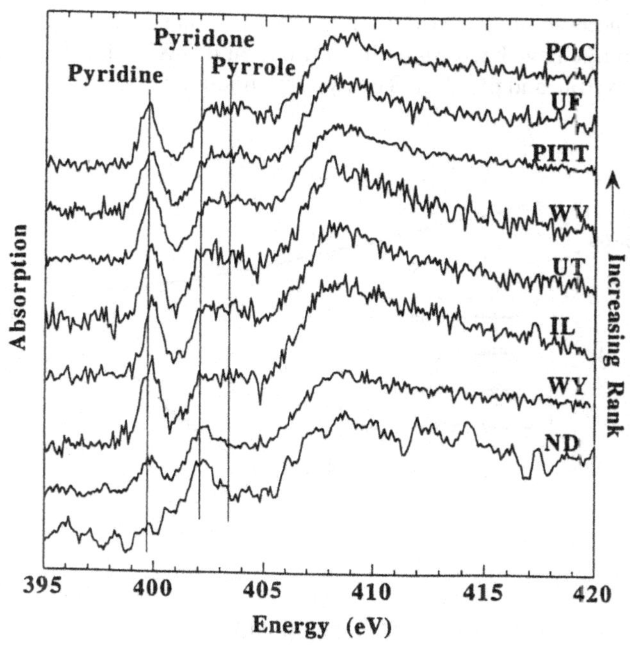

Figure 28. Nitrogen XANES spectra for the Argonne Premium Coals, from high rank (top) to low rank (bottom).[68] Pyridine, pyridone and pyrrole features are readily identified in the data. For the two low rank coals, the pyridine fraction drops to very low values, the pyridine peak at ~400 eV becomes very small; while the pyridone fraction becomes quite large as seen by the large peak at 402 eV. Low rank coals have large quantities of pyridone, but small quantities of pyridine. Evidently, with maturation, pyridone is converted to pyridine. The XANES pyridinium peak overlaps with the XANES pyridone peak; thus, pyridinium is included in the pyridone fraction. However, as discussed in the text, pyridinium is evidently present in smaller quantities than pyridone for lowest rank coals.

Table 6. Nitrogen speciation (%) in Argonne Premium Coals

Sample	Pyridine	Pyridone[a]	Pyrrole	Aromatic amine
POC	18	8	66	8
UF	18	8	66	8
PITT	20	6	65	9
WV	20	16	55	9
UT	17	15	60	8
IL	20	19	54	7
WY	10	29	51	10
ND	2	42	50	6

[a]Includes pyridinium[50].

Comparison of the pyridine resonance at 399.5 eV with the pyridone resonance 402 eV illustrates the substantial compositional differences in these two coals.[68]

The effect of aromatic amines on the coal spectra is somewhat difficult to display in the raw data because the coals appear to have relatively small and invariant fractions of aromatic amines. Thus, the assignment of the aromatic amine fraction in coals by XANES is somewhat uncertain. Nevertheless, the effect of aromatic amines can be observed when comparing coal spectra with spectra of a related system, petroleum asphaltene. Figure 31 shows on an expanded scale, the spectra of a high rank coal and a petroleum asphaltene. The petroleum asphaltene shows a shallow spectral valley at 405 eV while the coal shows almost no valley.[68] The aromatic amines exhibit a peak or, at least, a large signal increase at 405 eV (see fig. 26). Coals and petroleum asphaltenes are similar with regard to their high concentrations of varied pyrroles and pyridines. However, coals evidently also have an aromatic amine fraction which produces absorption at 405 eV, thereby "filling in" the spectral valley at this energy. Aromatic amines are known constituents in many coal derived materials.[69]

These XANES results on coals can be compared to XPS results for coals; in one case for the same set of coals. Early XPS experiments on coals showed the importance of pyrrolic and pyridinic structures, with the former in higher concentration.[70] Pyrolysis leads to an increase in the pyridine fraction in the char.[70] Coals spanning the bituminous to anthracite range have been studied; the fitting of this XPS data was restricted to two types of nitrogen, pyrrole and pyridine. The pyridine fraction varied from 14% to 35% and generally increased

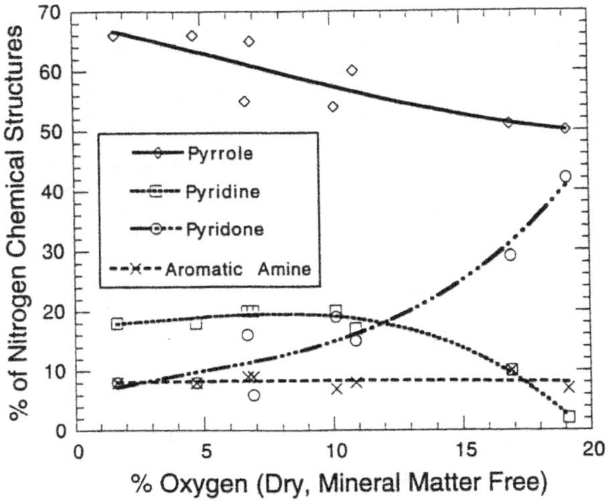

Figure 29. The fraction of nitrogen constituents in coals vs the coal oxygen content.[68] The oxygen content of coals decreases with increasing rank. The pyrrole fraction dominates all coals. Pyridone, an oxygen-containing pyridine derivative, is present in large quantities only for low rank coals. Note the dramatic loss of pyridone and the growth of pyridine with loss of oxygen, particularly for low rank coals. Aromatic amines are always present in small quantities.

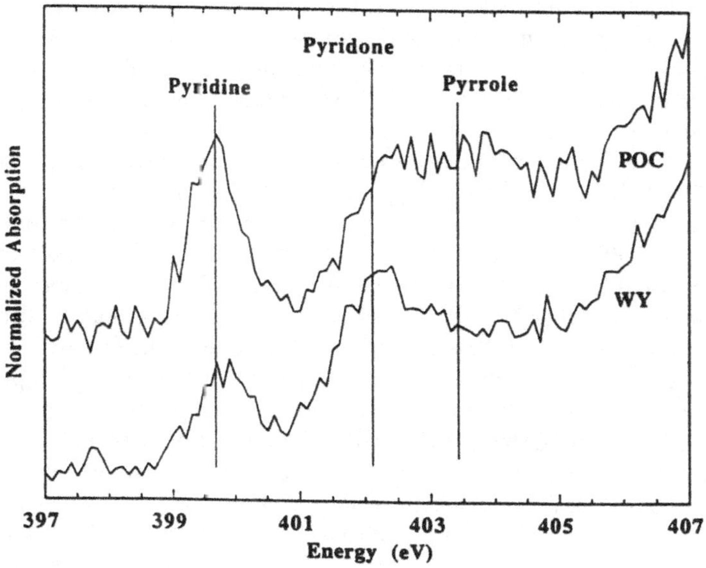

Figure 30. Nitrogen XANES spectra of a high rank (POC) and a low rank (WY) coal plotted with the same normalization.[68] The large pyridine peak and small pyridone peak of the high rank coal contrasts the small pyridine peak and large pyridone peak of the low rank coal. These observations are obtained directly from the data without the need of complicated data interpretation.

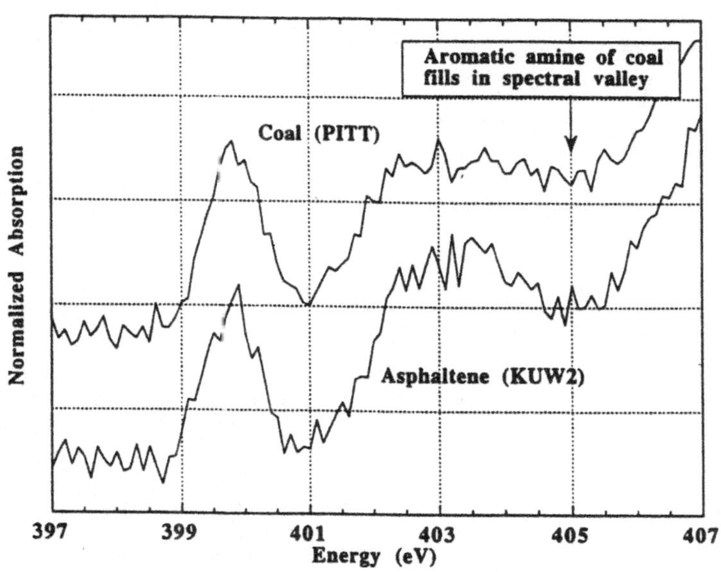

Figure 31. The nitrogen XANES spectra of a coal and an asphaltene showing the effect of the aromatic amine component of coals.[68] A small spectral valley at 405 eV is observed for the asphaltene. The aromatic amine component of the coal is thought to "fill in" this spectral valley in the coal spectrum.

with rank, except for the highest rank coals. For one coal with a pyridine fraction of 15%, the pyridine in the separate macerals ranged form 6% to 10%. It is not clear why all the maceral fractions contain pyridine fractions less than the whole coal. XPS studies have been extended to include other coals and coal derived products. Again, pyrrole and pyridine are the main components of these materials.[71]

More recently, nitrogen XPS analyses of coal have been revisited,[72] using the same eight coals which have been studied by XANES.[68] Pyrrole, pyridine and pyridinium complexes were used to analyze the data. Figure 32 shows the XPS spectra for three of these coal samples. The XPS data were collected for pyridone; the pyridone peak was found to be too close to the pyrrole peak to allow for the determination of the pyridone quantity. This work corroborates the earlier XPS and XANES findings that the primary forms of nitrogen in coals are pyrrole and, to a lesser extent, pyridine. The largest values of pyridinium nitrogen were obtained for the lowest rank coals to be 16%; with the percentage dropping to 3% for high rank coals. Thus, with maturation, pyridinium nitrogen is converted to pyridine. This is similar to the finding in XANES that with maturation, pyridone is converted to pyridine.

A XANES study was directed to look for the pyridinium component of coal.[50] The XANES pyridinium π* peak was found to be unresolvable from the pyridone peak, although at a slightly lower energy. In other words, both pyridine derivatives, pyridone and pyridinium yield similar XANES spectra. The XANES and XPS techniques are thus found to be complimentary. XPS methods are able to resolve pyridinium nitrogen, but not pyridone,

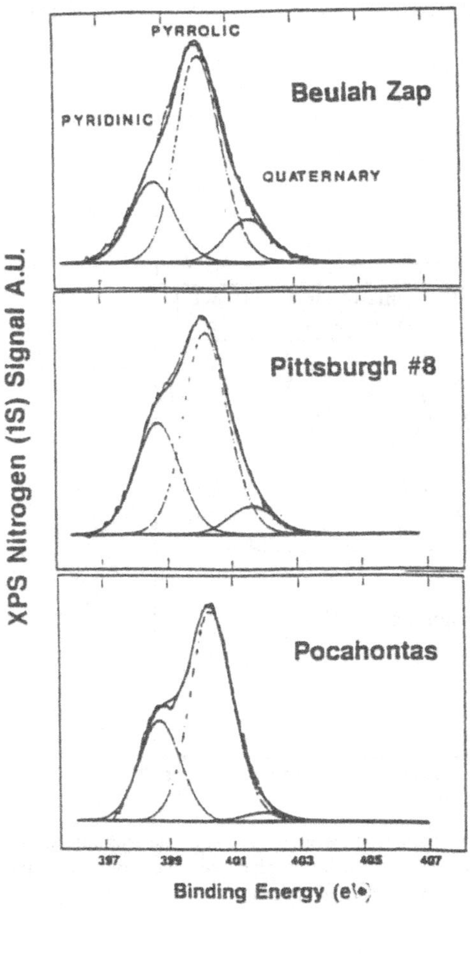

Figure 32. The nitrogen XPS spectra of three Argonne Premium Coals, Beulah-Zap (ND), Pittsburgh #8 (PITT), and Pocahontas (POC), showing the dominance of pyrrolic nitrogen, the large fraction of pyridinic nitrogen and the contribution of pyridinium nitrogen.[72] The XPS pyridone peak overlaps with the XPS pyrrole peak and cannot be fit separately. The XANES and XPS data on the same set of eight coals can be compared directly. General agreement with the two studies is found. Specific disagreements regarding low rank coals, and absolute quantities are discussed in the text.

while XANES methods can resolve the sum of both pyridone and pyridinium nitrogen. The XANES study found 42% pyridone in the lowest rank coal and 29% pyridone in the second lowest rank coal.[68] The XPS study found 16% pyridinium nitrogen in the lowest rank coals and 15% in the second lowest rank coal.[72] Thus, the conclusion from the combined results is that lowest rank coals contain significant pyridone as well as some amount of pyridinium nitrogen.[72] Both of these pyridine derivatives decrease in high rank coals.

The XANES and XPS data sets on the same coals do show some discrepancies. For instance, in XPS, the pyridone peak is closer to the pyrrole peak than to the pyridinium peak. Thus, it would appear that the majority of the pyridone component would be assigned to pyrrole in the XPS study. However, the sum of the pyrrole plus pyridone from the XANES study does not add to the pyrrole quantity found in XPS. Some of the variation could simply be due to uncertainties in the determinations. XPS analyses suffer from overlapping peaks giving rise to uncertainties in compositional analyses. For nitrogen XANES analyses, the pyridine peak is isolated and easily resolved. However, the pyridone and pyrrole peaks are close yielding some uncertainties.

We have already discussed the difference between XANES and XPS classifications of metalloporphyrins where XANES yields equal quantities of pyrrolic and pyridinic nitrogen while XPS analyses yield predominantly pyridinic nitrogen. Thus, perhaps some components of coals produce larger pyridine estimates from XPS than from XANES. Another possible source of difference between the XANES and XPS results corresponds to the very different depth of investigation in the sample. Nitrogen XANES using x-ray fluorescence detection has a depth of investigation of approximately 1/3 micron into the sample. Nitrogen XANES with electron yield detection has a smaller depth of investigation of about 10 nm. XPS requires electrons to leave the surface with no inelastic collisions; the depth of penetration for XPS is very small, 0.5-2 nm. Thus, XPS is really a surface measurement. The difficulties associated with the very small depth of investigation in XPS studies of coals have been investigated primarily by comparing surface to bulk concentrations of different elements.[71] If surface and bulk concentrations are not the same, then the XPS results may not be directly applicable to the bulk sample. For coals, some surface depletion of nitrogen has been found, particularly for high nitrogen coals.[71,73] Surface depletion of nitrogen was found for some of the Argonne Premium Coals.[72] Perhaps some of the differences between the XANES results and XPS results are associated with surface effects, particularly for chemical species involving oxygen, such as pyridones.

SUMMARY

XANES spectroscopy has been shown to be very effective in the analysis of molecular structures of sulfur and nitrogen in high molecular weight, complex mixtures such as asphaltenes, kerogens, bitumens, and coals. The large separation of resonances in XANES spectroscopy of various model compounds allows for the inclusion of several different molecular structures in the analyses and lends credence to the corresponding results. The entire sample is analyzed without the requirement of separation, and the normalization of the constituent chemical structures is readily accomplished. The relatively small mass fractions of heteroatoms in these materials is counterbalanced by the element specific nature of XANES methods.

The dependence of the energy of the 1s-3p XANES sulfur peak on the sulfur chemical class provides a straightforward and reliable method of analysis. Different oxidation states of sulfur are readily identified, and most importantly, thiophenes and sulfides are resolvable. All sulfur of asphaltenes is organic, while coals also contain appreciable quantities of pyrite, and the lowest rank coals contain sulfate. Thiophenes dominate the sulfur fraction of asphaltenes and the organic sulfur of coals. For many natural materials, the thiophene fraction increases with greater thermal maturity. Sulfides are also found in appreciable quantity. In asphaltenes, reduced forms of sulfur predominate, although oxidized sulfur is sometimes found, virtually all in the sulfoxide form. Coals do not contain significant organic, oxidized sulfur. It appears that some of the asphaltene sulfides are susceptible to oxidation and can convert to the sulfoxide, further suggesting that the sulfoxides which are present in asphaltenes are not aromatic. Some of this oxidation may be subsequent to the extraction of the crude oil from the ground. Similar observations regarding sulfide lability apply to the organic sulfur component of coal. The finding that both aromatic and alkyl sulfur are present in asphaltenes is similar to asphaltene carbon where significant quantities of both aromatic and saturated forms of carbon are found.

Nitrogen XANES studies of natural materials are greatly enhanced by the large separation of the 1s-π^* peaks of pyrrolic (nonbasic) and pyridinic (basic) nitrogen. Even complex aromatic rings systems which contain multiple nitrogen atoms, such as porphyrins and DNA bases, are found to be amenable to this classification for XANES analyses. Also, pyridones, aromatic amines and saturated amines produce spectral signatures, allowing for their identification and quantification. Virtually all of the nitrogen found in asphaltenes is contained within aromatic rings and is classified as pyrrolic or pyridinic. Essentially no saturated amine and, at most, only small quantities of aromatic amines are suggested by the data. In related materials such as kerogens, bitumens and coals, pyridones, and small quantities of pyridinium compounds are found to be present. With increasing maturation of coal, and with the conversion of kerogen in part to asphaltene, these oxidized pyridine analogues are lost, probably by conversion to pyridine. In addition, small quantities of aromatic amine are found in the coals, but saturated amines are absent.

XANES methods especially when combined with complementary methods such as XPS provide great insight into the molecular structures of asphaltenes, coals and related materials. The detailed analyses of sulfur and nitrogen have proven to be very fruitful. In the future, XANES studies of fossil fuel samples utilizing third generation synchrotron light sources with wigglers and undulators should provide much higher x-ray flux needed to improve significantly the signal to noise ratio, particularly for nitrogen. The resulting enhanced spectra should confirm the conclusions of existing XANES studies as well as provide greater precision for sulfur and especially nitrogen speciation. Perhaps the most promising, if challenging, new technique for heteroatom studies in fossil fuel samples is high resolution measurement of x-ray fluorescence. Emission x-ray fluorescence spectra are very sensitive, not only to the chemical function excited, but also to the particular resonance excited. Combining x-ray emission spectroscopy to the methods of x-ray excitation spectroscopy should yield a tremendous advance in heteroatom analysis.

ACKNOWLEDGMENTS

The author gratefully acknowledges Professor S.P. Cramer of U.C. Davis and of LBL for many illuminating and insightful discussions as well as for his many important suggestions upon his critical review of this manuscript. The author is also greatly indebted to Professor Sudipa Mitra-Kirtley of Rose-Hulman Institute and Dr. Jie Chen of U.C. Davis for many stimulating scientific discussions and, further, for their much appreciated reviews of this manuscript.

REFERENCES

1. G.V. Chilingarian, T.F. Yen, (Ed.), "Bitumens, Asphalts and Tar Sands," Elsevier Pub. Co. Amsterdam (1978)
2. J.W. Bunger, N.C. Li, (Ed.), "Chemistry of Asphaltenes," Amer. Chem. Soc., Wash D.C. (1981)
3. J.G. Speight, "The Chemistry and Technology of Petroleum," Marcel Dekker, Inc. New York (1980)
4. B.P. Tissot, D.H. Welte, "Petroleum Formation and Occurrence," Springer Verlag, Berlin (1984)
5. T.F. Yen, (Ed.) "The Role of Trace Metals in Petroleum," Ann Arbor Science Pub Inc., Michigan (1975)
6. G. Margaritondo, "Introduction to Synchrotron Radiation," Oxford University Press, New York (1988)
7. D.E. Eastman, Y. Farge, (Ed.), "Handbook on Synchrotron Radiation," North-Holland Pub. Co. Amsterdam (1983)
8. J. Stohr, "NEXAFS Spectroscopy," Springer-Verlag, Berlin (1992)
9. J. Goulon, C. Goulon-Ginet, R Cortes, J.M. Dubois, J. Phys. 43, 539 (1982)
10. P.A. Lee, P.H. Citrin, P. Eisenberger, B.M. Kincaid, Rev. Mod. Phys., 53, 769 (1981)
11. T.A. Tyson, K.O. Hodgson, C.R. Natoli, M. Benfatto, Phys. Rev B, 46, 5997 (1994)
12. J.L. Dehmer, J. Chem. Phys., 56, 4469 (1972)
13. H. Nakamatsu, T. Mukoyama, H. Adachi, Chem. Phys. 143, 221 (1990)
14. T.A. Ferret, D.W. Lindle, P.A. Heimann, H.G. Kerkhoff, U.E. Becker, D.A. Shirley, Phys. Rev. B 34, 1916 (1986)
15. S.P. Cramer "Extended X-Ray Absorption Fine Structure," D. Konigsberger, R. Prins, (Ed.) Plenum, New York (1988)
16. J.L. Dehmer, D. Dill, J. Chem. Phys., 65, 5327 (1976)
17. E.A. Stern, S.M. Heald, in "Handbook on Synchrotron Radiation," D.E. Eastman, Y. Farge, (Ed.), North-Holland Pub. Co. Amsterdam (1983)
18. S.S. Hasnian, J.R. Helliwell, H. Kamitsubo, J. Synchrotron Rad., 1, 1 (1994)
19. C.T. Chen, Nucl. Instrum. Methods Phys. Res., A256, 595 (1989)
20. C.T. Chen, F. Sette, Rev. Sci. Instrum. 60, 1616 (1989)
21. C.T. Chen, Rev. Sci. Instrum. 63, 1299 (1992)
22. Z. Hussain, E. Umbach, D.A. Shirely, J. Stohr, J. Feldhaus, Nucl. Instrum. Methods, 195, 115 (1982)
23. C.L. Spiro, J. Wong, F.W. Lytle R.B. Greegor, D.H. Maylotte, S.H. Lamson, Science, 226, 48 (1984)
24. B. Hedman, P. Frank, J.E. Penner-Hahn, A.L. Roe, K.O. Hodgson, R.M.K. Carlson, G. Brown, J. Cerino, and R. Hettel. Nucl. Instrum. Methods Phys. Res., A246, 797 (1986).
25. G.N. George, M.L. Gorbaty, J. Amer. Chem. Soc., 111, 3182, (1989)
26. G.P. Huffman, F.E. Huggins, S. Mitra, N. Shah, R.J. Pugmire, B. Davis, F.W. Lytle, R.B. Greegor, Energy & Fules 3, 200, (1989)
27. G.S. Waldo, R.M.K. Carlson, J.M. Moldowan, K.E. Peters, J.E. Penner-Hahn, Geochim. Cosmochim. Acta, 55, 801 (1991)
28. S.P. Cramer, F.M.F. deGroot, C.T. Chen, F. Sette, C.A. Kipke, D.M. Eichhorn, M.K. Chan, W.H. Armstrong, E. Libby, G. Christou, S. Brooker, V. McKee, O.C. Mullins, J.C. Fuggle, J. Amer. Chem. Soc., 113, 7937 (1991)
29. G.P. Huffman, S. Mitra, F.E. Huggins, N. Shah, S. Vaidya, F. Lu, Energy & Fuels, 5, 574 (1992)
30. G.S. Waldo, O.C. Mullins, J.E. Penner-Hahn, S.P. Cramer, Fuel, 71, 53 (1992)
31. J.-M. Ruiz, B.M. Carden, L.J. Lena, E.-J. Vincent, J.-C. Escalier, Anal. Chem. 54, 688 (1982)
32. M.L. Gorbaty, G.N. George, S.R. Kelemen, Fuel 69, 1065 (1990)
33. G.P. Huffman, F.E. Huggins, F.E. Francis, S. Mitra, N. Shah, "Process. Util. High-Sulfur Coals," R. Markuszewski, T.D. Wheelock, (Ed.), Elsevier, New York (1990)
34. R.C. Neaval, "Chemistry of Coal Utilization," M.A. Elliott, (Ed.) John Wiley & Sons, New York (1981)
35. R.M. Davidson, "Coal Science," I.G. Dryden, (Ed.) Academic Press, New York, Vol. 1 (1982)
36. M.W. Haenel, Fuel, 71, 1213 (1992)
37. M.L. Gorbaty, G.N. George, S.R. Kelemen, Fuel 69, 945 (1990)
38. S.R. Kelemen, G.N. George, M.L. Gorbaty Fuel 69, 939 (1990)
39. K.S. Vorres, Energy & Fuels, 4, 420 (1990)
40. L.M. Stock, R. Wolny, B. Bal, Energy & Fuels, 3, 651 (1989)
41. A.G. Douglas, B.J. Mair, Science, 147, 499 (1965)
42. M. Kasrai, J.R. Brown, G.M. Bancroft, K.H. Tran, J.-M. Chen, Fuel, 69, 411 (1990)
43. S.P. Cramer, O. Tench, M. Yocum, H. Kraner, L. Rogers, V. Radek, O.C. Mullins, S. Rescia, "X-ray Absorption Fine Structure -Proceedings of the 6th Internl. XAFS Conf." S.S. Hasnian (Ed.), Ellis Horwood, Chichester 640 (1991)
44. C.T. Chen, Y. Ma, F. Sette, Phys, Rev. A, 40, 6737 (1989)

45. J. Stohr, R. Jaeger, Phys. Rev. B26, 4111 (1982)
46. J.L. Dehmer, D. Dill, Phys. Rev. Lett. 35, 313 (1975),
47. J. Stohr, J.L. Gland, W. Eberhardt, D. Outka, R.J. Madix, F. Sette, R.J. Koestner, U. Doebler, Phys. Rev. Lett., 51, 2414, (1983)
48. S. Mitra-Kirtley, O.C. Mullins, J. van Elp, S.P. Cramer, Fuel, 72, 133 (1993)
49. S. Mitra-Kirtley, O.C. Mullins, J. van Elp, S.J. George, J. Chen, S.P. Cramer, J. Amer. Chem. Soc., 115, 252 (1993)
50. O.C. Mullins, S. Mitra-Kirtley, S.P. Cramer, private communication
51. M. Gouterman, "The Porphyrins," D. Dolphin, (Ed.), Academic Press, New York (1978)
52. S. Mitra-Kirtley, O.C. Mullins, J. Chen, J. van Elp, S. George, C.T. Chen, T. O'Halloran, S.P. Cramer, Biochim. Biophys. Acta, 1132, 249 (1992)
53. W. Saenger, "Principles of Nucleic Acid Structure," Springer-Verlag, New York (1984)
54. S. Mitra-Kirtley, O.C. Mullins, J.F. Branthaver, S.P. Cramer, Energy & Fuels, 7, 1128 (1993)
55. A.L. Johnson, E.L. Mutterties, J. Stohr, F. Sette, J. Phys. Chem., 89, 4075 (1985)
56. P. Skytt, J. Guo, N. Wassdahl, J. Nordgren, submitted Phys. Rev. Lett.
57. B. Durand, (Ed.) "Kerogen: Insoluble Organic Matter from Sedimentary Rocks," Editions Technip., Paris (1978)
58. E.C. Copelin, Anal. Chem. 36, 2274 (1964)
59. J.C. Petersen, R.V. Barbour, S.M. Dorrence, F.A. Barbour, R.V. Helm, Anal. Chem., 43, 1491 (1971)
60. S.P. Preece, J.F. Branthaver, S.-S. Kim, Div. Fuel Chem., Amer. Chem. Soc. Preprints, 37, 1342 (1992)
61. J.C. Petersen, J. Phys. Chem., 75, 1129 (1971)
62. A. Wilhelms, R.L. Pateience, S.R. Larter, S. Jorgensen, Geochim. Cosmoschim. Acta, 56, 3745 (1992)
63. M.Z. Zeller, R.G. Hayes, J. Amer. Chem. Soc. 95, 3885 (1973)
64. P.G. Gassman, A. Ghosh, J. Almlof, J. Amer. Chem. Soc., 114, 9990 (1992)
65. S. Muralidharan, R.G. Hayes, J. Amer. Chem. Soc., 102, 5107 (1980)
66. L.E. Hamilton, P.A. Sherwood, B.M. Reagan, Appl. Spectrosc., 47, 139 (1993)
67. Y. Niwa, H. Kobayashi, T. Tsuchiya, Inorg. Chem. 13, 2891 (1974)
68. O.C. Mullins, S. Mitra-Kirtley, J. van Elp, S.P. Cramer, Appl. Spectrosc. 47, 1268 (1993)
69. D.W. Later, M.L. Lee, B.W. Wilson, Anal. Chem. 54, 117 (1982)
70. R.B. Jones, C.B. McCourt, P. Swift, "Proceedings Internl. Conf. on Coal Science, Dusseldorf," Verlag Gluckauf, Essen 657 (1981)
71. S. Wallace, K.D. Bartle, D.L. Perry, Fuel, 68, 1450 (1989)
72. S.R. Kelemen, M.L. Gorbarty, P.J. Kwiatek, Energy & Fuels, 8, 896 (1994)
73. D.L. Perry, A. Grint, Fuel, 62, 1029 (1983)

Chapter III

SOLUBILITY AND PHASE BEHAVIOR OF ASPHALTENES IN HYDROCARBON MEDIA

Roberto Cimino, Sebastiano Correra, Alberto Del Bianco, and Thomas P. Lockhart

Eniricerche S.p.A.
20097 San Donato Milanese, Italy

INTRODUCTION

Several decades of experimental and theoretical investigation notwithstanding, only a partial description of the physical state of asphaltenes in crude oil and hydrocarbon solvents and of their tendency to undergo phase separation has been achieved. The strong interest in developing a better understanding of the solution behavior of asphaltenes is motivated by the impact of their phase separation on the production, transportation, refining, and utilization of crude oil. The tendency of asphaltenes to sediment and flocculate during storage and transportation, for example, is a serious economic issue. Less familiar, perhaps, are the production problems that can be caused by the formation of asphaltene deposits within the reservoir or wellbore.

Our own interest in asphaltene phase behavior was stimulated when asphaltene deposition within the reservoir rock was identified as the cause of the sharp productivity decline of the wells in a major Italian oilfield.[1] A remedial treatment was quickly developed which involved soaking the formation surrounding the afflicted wells with a large volume of aromatic solvent in order to dissolve the asphaltene deposits. Though it was possible by this means to recover much of the wells' productivity, the relative inefficiency and significant cost of these treatments stimulated the organization of a research program focused on (1) identifying a more effective hydrocarbon solvent and (2) developing a reliable physical model that would serve to optimize production from the afflicted reservoir. It was expected, of course, that the thermodynamic model, if valid and general, would be applicable to other crude oils and reservoirs as well as to asphaltene problems in other phases of the petroleum cycle.

It quickly became clear to us that the successful realization of these objectives depended on the development of a more accurate description of the solution behavior of asphaltenes in hydrocarbon media. As we will show below, key elements that had to be taken into account in order to achieve this insight include the compositional heterogeneity of asphaltenes, their tendency to form complex, aggregated structures in hydrocarbon solvents, and the nature of their interaction with the surrounding hydrocarbon medium. In this Chapter

Asphaltenes: Fundamentals and Applications, Edited by E. Y. Sheu and O. C. Mullins
Plenum Press, New York, 1995

we critically examine the current understanding of the chemical nature of the asphaltenes and of their physical state in hydrocarbon solutions, including the approaches that have been pursued in modelling their behavior in hydrocarbon media. We also describe the experimental and theoretical approaches that have been brought to the study asphaltene solution behavior in our laboratories. These are based above all on an appreciation of the compositional heterogeneity of the asphaltene component of crude oils and on the decision to treat the asphaltenes as lyophilic (*i.e.*, solvated) colloids. Parting from these ideas, we have constructed and validated a conceptual framework that is capable of rationalizing the solution behavior of the asphaltenes. This has been used, on the one hand, to identify more powerful solvents for the remedial treatment of reservoirs in which asphaltene deposition constitutes a problem for production (see Part I) and, on the other, to develop a useful, predictive theoretical model (Part II).

PART I. ASPHALTENE FRACTIONATION AND SOLUBILITY STUDIES

The Functional Definition of Asphaltenes

Petroleum is an extremely complex mixture of hydrocarbons whose composition and physical-chemical properties vary continuously from the simplest paraffinic structures (hydrocarbon gases) to macromolecules composed of many tens of carbon atoms, heteroatoms, and metals.[2] In view of this fact, the methods developed for the characterization of petroleum first simplify the mixture by separating it, as far as possible, into fractions with homogeneous physical-chemical properties. The primary fractionation is by distillation, which effects separation on the basis of volatility. Once the distillation fractions have been obtained, they can be further fractionated into homogeneous classes of compounds according to their polarity.

As for the distillation residues, generally one operates so as to separate them into four classes of compounds: saturates, aromatics, resins, and asphaltenes. The fractionation methods employed rely on selective precipitation (for the asphaltenes) and/or chromatographic techniques.[3-15] Of the four classes of compounds, only the saturates are distinguishable from the rest of the hydrocarbons in the mixture on the basis of objective criteria. The absence of π -bonds in the constituents of this class allows them to be easily differentiated and therefore separated from the aromatic component by virtue of the difference in their polarities. The remainder of the residue is composed of aromatic and heteroaromatic compounds of varying degrees of condensation, alkyl substitution, and functionalization. This mixture constitutes a *compositional continuum* with respect to both molecular weight and polarity.[2,3] Distinctions between the aromatic, resin, and asphaltene fractions, therefore, are inherently arbitrary and must be made in the light of the conditions employed to effect their separation. This point is illustrated by the precipitation of the asphaltenes, whose composition is strongly influenced by the nature of the precipitating solvent and the operating conditions. It follows from these considerations that standardized analytical conditions must be defined in order to compare the asphaltene content of different crude oils.[16-18]

The term "asphaltene," then, identifies a *solubility class* which is distinguishable from the remainder of the crude oil (de-asphalted oil, "DAO") in being insoluble in the precipitating solvent under the operating conditions employed, but which may be composed of molecules that differ significantly in their chemical characteristics (*e.g.*, molecular weight, aromaticity, functional groups, etc.). Conceptually, the separation of the asphaltenes by precipitation can be likened to the fractionation of the volatile components of petroleum by

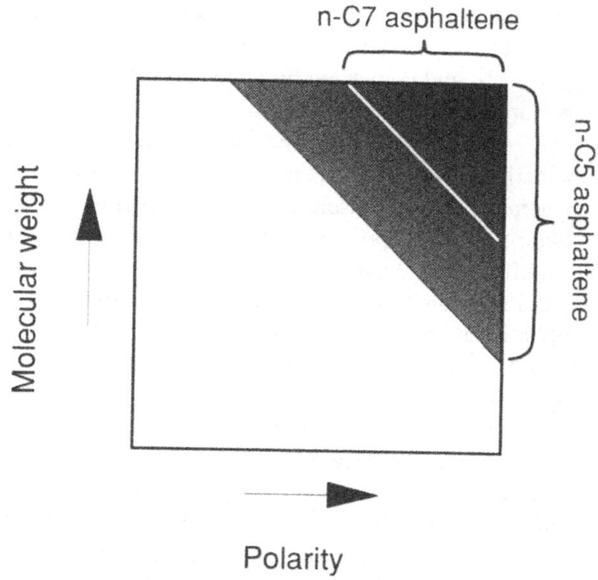

Figure 1. Hypothetical diagram representing the molecular characteristics of the asphaltenes precipitated from petroleum by n-alkane solvent addition (from ref. 20).

distillation: in both cases separation is effected with reference to an arbitrarily defined physical-chemical cut-off point (*i.e.*, boiling point or solvent quality).[19] This consideration has been illustrated by Long,[20] who employed a two-dimensional graph to represent the range in molecular weight and polarity of the molecules comprising a given petroleum, and diagonal lines to indicate, for each precipitating solvent, the characteristics of the asphaltene fraction unstable with respect to precipitation (Figure 1).

Just as the asphaltene fraction of a given crude oil contains a distribution of chemical species, there is significant variation in the average chemical characteristics (aromaticity, degree and nature of alkyl substitution, and heteroatom and metal content) of asphaltene fractions isolated from different oils.[21-27] This is shown in Table 1, which indicates the range of compositions found for n-C7 (n-heptane) asphaltenes precipitated from different crude oils.

In this regard, NMR methods[28-29] have proven extremely useful for defining the average molecular parameters of asphaltene fractions. Through the application of ^{13}C-NMR,[30-31] Distorsionless Enhancement by Polarization Transfer (DEPT),[32-33] Gated Spin Echo (GASPE),[34-35] and 2D-NMR,[36-37] the hydrocarbon skeleton can be described to a high

Table 1. Range of compositions found for asphaltenes precipitated from crude oils with n-heptane

Yield of precipitate (wt.%)	Up to 30
H/C ratio	0.8–1.4
S (wt.%)	0.5–10.0
N (wt.%)	0.6–2.6
O (wt.%)	0.3–4.8
fa[1]	0.45–0.70
n[2]	4–7

[1]Aromaticity factor (fraction of aromatic carbon) from ^{13}C-NMR.

[2]Number of alkyl C atoms per side chain.

degree of detail. The picture of the asphaltenes deduced from these studies is that of aromatic polycyclic clusters variably substituted with alkyl chains that may be quite long (up to C10-C12), and connected by alkyl and heteroatom bridges. The degree of condensation of each aromatic cluster may be more or less elevated, but generally does not exceed 5-6 rings.[38]

The presence of functional groups contributes strongly to the physical-chemical properties of the asphaltenes.[2,39,40] Sulfur, by far the predominant heteroatom, occurs principally in condensed, thiophenic structures (benzo-, dibenzo- and naphthobenzothiophene), but is also present in aliphatic (sulfides and disulfides) and oxidized forms.[41-45] Often the sulfides and disulfides are present as bridges between hydrocarbon clusters.[46] Nitrogen-containing functional groups have been shown to be both basic in nature (in the form of aliphatic and aromatic amines, particularly primary amines and pyridine derivatives) and neutral (as indoles, carbazoles, and amides as well as porphyrinic nitrogen[47-49]). The application of X-ray Absorption Near-Edge Structure (XANES) spectroscopy to the study of asphaltenes has led to the conclusion that a large portion of the nitrogen present is aromatic, in pyrrolic rather than pyridinic form.[50] Oxygen is present predominantly in the form of acidic functional groups (carboxylic acids and phenols), though ketonic and condensed derivatives such as furan have also been identified.[51,52]

The organometallic compounds present in petroleum contain predominantly the metals Ni, V, and Fe. These compounds are concentrated in the heaviest part of the crude oil and, not surprisingly, the asphaltene fractions are rich in these metals. A significant part of this organometallic component is porphyrinic in nature[53-56] though many other structures, not yet better identified, are also present.[57-59]

The tendency of the asphaltenes to form aggregates in hydrocarbon solution is one of their most characteristic features (see below) and complicates the determination of their molecular weight. The average molecular weights measured by means of vapor pressure osmometry (VPO) or size exclusion chromatography (SEC) are significantly influenced by the conditions of the analysis (temperature, asphaltene concentration, solvent polarity).[23,60-70] For this reason, molecular weights reported in the literature have relative value only and can be far removed from the values for the single, unassociated molecules.

Speight showed that VPO molecular weights vary significantly in solvents with different dielectric constants.[23] This point was confirmed by Al Jarrah *et al.*, who carried out systematic VPO measurements on the n-C6 asphaltenes from an Iraqi crude.[66] They found values of the number average molecular weight (M_n) ranging between 935 and 16,840 amu for the solvents nitrobenzene, tetrahydrofuran (THF), chlorobenzene, and benzene, and decreasing from 13,900 to 6330 in THF as the temperature was raised from 37° to 60°C. Similarly, Chung *et al.*[68] have reported on the influence of the asphaltene concentration on the values of molecular weight measured, and pointed out the importance of working at as high a dilution as possible in order to measure a molecular weight that at least approaches that of the unassociated asphaltene monomers. Conditions allowing the determination of the "true" molecular weight of the asphaltene component (that is, the mean molecular weight of the asphaltene molecules in the absence of associative interactions) have not yet been established. In comparison to VPO, SEC offers the advantage of giving the molecular weight distribution rather than a single value for M_n. In this case too, however, the conditions of analysis and the choice of molecular weight standards influence the result obtained, so that only relative comparisons can be made.

By means of Small-Angle Neutron Scattering (SANS) studies, Overfield[71] and Ravey[72] and their coworkers have recently confirmed directly the existence of aggregates in oil and have described their physical dimensions and shape as a function of the solvent used. Sheu *et al.*[73] employed surface tension measurements in order to study the self-association of asphaltenes in pyridine and nitrobenzene. A discontinuous transition in the surface tension as a function of asphaltene concentration was interpreted as the critical asphaltene concentration above which self-association occurs.

In the past, efforts were made to describe the asphaltene fraction in terms of a single, representative asphaltene "molecule" incorporating, in the correct proportions, all of the chemical and elemental constituents known to be present in a given asphaltenic matrix. An example of two such asphaltene "molecules," each representing the asphaltene fraction of a crude oil, are shown in Figure 2. While this approach gives an idea of the structural complexity of the compounds comprising the asphaltene component, it obscures the highly differentiated chemical nature of the molecules in this petroleum fraction (*cf.* Long's

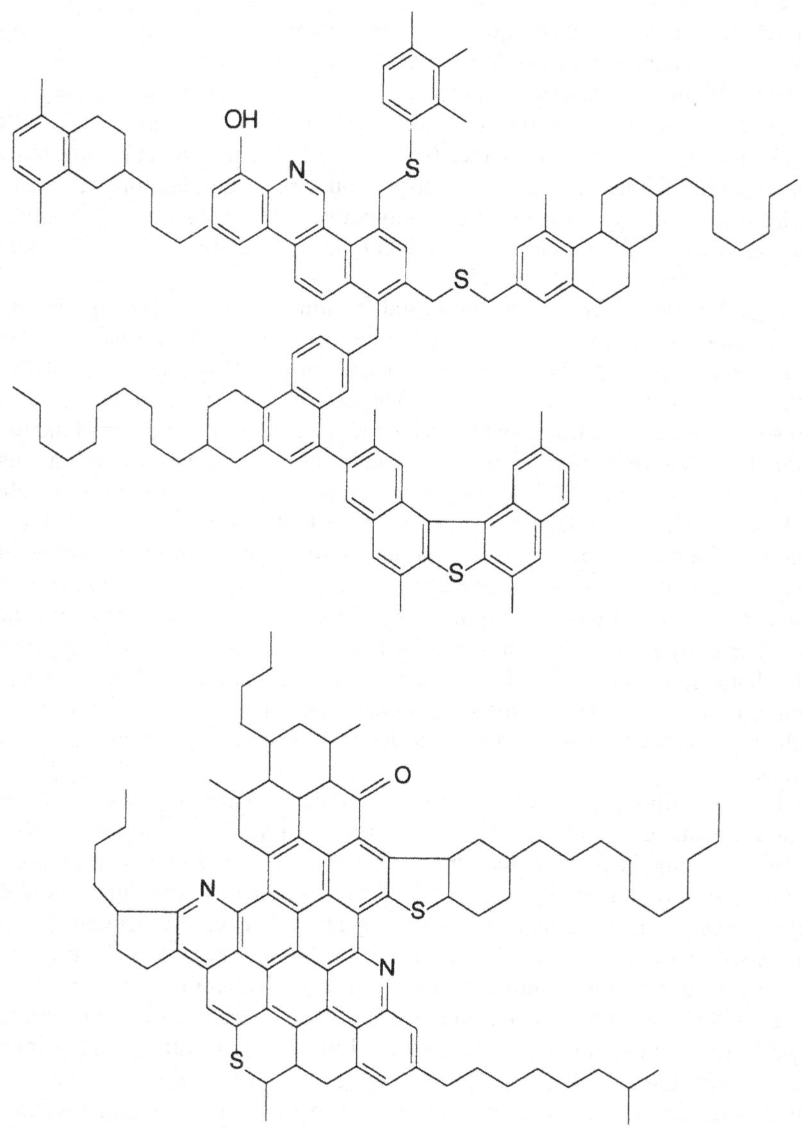

Figure 2. Hypothetical structures for asphaltenes of different origins (ref. 24; reproduced with permission of the publisher).

description of the asphaltenes[20] represented in Figure 1). The idealized molecular structures defined by this method, consequently, are of little utility for inferring the chemistry and physical chemical properties of the asphaltenes.

Asphaltene Fractionation

In analogy with the fractionation of the volatile components of petroleum by distillation, the non volatile residue (550+°C) can be further fractionated into asphaltenes and maltenes by treatment with solvents (or solvent mixtures) capable of solubilizing selectively some of the chemical constituents of the residue. In this case, the criteria for fractionation of the residue (*e.g.*, molecular weight, polarity, number or type of functional groups, etc., and their various combinations) is a function of the characteristics of the solvent chosen: both the quantity and the quality of the solubilized material will depend on the physical-chemical characteristics of the solvent, or solvent mixture.

It is useful, in discussing solvation phenomena, to introduce the solubility parameter, δ, which provides a measure of the work that must be done to overcome the interaction between the molecules of a pure substance (see below).[74] The parameter δ takes into account the effect of three distinct contributions: dispersion forces, dipolar forces, and hydrogen bonding. In order to assure efficient interaction between two substrates, not only must the solubility parameters of the solute and solvent be similar, but the single components contributing to δ must be similar as well.

Speight has described the solvent-based fractionation of the asphaltene component of petroleum and showed that is it possible to obtain asphaltene fractions characterized by different degrees of aromaticity or heteroatom content by using benzene/pentane or benzene/methanol mixtures in variable ratios.[16] The use of solvent mixtures of a polar and a non-polar solvent in order to fractionate an asphaltene sample, however, will tend to "direct" the fractionation by introducing polar forces and hydrogen bonding, as well as dispersion forces, as factors determining which components of the asphaltene sample are soluble in the mixture. In contrast, the exclusive use of saturated hydrocarbons, whose solvent power derives entirely from the dispersion forces, should in principal permit the fractionation to be carried out with variation of only a single component of the solubility parameter. It is known that the quantity of asphaltenes precipitated by linear paraffins decreases with increasing carbon number until a limiting value is reached above n-C7 to n-C9.[19] The cycloparaffins, on the other hand, have better solvating power toward the asphaltenes and often fail to induce their precipitation from crude oil. This suggested to us that it might be possible to fractionate the asphaltene component of an oil in a controlled fashion by using mixtures of cyclohexane and n-alkanes.

Indeed, we find experimentally that the dilution of petroleum with an excess of cyclohexane/n-pentane mixtures provokes the precipitation of a portion of the asphaltenic material. By removing the precipitated asphaltenes and evaporating the solvent, it is possible to carry out a successive partial precipitation of the asphaltenes remaining in the crude oil using a cyclohexane/n-pentane mixture containing a higher level of n-pentane. By operating such that the value δ of the anti-solvent mixture is decreased progressively, the n-C5-insoluble component can be divided into a number of fractions of similar size.

The results of a series of experiments of this type, carried out on three crude oils of markedly different compositional characteristics (Table 2), are summarized in Figures 3 and 4. In these figures, the cumulative weight percent of the asphaltenes precipitated (referred to the quantity of asphaltenes precipitated by pure n-pentane) is plotted against the volume % of n-pentane in the cyclohexane/n-pentane mixture. Figure 3 reports the results of fractionation experiments carried out on the n-C5 asphaltene component of the heavy, aromatic crude oil #1 and the medium-heavy crude #2. Whereas the asphaltenes in oil #1

Table 2. Summary of the properties of crude oils #1-3

	Oil # 1	Oil # 2	Oil # 3
API gravity	7.80	18.60	41.20
Sulfur content (%)	7.80	4.60	0.30
n-C5 asphaltenes (%)	22.70	18.70	0.90
Distillation fractions (% by weight)			
C5–200°C	10.70	17.00	31.10
200°–350°C	7.00	17.60	33.50
Atm. residue 350°C +	82.30	65.40	35.40

were fully soluble in pure cyclohexane, 40% of the total asphaltenic material in oil #2 precipitated upon dilution with this solvent. This pronounced difference in their n-pentane precipitation thresholds underscores the point that the solubility properties of the asphaltene fractions of different crude oils can vary markedly.

The fractionation results for the lighter crude oil #3 (Figure 4) require closer examination. A residue, composed of roughly equal parts of asphaltenic material and C30-C46 waxes, spontaneously precipitates from this oil at ambient conditions. In cyclohexane/n-pentane fractionation experiments carried out on oil #3 from which the residue had been removed, asphaltene precipitation first occurred when the volume of n-pentane in the solvent mixture reached 60%. Cyclohexane/n-pentane fractionations were also carried out on the entire asphaltene component of the oil by dissolving the residue and the n-C5 asphaltene fraction together in a small amount of tetralin. The fractionation behavior for the two samples is nearly coincident for levels of n-pentane in the solvent mixture of 60% or more (Figure 4). The subfractions generated from the entire asphaltene component at n-pentane levels from 0 to 60%, therefore, must arise from the residue alone. It is evident from the fractionation behavior of the entire asphaltene sample that the residue constitutes a single, continuous compositional distribution with the rest of the soluble asphaltenic material. This result demonstrates that the formation of asphaltenic deposits can involve only a subset of the entire asphaltene component or, conversely, that the solubility characteristics of the deposit can differ from those of the rest of the asphaltene fraction. We will return to this point below.

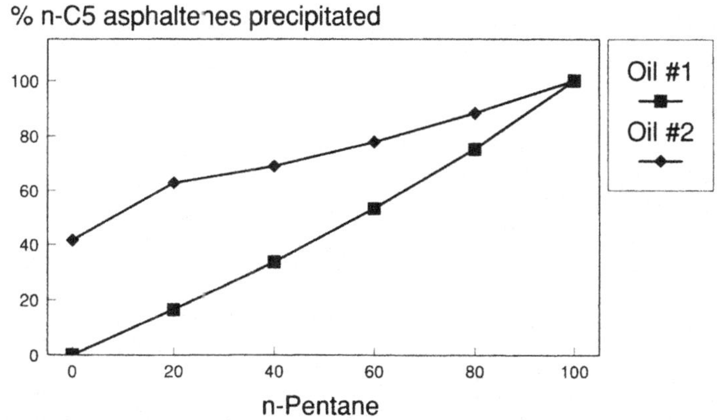

Figure 3. Fractionation of the asphaltenic component of crude oils #1 and #2 as a function of the vol % n-pentane in cyclohexane/n-pentane mixtures.

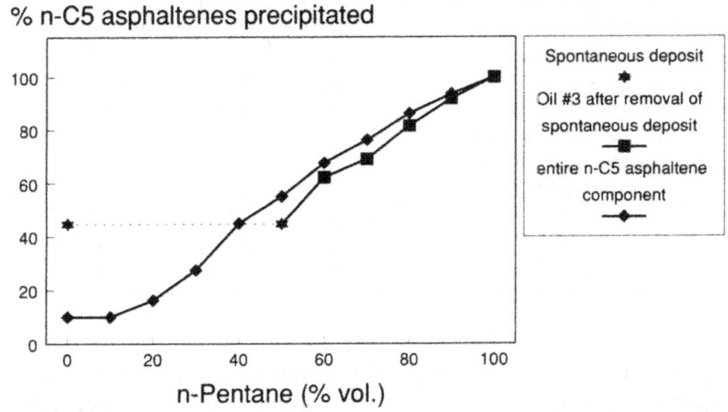

Figure 4. Fractionation of the asphaltenic component of crude oil #3 as a function of the vol % n-pentane in cyclohexane/n-pentane mixtures.

The asphaltene samples generated from the three oils by the fractionation procedure were characterized by elemental analysis, SEC, and NMR (Table 3). For all three crude oils, the repeated cycles of precipitation generated asphaltenic subfractions significantly differentiated with respect to their aromaticity (Figure 5) and heteroatom content, nitrogen in particular (Figure 6). Given the high percentage of nitrogen present in the form of the polar pyrrolic and pyridinic functionalities,[50] we expect the polarity of the precipitated asphaltenes to correlate with their nitrogen content. The variations with respect to the average molecular weight, on the other hand are much less substantial and more difficult to interpret.

It would appear, therefore, that the tendency of asphaltenes to precipitate *from a given crude oil* on account of a variation in the δ value of the precipitating solvent mixture

Table 3. Characterization of asphaltenic deposits and asphaltene subfractions*

	Oil #1				Oil #2				Oil #3			
	wt.%	M_w	N(%)	*fa*	wt.%	M_w	N(%)	*fa*	wt.%	M_w	N(%)	*fa*
Well deposit									—	1,600.00	0.81	0.75
Spontaneous deposit									0.39	1,300	0.84	0.74
Asphaltene fraction†												
100:0					7.75	1,700	0.95	0.60				
80:20	3.76	3,100	1.18	0.54	3.94	1,700	0.93	0.58				
60:40	3.90	3,300	1.19	0.50	1.71	1,800	0.87	0.57				
50:50												
40:60	4.45	3,300	1.09	0.48	1.66	1,900.00	0.84	n.d.	0.15	1,300.00	0.69	0.63
30:70									0.06	1,200	0.67	0.63
20:80	4.98	3,100	0.96	0.46	1.93	2,000	0.79	0.53	0.11	1,200	0.66	0.59
10:90									0.09	1,100	n.d.	0.58
0:100	5.64	2,800	0.84	0.43	1.67	2,000	0.69	0.52	0.07	1,300	0.65	0.55

*Abbreviations: wt.%, weight % of precipitated asphaltene with respect to oil; M_w, weight average molecular weight; N(%), weight % of nitrogen in the asphaltene fraction; *fa*, the aromaticity factor (see text).
†Cyclohexane : n-C5 ratio.

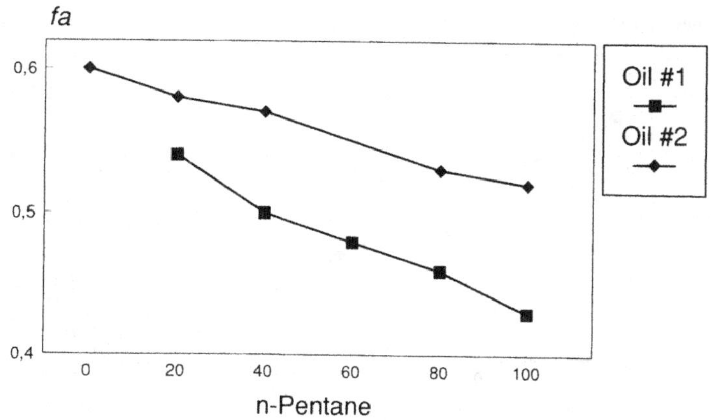

Figure 5. Aromaticity factor (*fa*) of the asphaltene subfractions obtained by selective precipitation.

is more closely related to the aromaticity and polarity of the asphaltenes than to their dimensions. However, a comparison of the characteristics of the asphaltene subfractions precipitated from the three crude oils at the same cyclohexane/n-pentane solvent composition (recall that, by virtue of the progressive fractionation procedure, these subfractions contain asphaltene molecules possessing a narrow range of solubility properties) provides us with a broader perspective on the interplay between the physical-chemical attributes that determine asphaltene solubility. From the results reported in Table 3, we see that the same diluent mixture that provokes the precipitation of high molecular weight asphaltenes of low aromaticity from crude oil #1 precipitates more highly aromatic and polar asphaltenes of lower molecular weight from crude oil #3. We conclude, therefore, that molecular weight, together with aromaticity and polarity, combine to determine asphaltene solubility in hydrocarbon media. Further support for this conclusion is provided in Figure 7, which shows the good correlation between the molecular weight and aromaticity factor for the n-C5 to n-C7 asphaltene fractions (that is, the portion of the asphaltenic component recovered by n-C5 precipitation from an oil previously de-asphalted with n-heptane) for eight crude oils.

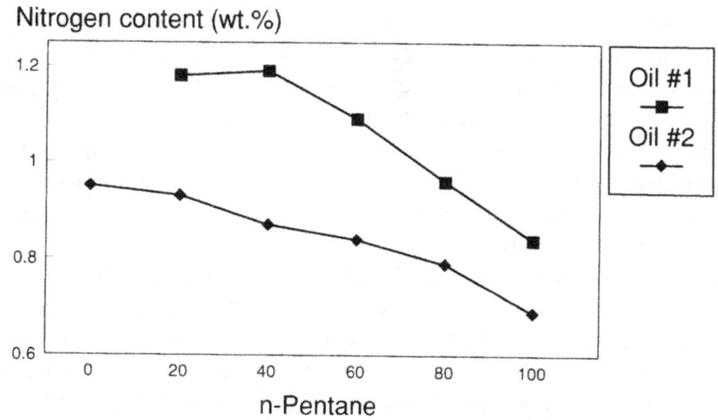

Figure 6. Nitrogen content of the asphaltene subfractions obtained by selective precipitation.

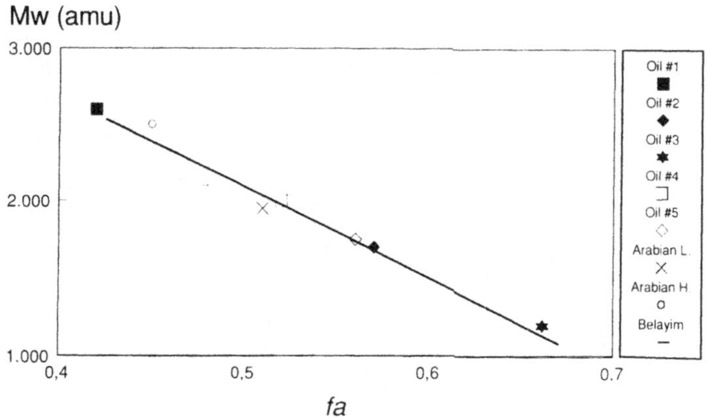

Figure 7. Relationship between molecular weight (M_w) and aromaticity (*fa*) for n-C5/n-C7 asphaltene subfractions of different origins.

Previous studies have also suggested that asphaltene size (molecular weight) and polarity are important factors influencing solubility.[75,76] Snape and Bartle, for example, identified molecular weight, the concentration of acidic hydroxyl groups, and the ratio of internal aromatic carbon to the total aromatic carbon, as important to solubility, and related these parameters to an empirical "solubility index."[75]

It is also of interest to compare the composition of the asphaltene field deposits formed from crude oil #3 with the fractions generated by the selective precipitation procedure developed (Table 3). The aromaticity, nitrogen content, and molecular weight of the asphaltene deposit are equal to, or greater than, those of the spontaneous residue formed under ambient conditions, and considerably removed from those of the asphaltenes stable in the crude oil. This result again underscores the fact that asphaltene phase separation and deposit formation in the field (*i.e.*, the problems that we are interested in addressing) may involve only a portion of the asphaltene fraction generated from the crude oil in the laboratory by n-C5 precipitation.

In this section we have seen that it is possible to fractionate the asphaltene component of a crude oil into distinct subfractions of varying solubility characteristics by means of their selective precipitation with organic solvents. The chemical characterization of the asphaltene fractions thus generated identifies polarity and aromaticity as the principal factors defining the heterogeneity of the asphaltene molecules within the n-C5 insoluble fraction. A comparison of the characteristics of narrow asphaltene subfractions obtained from different crude oils indicates that molecular weight (which can vary significantly for the asphaltene component of different crude oils) and aromaticity, together with the polarity, combine to determine solubility. Although attempts to further define the relationship between asphaltene structure and composition and solubility are still at a primitive stage, the fractionation approach described above appears to be a promising path toward this goal.

The fractionation studies provide direct and unambiguous evidence of the fact that the asphaltene component of a crude oil is composed of a broad range of molecular species, *i.e.*, a compositional continuum. We will show in the following sections how the fact that the different molecular species comprising the asphaltenic component of a crude oil can display a range of solubility behavior has important consequences for the analysis and resolution of production problems arising from asphaltene deposition and for modelling the flocculation process.

Asphaltene Solvent Studies

As noted in the introduction, petroleum production can be seriously compromised by the formation of asphaltene deposits within the pores of the reservoir rock (formation damage), in the wellbore, or in the surface facilities. Aromatic solvents such as toluene and xylene or light petroleum distillates are normally used in the petroleum industry to remove these deposits.[77-81] We will show in this section that the effectiveness of such solvent treatments depends strongly on both the chemical characteristics of the solvent and on the specific nature of the petroleum deposits.

Screening tests carried out with the scope of selecting the best solvent for treating a particular problem have typically aimed at determining the "asphaltene solubility" in the solvents under examination.[78,82,83] We believe that there are two important limitations to the solvent tests as they are commonly performed. First, (n-C5 or n-C7) asphaltene samples precipitated from the crude oil in the laboratory are often used. In light of the heterogeneity of the asphaltene component and the results for oil #3 discussed above, it is clear that in many (perhaps most) cases precipitation in the field will involve only a portion (the least soluble fraction) of the conventionally defined asphaltene fraction. Although the structural and physical-chemical differences between the asphaltenic deposit and the entire asphaltene component of the oil will presumably depend on the nature of the destabilization mechanism leading to asphaltene flocculation in the field, it would appear reasonable to expect that, in general, asphaltene deposits will be composed of the more highly aromatic and polar portion of the asphaltene fraction of the crude oil. For this reason, it is essential that solvent evaluation studies be carried out on actual asphaltene deposits, where possible, or on the least soluble portion of the asphaltene component, if generated by precipitation from the crude oil in the laboratory. Our studies show, in fact, that in some cases typical asphaltene solvents such as toluene fail to dissolve a significant portion of asphaltenic field deposits (see below).

Secondly, it is common practice to determine the "solubility" of asphaltene deposits from experiments in which a fixed mass of asphaltene is placed in contact with a fixed volume of solvent at a single deposit-to-solvent ratio. The "solubility" (in g of asphaltene solubilized per mL of solvent) of the deposit is calculated from the amount of sample dissolved in the volume of solvent used. Implicitly, this method assumes that the asphaltene deposit is uniform with respect to its solubility: *i.e.*, if one-half of an asphaltene sample dissolves in a given volume of solvent, then it is expected that the entire sample will dissolve at twice the solvent/deposit ratio. However, as we saw above in the fractionation studies, the asphaltene component of an oil is comprised of molecular species possessing a wide range of chemical and structural characteristics and the deposits are likely to be concentrated in the most polar and aromatic portion with respect to the n-C5 or n-C7 asphaltene fraction. It is our experience, in fact, that a portion of the asphaltenic deposits recovered from the field may be insoluble in any volume of the solvent under study.

These considerations suggest that the traditional method of asphaltene solvent testing may provide incomplete or even misleading results. We prefer to evaluate the solubilizing characteristics of solvents toward asphaltenic deposits in terms of a *"dissolution curve"* describing the solubility behavior over a wide range of dilution. In this approach, the affinity of a solvent toward a given sample is established on the basis of two indices: *Smax*, the maximum solubilization of the sample reached (in weight percent), independent of the volume of solvent used, and *Vmin*, the minimum volume of solvent required to reach Smax. Figure 8 illustrates the hypothetical behavior of three solvents which differ in their minimum volume required to achieve maximum asphaltene solubilization (*cf.* Vmin of solvent 1 *versus* solvent 2) and in the maximum degree of solubilization of the sample (*cf.* Smax of solvent 3 *versus* solvents 1 and 2). The evaluation of solvents by the method proposed is completed by information on their relative *kinetics of solubilization* of the asphaltene sample; that is,

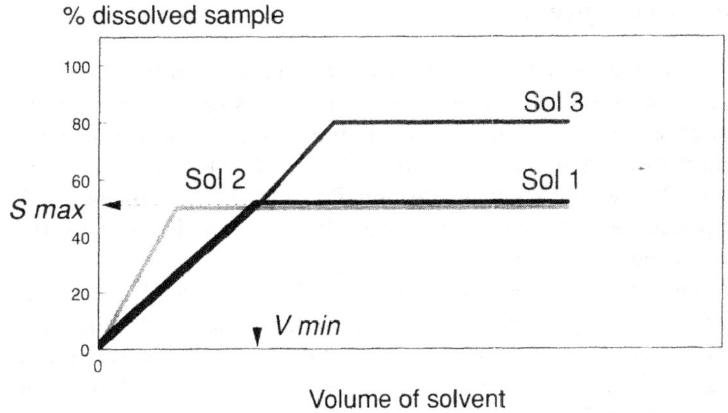

Figure 8. Hypothetical asphaltene deposit dissolution curves describing the solubility behavior for three different solvents (Sol 1-Sol 3).

the time necessary to reach thermodynamic equilibrium at any point along the dissolution curves shown in Figure 8.

Parting from these considerations, we carried out a systematic study of the solvent capacity of the hydrocarbon solvents normally employed by the industry for field treatments, using an asphaltene deposit from crude oil #3 recovered from the field. Some chemical properties of the sample were given in Table 3. For each solvent, Smax was determined at room temperature as a function of the sample-to-solvent ratio for the powdered deposit at "thermodynamic equilibrium" (*i.e.*, after sonication and 12 h agitation). The solubilization of the asphaltenes was calculated by difference from the weight of undissolved residue and is expressed as a value in percent by weight. Asphaltene dissolution kinetics were evaluated using compressed disks prepared from the powdered deposit according to a standardized procedure. The disks were suspended in a fixed amount of solvent (>Vmin) in a glass autoclave and maintained at constant temperature under mild agitation. The concentration of asphaltene dissolved during the test run was determined on filtered aliquots of the solution by means of a UV-vis spectroscopic method.[84]

A preliminary screening using pure model solvents showed that the solubility curves of aromatic hydrocarbon solvents can differ markedly. Thus, whereas alkylbenzenes dissolved at most 50% of the sample, the higher-condensed aromatic hydrocarbons tetralin and 1-methylnaphthalene gave more than 90% dissolution (Figure 9). The higher-condensed aromatic hydrocarbons also gave markedly faster dissolution kinetics. These results show that even relatively small variations in the solubility parameter significantly influences the performance of hydrocarbon solvents (*cf.* the values of δ for toluene and tetralin, 18.0 and 19.6 $MPa^{1/2}$, respectively). Studies with toluene showed that the maximum solvent capacity was relatively insensitive to temperature over the range 20°-150°C, although the rate of dissolution was enhanced significantly.[84]

The finding that monocyclic aromatic solvents such as toluene and xylene gave less than 50% dissolution, irrespective of the temperature, contact time, or solvent-to-deposit ratio, suggested that their use in field treatments is far from optimal. The use of pure condensed aromatic hydrocarbons in their place, however, is not economically feasible. We sought, therefore, to identify low-cost refinery streams whose composition and performance would best approach that of the condensed aromatics.

Distilled hydrocarbon mixtures from fossil fuels are normally composed of two main classes of compounds: saturates and aromatics. Within the boiling range from about 80° to

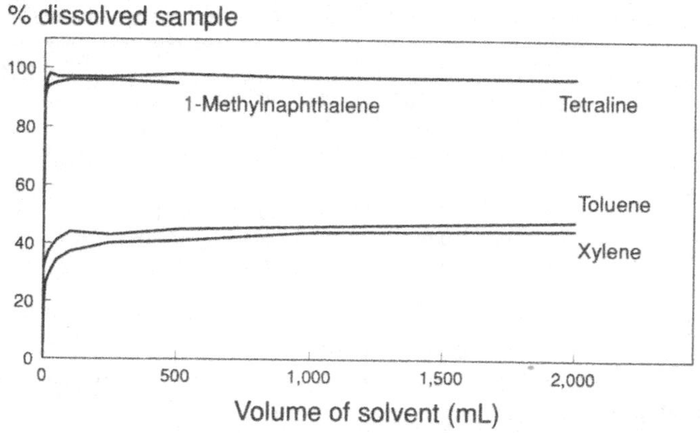

Figure 9. Dissolution curves of 100 mg samples of the asphaltene deposit from oil #3 in model solvents at room temperature.

350°C (which, for physical-chemical reasons, is the range of interest for use in oilwell treatments), the aromatic fraction is comprised of molecules of varying degrees of condensation. These range from the alkylbenzenes (z = -6) to tri-condensed aromatic hydrocarbons (z = -18), where the z number (a conventional procedure of classifying hydrocarbons) is referred to the formula C_nH_{2n+z}. Though the asphaltene dissolution capacity of these streams will evidently be related to the solvent composition, prediction of the solvency characteristics of a given product on the basis of its molecular composition is not simple. We observed, for example, that the asphaltene dissolution capacity of three refinery-derived gasoils (S1-3) was markedly different, in spite of their quite similar hydrocarbon distributions (Table 4).

In order to better understand the relationship between composition and the dissolving power of complex hydrocarbon mixtures, we carried out dissolution studies using solvents composed of variable ratios of n-hexadecane (cetane), toluene, and 1-methylnaphthalene, which were chosen as models of the three principal pseudo-components, saturates, alkylbenzenes (z = -6), and condensed aromatic hydrocarbons (z < -6), respectively, present in the available hydrocarbon streams. The levels of maximum dissolution of the asphaltenic deposit (Smax) achieved with different mixtures of the three solvents are presented in the ternary diagram shown in Figure 10. Regions of iso-solubility emerged clearly from such studies. Most important for our purposes were those hydrocarbon blends having a solvent capacity

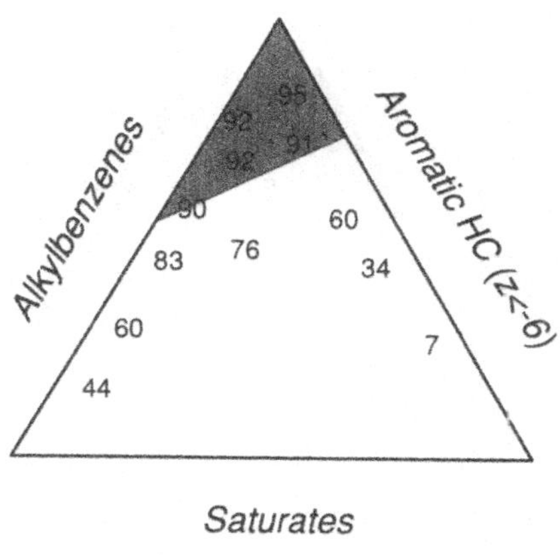

Figure 10. Ternary diagram reporting asphaltene deposit solubilization in different hydrocarbon mixtures (from ref.85; reproduced with permission of the publisher).

Table 4. Characteristics of asphaltene deposit-removing solvents*

	S1	S2	S3	S4	S5	Tol.	S. 150	S. 200
Density at 15°C	0. 979	0. 963	0. 988	0. 998	1. 041	0. 867	0. 902	1.00
Viscosity at 38°C, cSt.	6.88	6.36	7.57	4.07	2.51			
Flash point, °C	116.00	126.00	128.00	120.00	112.00	4.00		
Elemental analysis, wt%								
C	87.40	87.80	86.90	90.70	90.80			
H	9.10	9.90	9.50	8.50	6.50			
N	< 0.3	< 0.3	< 0.3	< 0.3	0.40			
S	1.70	1.80	2.50	0.20	—			
LC separation, wt %								
Saturates	**19.20**	**26.00**	**19.90**	—	—	—	**1.10**	—
Aromatics	80.80	74.00	80.10	100.00	100.00	100.00	98.90	100.00
HC distribution via MS, wt %								
Paraffins	11.50	15.20	10.90	—	—	—	1.10	—
Naphthenes	7.70	10.80	9.00	—	—	—	—	—
Alkylbenzenes	**3.60**	**4.70**	**9.50**	**6.80**	**1.30**	**100.00**	**77.50**	**11.40**
Naphthenobenzenes	6.10	7.40	9.70	13.90	4.10	—	10.90	7.30
Diaromatics	51.50	41.10	40.10	77.60	90.20	—	10.50	81.00
Polyaromatics	19.60	20.80	20.80	1.70	4.40	—	—	0.30
Total aromatics z < -6	**77.20**	**69.30**	**70.60**	**93.20**	**98.70**	—	**21.40**	**88.60**
Solubility of oil #3 well deposit (Smax)	**91.00**	**56.00**	**70.00**	**95.00**	**95.00**	**40.00**	**46.00**	**95.00**

*S1–S5 refer to industrial streams of different origin; S.150 and S.200 indicate Solvesso™ 150 and 200. LC = liquid chromatography; HC = hydrocarbon; MS = mass spectrometry.

for the asphaltene deposit higher than 90% (shaded region in Figure 10). Interestingly, we found that there is a rapid decline in the asphaltene solubilizing capacity for solvent compositions falling outside the region indicated.

Results for a number of industrial hydrocarbon streams (Table 4) confirm the excellent dissolution capacity of those whose composition, defined in terms of their saturate, alkybenzene, and condensed aromatic hydrocarbons content, falls within the region defined by the shaded triangle of Figure 11; such solvents can therefore be qualified as potential candidates for wellbore cleanup operations. A low-cost, industrial solvent identified on the basis of these criteria has been used to remove the asphaltene deposits in the Italian reservoir mentioned in the Introduction.[85] Analyses on the solvent recovered from the reservoir after the treatments show consistently higher (by a factor of 2 to 3) asphaltene concentrations than

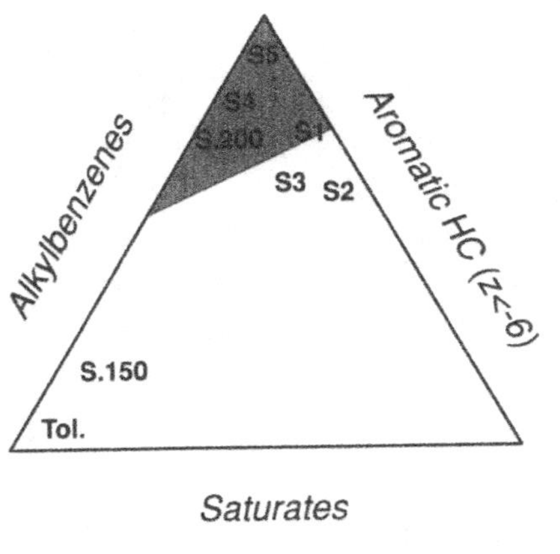

Figure 11. Ternary diagram indicating composition of solvents evaluated for asphaltene deposit solubilization (see Table 4).

when toluene is used. Yet more important, from an economic point of view, is the fact that the length of the shut-in period (or reservoir "soaking" time) required to achieve satisfactory asphaltene deposit removal is reduced from greater than 20 hours to just 30 minutes.

PART II. MODELLING OF ASPHALTENE SOLUTION BEHAVIOR

Asphaltene Solution Behavior and Prior Modelling Approaches

Descriptive Models of Asphaltenes in Solution. As noted above, the pronounced tendency of asphaltenes to form aggregates in hydrocarbon solution is one of their most characteristic traits. Thus, while a number of experimental methods indicate that isolated asphaltene "monomers" have an average molecular weight of the order of 1000 g/mol, numerous methods including VPO, SEC, and scattering techniques (SAXS, SANS) show that they associate spontaneously in most hydrocarbon media, arriving at empirical molecular weights of 10,000 g/mol or more (VPO) and physical dimensions from 10-100 nm (SAXS, SANS). These aggregates are commonly referred to as micelles in the literature, based on an analogy drawn early on[86-89] with the behavior of surfactant molecules in solution.

Under unfavorable solvent conditions the asphaltene micelles are prone to further aggregation into clusters that are unstable with respect to precipitation. The generally accepted picture of the physical behavior of asphaltenes in hydrocarbon media is illustrated schematically in eq.1 below.[90]

asphaltene
molecules <=====> micelles <=====> micelle clusters/floccs (1)

The asphaltenes appear to display the same qualitative behavior both in petroleum and in organic solvents.[71,72] As discussed in the preceding section, studies on asphaltenes dissolved in hydrocarbon solvents have shown that the dimensions of the micelles depend strongly on the nature of the solvent, the asphaltene concentration, and the temperature.

On the basis of their physical size, the micellar aggregates formed in hydrocarbon media are correctly defined as colloids.[86-89,91] Considerable effort has been dedicated to understanding the physical nature of the interaction of the asphaltene colloids with the hydrocarbon medium and, in particular, the basis for their stability in crude oil under reservoir conditions. Empirical observations indicate that the resins play an important role in stabilizing asphaltenes in crude oil.[27,89] The resins differ from the latter in having a higher affinity for the petroleum medium. VPO and scattering studies[92,93] on isolated resin fractions show that they too tend to self-associate in organic solvents. Finally, as the discussion in Part I suggests, the resin and asphaltene fractions are probably best viewed as two "adjacent" portions of a single broad compositional continuum that contains the polar, aromatic, and higher molecular weight components of petroleum.

It will be useful to our discussion to digress for a moment in order to introduce some basic considerations on colloidal systems.[91] First of all, colloids dispersed in a fluid media can be classified as one of two fundamentally distinct types, where the critical distinction regards the number of phases present. *Lyophobic* colloids are not solvated by the liquid medium, and their dispersions are *two-phase* systems. *Lyophilic* colloids are *solvated* by the surrounding medium, with which they constitute a one-phase system. Hence, lyophilic dispersions are true solutions. High molecular weight polymers and large proteins dissolved in aqueous or hydrocarbon media are familiar examples of lyophilic colloidal dispersions. The case of the micellar structures formed by surfactant molecules in aqueous and hydrocarbon media is more complicated and, in view of the analogy drawn with asphaltene colloids, is worthwhile considering in further detail.

Surfactant micelles may belong to either the lyophilic or lyophobic class of colloids. When the surfactant aggregation number is high and the polydispersity (*i.e.*, the range of variation in the number of surfactant monomers present in the micelles) is low, then the micelles are best described as a separate phase. When the aggregation number is low (as is generally the case in hydrocarbon solvents) and the polydispersity is high, then the micelles are most accurately represented as solvated, lyophilic colloids.[91] In both cases, however, the micellar structures formed by surfactants are thermodynamically stable in the medium, and do not spontaneously form clusters or higher aggregates. Evidently, therefore, neither lyophilic nor lyophobic micellar colloids require the presence of other stabilizing agents (*e.g.*, dispersants) in order to resist flocculation.

While the formation of asphaltene aggregates in hydrocarbon solvents and crude oil has been likened to the micellization of surfactants, most discussions of the behavior and stability of asphaltene solutions in which the micellar model has been invoked have not been faithful to the principal elements of surfactant micellization described above. This is immediately evident if we consider the so-called "colloidal model" (referred to as the *steric-stabilization model* in this Chapter) which has dominated the thinking of researchers in the field and has served as the basis for virtually all attempts to rationalize the phase-stability of asphaltene solutions. This model, which was well developed by the early 1940's,[87-89] holds that the asphaltene micelles are composed of an insoluble, aromatic core onto the surface of which resin molecules adsorb, thereby providing a steric stabilization against their flocculation and precipitation.[19,27,92,94,95] More recently, the phase separation of asphaltenes upon the addition of nonpolar solvents to the crude oil has been further rationalized[96-99] in terms of the following, explicit mechanism: the reduction of the polarity of the hydrocarbon medium, which is presumed to make it a better solvent for the resins, provokes the dissociation of the peptizing resin molecules from the surface of the solid asphaltene particles, destabilizing the colloids with respect to flocculation. In light of the previous discussion, we see immediately that asphaltene micelles are assumed to be lyophobic dispersions inherently unstable with respect to flocculation. Also, a crucial role—peptization or steric stabilization—has been postulated for the resins, which finds no analogy with the case of surfactant micelles.

One might expect there to be a compelling motive for departing so radically from the surfactant micelle model. The role postulated for the resins, however, is not supported by direct experimental observation (*cf.* refs. 100, 101). In fact, a critical assumption of this model— that resins are physically present at the interface between the asphaltene micelles and the crude oil— has yet to be demonstrated. The steric-stabilization model will be further evaluated below.

An alternative explanation for the stabilizing influence of the resins, and for the stability of asphaltenes in hydrocarbon media in general, is that the asphaltenes are lyophilic colloids solvated by the surrounding medium. This view implies that the asphaltene monomers and micelles are in thermodynamic equilibrium, where the degree of association is determined by the relative energies of solvation of the monomers and the micelles, and the entropy and enthalpy changes associated with the self-association of the asphaltene monomers. According to this model, the stabilization of the asphaltene micelles by the resin fraction derives from the contribution of the latter to the solvent power (polarity and aromaticity) of the medium. Analogously, destabilization of the asphaltenes leading to their phase separation occurs when the solvating power of the medium toward the asphaltene monomers and micelles is reduced to the point where they are no longer fully soluble.

Relatively few previous studies have explicitly discussed or adopted this point of view to account for asphaltene solution behavior.[64,72,93,101] Noteworthy among these are

SAXS and SANS scattering studies carried out by Espinat and co-workers,[72,93] who observed that the size of the asphaltene "micelles" varies as a function of the temperature, asphaltene concentration, and the identity of the solvent in the same way as found in earlier VPO studies. These studies establish that the self-association of the asphaltenes is reversible, and the authors concluded that: "the sizes determined by neutron scattering in the solution state result from a true thermodynamic equilibrium between 'primary' particles and reversible aggregates, the equilibrium constant depending on some property of the solvent." The principal elements of this study have been confirmed in a SANS study on asphaltenes in toluene reported by Sheu *et al.*,[102] who further defined the polydispersity in the dimensions of the asphaltene aggregates in solution. Studies on the aggregation behavior of asphaltenes in crude oil were reported recently by Carnahan and coworkers.[103] Although interpreted in terms of the customary "stabilized colloid model," the experimental results produced in the latter two studies establish the similarity of the asphaltene micellization process in crude oil and in solutions of isolated asphaltene fractions dispersed in organic solvents.

The critical distinction between the lyophilic colloid and steric stabilization models lies in the issue of whether the asphaltene colloids are solvated or suspended in hydrocarbon media. Curiously, over the past 30 years, most experimental studies on the physical state of asphaltenes in hydrocarbon media have focused only on determining the size of the asphaltene colloids in hydrocarbon media. It should be clear from the preceding discussion that such studies provide little if any insight into the fundamental question of the *nature of the interaction* of asphaltene colloids with the surrounding media. As we will see in the next section, these issues have extremely important implications for attempts to develop physical models of the phase behavior of the asphaltenes. Therefore, later in this Part we will critically re-examine some of the evidence in the literature in order to better evaluate the two conflicting explanations (solubilization *versus* steric stabilization) for asphaltene stability in hydrocarbon media.

Physical Models of Asphaltene Stability. Both the steric ("colloidal") stabilization and solvation hypotheses to account for the stability of asphaltene solutions have been adopted in the physical models of asphaltene phase behavior reported in the literature. The prototype steric stabilization model developed by Leontaritis and Mansoori[94] adopted the picture of asphaltene stabilization proposed by Nellensteyn;[87] *i.e.*, that the insoluble, solid asphaltene particles are maintained in solution (peptized) by resins adsorbed onto the surface of the asphaltene particles.[95,104] In their "thermodynamic-colloidal" model of asphaltene precipitation the chemical potential of the resins in solution was expressed as:

$$\Delta\mu_2 = RT\,[\ln(\phi_2) - (r-1)(1-\phi_2) + r\,\chi\,(1-\phi_2)^2] \quad (2)$$

where $\Delta\mu_2$ is the difference between the chemical potentials of the resins in solution and in their reference state;
R is the gas constant;
T is the absolute temperature;
ϕ_2 is the volumetric fraction of the resins in solution;
χ is the interaction parameter between resins and solvent;
r is the ratio between the molar volumes of the resins and the solvent.

According to the physical assumptions of the model, $\Delta\mu_2^{CR}$ is the value of the chemical potential of the resins when the number of resin molecules adsorbed on the surface of the asphaltene micelles is the mininum required for their peptization. It follows that the asphaltene fraction will be stable to precipitation when $\Delta\mu_2 > \Delta\mu_2^{CR}$ and unstable when $\Delta\mu_2 < \Delta\mu_2^{CR}$. This approach has since been extended and further elaborated in papers by both

Leontaritis and Mansoori.[95,97] To date, this model has been employed for data regression, but its *predictive* power has not yet been demonstrated. In addition, the underlying physical assumptions of this model have not been experimentally verified, as we will discuss in the subsequent sections.

Almost all of the models invoking solvation as the basis for the stability of asphaltene solutions treat the system using the Flory theory[105] for polymer solutions. In this approach, first reported by Hirschberg *et al.*,[106] the asphaltenes are treated as solvated, *monodisperse* macromolecular solutes. The resin fraction is not considered explicitly in this approach, but is treated as an undifferentiated part of the solvent medium. By considering the asphaltene/oil solution as a pseudobinary system, Hirschberg *et al.* derived a simple expression for the free energy of mixing, ΔG_{mix}:

$$\Delta G_{mix} = RT\,[n_1 \ln(\phi_1) + n_2 \ln(\phi_2) + n_1\phi_2\chi] \tag{3}$$

where n_1 = number of moles of solvent (oil);
n_2 = number of moles of solute (asphaltenes);
ϕ_1 = volume fraction of solvent;
ϕ_2 = volume fraction of solute;
R = gas constant;
T = absolute temperature;
χ = Flory interaction parameter between solute and solvent.

The parameter χ can be written as:[107]

$$\chi = \frac{v_1}{RT}(\delta_2 - \delta_1)^2 \tag{4}$$

where v1 = solvent molar volume;
δ_1 = solvent solubility parameter;
δ_2 = solute solubility parameter.

The chemical potential of both the asphaltenes (μ_2) and the solvent (μ_1), are obtained by differentiation of (3) with respect to n_1 and n_2:

$$\mu_1 - \mu_{01} = RT\,[\ln(\phi_1) + (1-1/r)\,\phi_2 + \phi_2^{\,2}\chi] \tag{5}$$

$$\mu_2 - \mu_{02} = RT\,[\ln(\phi_2) - (r-1)(1-\phi_2) + (1-\phi_2)^2\, r\,\chi] \tag{6}$$

in which $r = v_2/v_1$
where v_2 is the solute molar volume, and
μ1 = chemical potential of the solvent;
μ01 = reference chemical potential of the solvent (ie., pure solvent);
μ2 = chemical potential of the solute;
μ02 = reference chemical potential of the solute (ie., pure solute).

When there is an equilibrium between two phases, one poor (I) and the other concentrated (II) in asphaltenes, the chemical potential of any component must be the same in both phases:

$$\Delta\mu_1^{\,I} = \Delta\mu_1^{\,II} \tag{7}$$

and

$$\Delta\mu_2^{I} = \Delta\mu_2^{II} \quad (8)$$

where $\Delta\mu_i = (\mu_i - \mu_{0i})$.

At this point, Hirschberg *et al.* introduce a critical assumption in order to simplify the problem: a pure asphaltenic pseudophase is assumed to separate from solution at the precipitation threshold (eq.9).

$$(\mu_2 - \mu_{02})^{II} = 0 \quad (9)$$

Further postulating that $\phi_2 << 1$ and using eq. 4, they obtained:

$$\ln(\phi_2) - (\frac{v_2}{v_1} - 1) + \frac{v_2}{RT}(\delta_2 - \delta_1)^2 = 0 \quad (10)$$

Although several groups have employed this model for interpolating experimental data, it has not yet proved its utility for predicting phase behavior.[108,109] Other models treating asphaltene dispersions as solvated systems have used the approach of Scott and Magat[110] for polydisperse polymer solutions in place of the Flory equation. Once again, however, the simplifying assumption is made that one of the two phases at the precipitation threshold is constituted by solvent-free, solid asphaltenes (eq. 9).[111-113] Although useful for simplifying the mathematics in both this and the Flory treatment employed by Hirschberg *et al.*, the assumption that asphaltene destabilization occurs with the formation of a pure asphaltenic phase puts the asphaltene solvation models into direct conflict with the usual treatment of polymer solutions,[105] from which they were inspired. We will show below that a straightforward alternative treatment is possible that remains faithful to the polymer theory upon which Hirschberg's approach is based and will demonstrate the utility of the model thus derived for predicting asphaltene phase behavior.

In addition to the two asphaltene solvation models described above, two groups have recently approached the problem by solving complete phase equilibrium systems.[114,115] In these studies the asphaltenes were treated as a single pseudo-component and, at least in principle, no simplifying assumptions were made. The use of these models, which contain too many parameters to be useful for predicting asphaltene phase behavior, has been limited thus far to the interpolation of experimental data.

A Critical View of the Steric Stabilization Model. As we have noted, the steric stabilization or "colloidal" model presently forms the basis of most discussions of asphaltene solution and phase behavior (*cf.* refs. 95-97,103). On the other hand, SAXS and SANS scattering studies of asphaltene aggregation[72,93,102,103,116] have provided strong evidence of the reversibility of the associative phenomena and other observations that are sharply inconsistent with this model. In fact, the dominance of the steric-stabilization model has been such that even the most familiar behavior displayed by asphaltenes in solution has not been discussed in the literature in terms of the thermodynamic model. In this section we will show that, upon re-examination, several previously published experimental results offer additional and compelling evidence in favor of the thermodynamic model.

Micellization Behavior of Asphaltenes in Organic Solvents. As described above, systematic VPO studies[24,64,66,68] have shown that the experimental molecular weights of asphaltenes dissolved in organic solvents are strongly influenced by the chemical nature of the solvent, the temperature, and the concentration of asphaltenes in solution. Speight noted[64] that such experimental results can be explained equally well by hypothesizing a change in the degree of solvation of the asphaltenes by the solvent. We strongly agree. In fact, we would emphasize that the variation of the empirically determined molecular weights of asphaltenes in solution with changes in solvent quality, temperature, and concentration, is characteristic of a solute/solvent system where the solute is present as an equilibrium distribution of monomeric and aggregated structures (*i.e.*, asphaltene monomers/micelles). In this context it may be helpful to note explicitly that lyophilic micellar colloids, which form true solutions, display normal colligative properties. From the response of the empirical asphaltene molecular weights to relatively modest modifications of the conditions, we can conclude that the energies of the asphaltene monomers and micelles must be fairly closely balanced (*cf.* also ref. 102).

The SAXS and SANS studies on the dimensions of associated asphaltene structures (micelles)[72,93,102,116] in organic solvents provide direct evidence that the equilibria measured do in fact derive from (reversible) changes in the degree of aggregation of the asphaltene molecules. We consider the results of a SANS experiment reported by Espinat and Ravey[93] to provide a critical test of the hypothesis that resins stabilize asphaltene aggregates by adsorbing onto their surface. They found that the addition of resins to an asphaltene solution in toluene did not lead to an increase in the asphaltene aggregate size, as predicted by the steric-stabilization model, but instead brought about a reduction in the physical dimensions of the asphaltene micelles. We note that this experimental result is identical to that observed when a good solvent is added to an asphaltene solution.

As mentioned above, one of the characteristics of lyophilic surfactant micelles is a relatively low aggregation number and a broad distribution of aggregates sizes. Experimental estimates of the number of monomers per asphaltene micelle, obtained in several SEC and scattering studies,[72,92,93] fall in the range from about 3 to 20. The clear evidence of a broad size distribution of the asphaltene aggregates obtained by Carnahan and coworkers led them to postulate[103] the existence of a "stepwise equilibrium model" similar to those which have been discussed in the case of lyophilic surfactant colloids.[91]

Stability of Asphaltene Micelles in Solution. Results of VPO and scattering studies have firmly established that isolated asphaltene fractions associate spontaneously in hydrocarbon solvents to form micelles comparable in size to those present in crude oil, and that they display the same general solution behavior. The asphaltene micelles formed in (good) organic solvents show every sign of being thermodynamically stable: they form spontaneously when the asphaltene solids are dispersed in the solvent; their dimensions remain unchanged over time;[117] they are stable indefinitely against precipitation.

Noting that these properties are displayed by solutions of asphaltenes in the absence of the resin fraction, we must conclude that *the presence of resins is not a necessary condition for the stabilization of asphaltene micelles against further aggregation in hydrocarbon media*. It follows that there is neither a logical nor a compelling motive for invoking steric-stabilization by resin molecules in order to account for the stability of the asphaltene fraction in crude oil.

The Reversibility of Asphaltene Precipitation. It has been claimed in support of the steric-stabilization model that the formation of asphaltene deposits is irreversible (*cf.* ref. 97). This conclusion appears to be based on anecdotal field observations; certainly, common experience is that asphaltenes can be precipitated and re-dissolved repeatedly and quantitatively from organic solvents. Obviously, it is critical to any meaningful evaluation of the reversibility of a phenomenon to establish directly that equilibrium has been effectively achieved under the experimental conditions. This is particularly true when dealing with dissolution phenomena of complex, high molecular weight materials, where there may exist a significant kinetic barrier to dissolution.

The results of a recent study[118] on the reversibility of asphaltene phase separation are of interest to this discussion. The experiments consisted in placing precipitated asphaltene fractions from a crude oil in contact with the de-asphalted crude (DAO) and evaluating the extent to which the asphaltenes re-dissolved. The authors found in two different experiments that 23 and 56%, respectively of the asphaltene deposits had re-dissolved at the "end of the experiment" (the time scale of the experiments and criteria for defining dissolution as complete were not better defined). Noteworthy is the conclusion of the study: "These tests show that the asphaltene dissolution-precipitation process is not thermodynamically reversible."[118] While this might be viewed as a case where the glass is seen to be either half-empty or half-full depending on the point of view of the observer, we would point out that the steric-stabilization model as commonly invoked[96-99] defines the asphaltenes as insoluble, solid particles. This model, thus, predicts that *none* should have re-dissolved in these experiments. On the other hand, it would not be surprising to find that a precipitated asphaltene matrix should dissolve slowly; the same behavior is generally observed for polymer systems, which are often very slow to re-establish equilibrium from the solid phase. We look at the "half-full glass" of asphaltenes (23-56% redissolved asphaltenes), therefore, as providing significant evidence against the current steric-stabilization model.

In concluding this section, we would point out that the range of behavior displayed by solutions of asphaltenes in hydrocarbon media is precisely that which would be expected for a solvated (lyophilic) colloidal system. On the other hand, we have shown that the steric-stabilization model for the stability of asphaltene solutions contains a number of unusual, and unsubstantiated elements; it is certainly difficult to account for the qualitatively similar aggregation behavior displayed by asphaltene fractions *free of resins* and the asphaltene fractions in crude oils or vacuum resids in terms of the steric-stabilization model.

Good scientific method [consider the criterion ("razor") of Occam, or Popper's contributions[119]] strongly counsels the adoption of the simplest among competing hypotheses capable of explaining a physical phenomenon, and the critical testing of models in order to establish the limits of their ability to describe and predict physical results. The accumulation of untested, *ad hoc* hypotheses in a physical model is usually a strong indication that the underlying model does not faithfully represent physical reality.[119] In view of these considerations and the results of the critical analysis of the steric-stabilization model presented in this Chapter, we suggest that the asphaltene monomers, and the micellar structures they form in hydrocarbon matrices, are best viewed as lyophilic (solvated) colloidal species.

Development and Applications of a New Thermodynamic Model

The above considerations encouraged us to pursue thermodynamic models to describe the solution behavior of the lyophilic asphaltene colloids. Since, as we will show, the analogy with polydisperse polymer systems remains the basis of our studies, it is useful to recognize explicitly not only the similarities but also the several significant differences between these two lyophilic colloidal systems. First, while polymers normally present a large distribution of molecular weights, for the asphaltenes this distribution is relatively narrow (the ratio M_w/M_n from SEC analysis is usually about 1.5).[92,117] On the other hand, the variation in the chemical characteristics of the different molecular weight fractions of a polymer is usually negligible, while the polydispersity of the asphaltenes is defined principally by the continuous variation in their chemical and structural properties. Finally, from the point of view of Flory's theory of polymer solutions,[105] one of the critical factors for polymer phase behavior is the ratio of the molar volume of the polymeric solute to that of the solvent. While for polymers this ratio is usually > 100-1000, for the asphaltene monomers it is of the order of 10. In terms of Flory's polymer solution theory, then, the differences between the lyophilic asphaltene and polymer colloids are numerous. We have opted to base our final judgement as to the *significance* of these differences on the predictive capability of the models of asphaltene phase behavior derived using polymer solution theory.

Cloud point measurements (in which the concentration of antisolvent necessary to induce phase separation of the polymer is determined) have often been employed in thermodynamic studies on polymer solutions.[120] Measurements essentially analogous to polymer cloud point determinations have been reported for asphaltene solutions as well.[121,122] In these asphaltene flocculation studies UV spectrophotometry was employed to detect the increase in optical density of asphaltene-solvent mixtures at the flocculation threshold during dilution with antisolvent. Previous use of this titration technique has been limited to studies on the affinity of solvents for asphaltene or asphaltic phases and for evaluating the influence of additives on asphaltene phase stability. We felt, however, that this procedure might also be useful for evaluating thermodynamic models of asphaltene phase behavior, and that it would be much simpler than the experimental methods conventionally employed for this purpose.[106,108] The cloud point studies can be carried out with model solvent mixtures as well as with crude oils, and with selected asphaltene samples, such as those generated by fractionation. This possibility of carrying out thermodynamic studies on well-defined and simplified systems held out the promise of achieving a less ambiguous evaluation of the competing models and, potentially, a more precise analysis of the molecular and structural parameters that control asphaltene phase behavior.

We have performed[123] systematic titration studies at ambient conditions on various asphaltenic fractions separated from crude oil #3 dissolved in several different hydrocarbon solvents. Typical results are shown in Figure 12, where the ratio of the weight of precipitant (antisolvent) at the precipitation threshold/weight of the asphaltenes, Y_{pt}, is plotted against the weight of diluent (solvent) at the precipitation threshold/weight of asphaltenes, X_{pt}. The linear dependence between these two variables is a characteristic feature of these plots, and indicates that *the asphaltene precipitation threshold is independent of the asphaltene concentration in the mixture and depends only on the quality of the solvating medium*, over the range of concentrations investigated. The same, unexpected result is found when data reported in earlier studies[121,122] are analyzed in the same way, suggesting that the behavior is general.

Hirschberg's model for asphaltene phase behavior has been assessed by interpolating the titration data using eq.10, where the asphaltene solubility parameter δ_2 has been used as an adjustable parameter and the asphaltene molar volume has been calculated from the

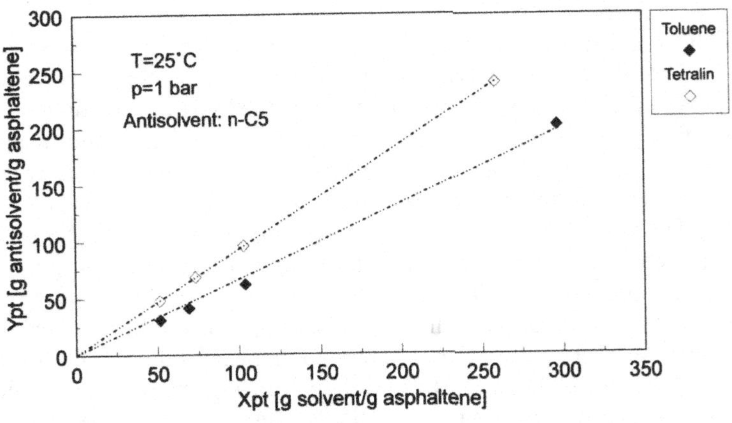

Figure 12. Precipitation threshold of n-C5 asphaltenes in model solvent mixtures (see text).

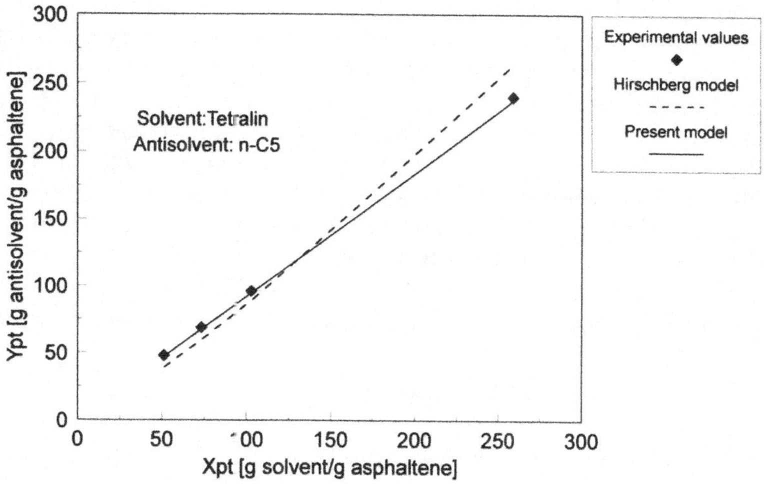

Figure 13. Comparison of the predictions of the n-C5 asphaltene precipitation threshold by the Hirschberg *et al.* and the proposed model.

molecular weight and the density of the asphaltene fraction (determined from SEC and wide angle X-ray scattering, respectively). The agreement between this model and the experimental data is poor (Figure 13): in particular, whereas the model predicts an exponential dependence of the precipitation threshold on the initial solute concentration (Y_{pt} and X_{pt}, respectively), a linear trend is found experimentally. This result led us to re-examine the application of the basic pseudo-binary formulation of Flory's theory for polymer solutions[105] to the problem of asphaltene phase stability.

In analogy with the conventional treatment of polymer phase equilibria,[124] the asphaltene titration experiment can be represented graphically in the χ/ϕ_2 plane as shown in Figure 14 (a value of 10 has been assigned to the ratio v_2/v_1). The continuous curve in the figure defines the boundary between one-phase and two-phase mixtures. As Tompa pointed out[105] for the

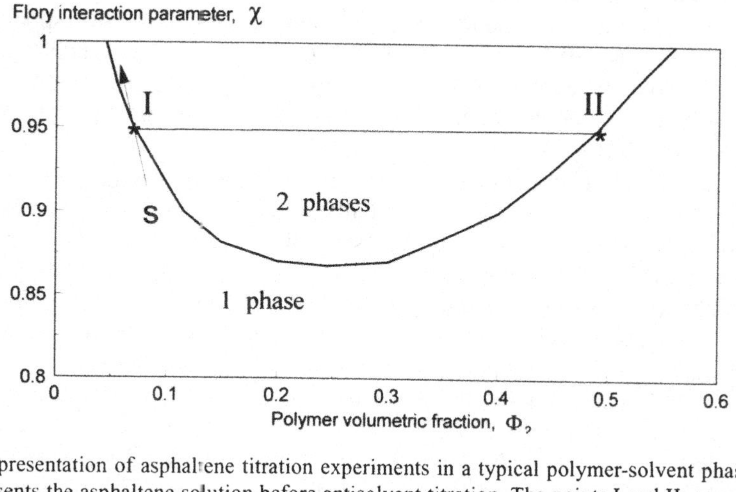

Figure 14. Representation of asphaltene titration experiments in a typical polymer-solvent phase envelope, where S represents the asphaltene solution before antisolvent titration. The points I and II, connected by the tie line, indicate the polymer-lean and polymer-rich phases, respectively, formed at the precipitation threshold.

flocculation of binary polymer solutions, “except near the critical point, one of the phases is practically pure solvent.” Thus, to a first approximation, polymer theory predicts that *pure solvent, rather than pure solute* separates from solution at the flocculation point (see point I in Figure 14). Recall that Hirschberg *et al.*, made precisely the opposite simplifying assumption in their treatment of asphaltene solutions, perhaps because of the historical steric-stabilization model, in which the asphaltenes are supposed to be a dispersed, insoluble, solid phase (*i.e.*, lyophobic colloids). In our development of a new asphaltene solution model below, we will be faithful to the analogy with the well-established polymer theory and hypothesize that one of the nucleating phases is pure solvent, rather than pure asphaltene solute.

Assuming that a liquid asphaltenic phase (containing solvent, point II in Figure 14) and a pure solvent phase (point I in Figure 14) nucleate during phase separation, we write:

$$\Delta\mu^{I}_{1} = (\mu_1 - \mu_{01})^{I} = 0 \tag{11}$$

From the equilibrium equation, $\Delta\mu_1{}^{I} = \Delta\mu_1{}^{II}$, and using Flory's theory to derive the chemical potentials involved (see the section “Physical Models of Asphaltene Stability” above), eq.12 is obtained:

$$\Delta\mu^{II}_{1} = \ln(1-\phi_2^{II}) + (1-1/r)\,\phi_2^{II} + \chi\,(\phi_2^{II})^2 = 0 \tag{12}$$

This equation relates the precipitation threshold to the interaction parameter χ [*i.e.*, the difference between δ_1 and δ_2 (eq. 4)], the molar volume ratio r, and the volumetric fraction of asphaltenes in phase II, $\phi_2{}^{II}$. Numerical simulations for typical values of χ and r show that $\phi_2{}^{II}$, which gives the concentration of asphaltenes in phase II, is almost insensitive to the initial asphaltene concentration, in accord with the experimental observations discussed above. In light of these considerations, $\phi_2{}^{II}$ will be taken to be constant in our treatment.

Examination of Figure 13 shows that eq.12 fits the experimental titration data much better than the model developed by Hirschberg. Another comparison of the two models was accomplished by simulating the variation in the flocculation threshold observed by Hotier and Robin[121] when a crude oil was titrated with n-paraffins of increasing carbon number. In these experiments a maximum in Y_{pt} for asphaltene precipitation was found for n-C7. Simulations of these experiments using eqs.10 and 12 were carried out by assuming reasonable values for the asphaltene and oil molar volumes (1500 and 150 cm^3/mol, respectively), their solubility parameters (23.5 and 20.4 $MPa^{1/2}$), the asphaltene volume fraction in the oil, ϕ_2 (0.005), and the volumetric fraction of asphaltenes in phase II, $\phi_2{}^{II}$ (0.8). Whereas eq.12 predicts the existence of a maximum in Y_{pt} (for n-C8) as found experimentally, a monotonic trend of the precipitation threshold versus the precipitant carbon number is predicted by eq.10 (Figure 15).

The steric-stabilization model of Leontaritis (eq.2) requires the presence of resins as an essential condition for the stability of asphaltene solutions. For this reason is it inherently unsuited for the analysis of experimental results obtained on model solutions of isolated asphaltene fractions (in fact, it incorrectly predicts that such solutions will be unstable under all conditions). It is, however, possible to evaluate this model by carrying out titration experiments on the crude oil, which contains both the asphaltene and resin components. As shown in Figure 16,[123] eq.2 fails to reproduce the linear trend between the experimental variables Y_{pt} and X_{pt} found for the stock-tank oil #3. Once again, the best fit to the experimental data is achieved with eq. 12.

As will be shown in the next section, “Applications of the Model,” important additional support for the new model is provided by the comparison between the predicted flocculation threshold of live oils and data available from the oilfield.

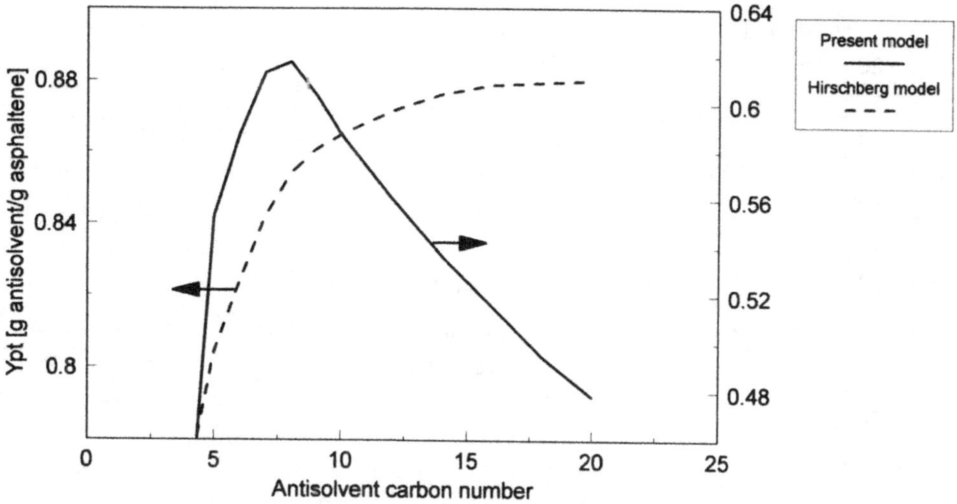

Figure 15. Simulation of the asphaltene precipitation threshold (Y_{pt}) versus the carbon number of the paraffinic precipitant.

With the objective of predicting asphaltene stability in live oils, it is helpful to express eq. 12 in a graphical form. As will be shown in the next section, such a representation provides a simpler, more immediate assessment of the potential stability of the asphaltenes in a given live oil as a function of the principal variables (pressure and composition) relevant to reservoir operations (pressure depletion, gas injection for pressure maintenance). Eq. 12 is an implicit function of the following parameters:

$$\Delta\mu_1{}^{I} = f(v_1, \delta_1, v_2, \delta_2, \phi_2{}^{II}) = 0 \tag{13}$$

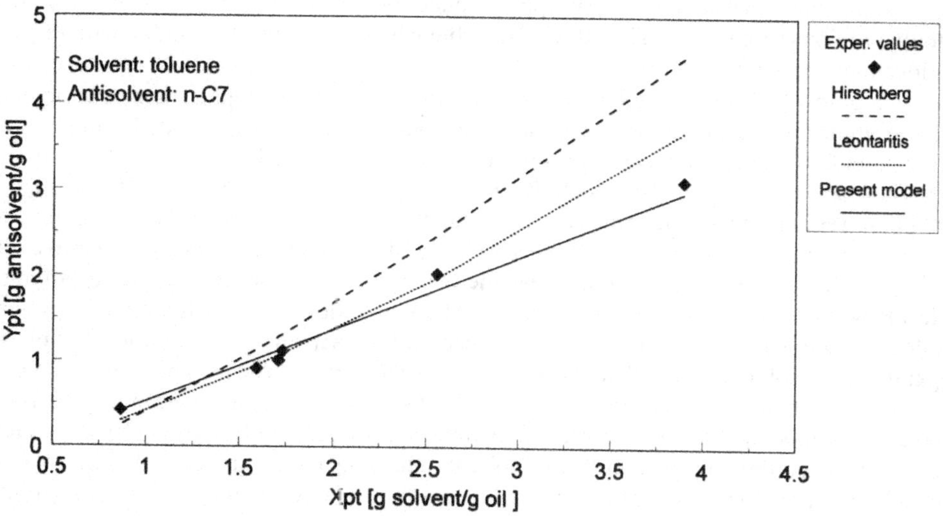

Figure 16. Predicted n-C5 asphaltene precipitation thresholds for stock-tank oil #3 using the Hirschberg *et al.*, Leontaritis, and proposed model (from ref. 123; reprinted with permission of the publisher).

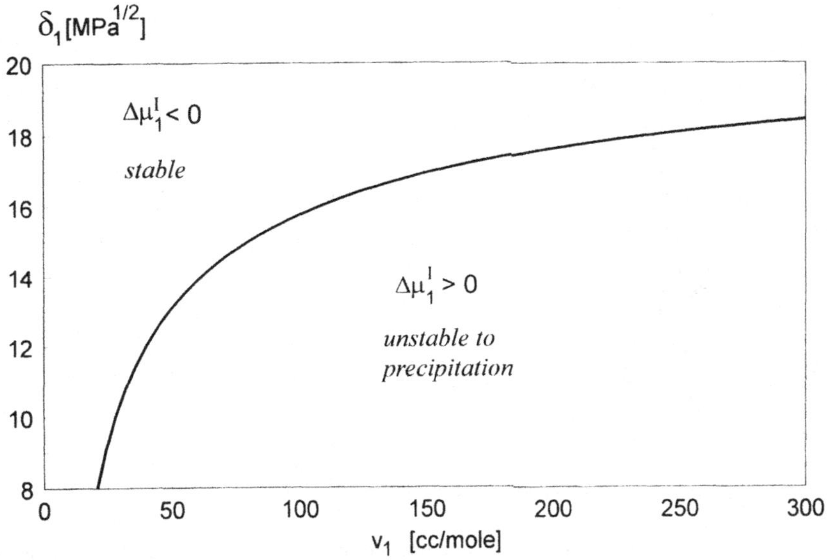

Figure 17. Asphaltene solution phase boundary represented in the v_1 *vs* δ_1 plane (from ref. 123; reprinted with permission of the publisher).

where parameters v_2, δ_2, and ϕ_2^{II} refer to the asphaltenes and v_1 and δ_1 to the oil. Given a live oil at a specified temperature and pressure (reservoir conditions, for instance), the parameters related to the asphaltenes, which are determined by their chemical nature and by the temperature and pressure, are constant. In these conditions, eq.12 is a function of the two variables related to the oil alone:

$$f(v_1, \delta_1) = 0 \tag{14}$$

This function describes a curve (phase boundary) on the plane (v_1, δ_1) which it divides into two regions (Figure 17). The oil will be stable where $\Delta\mu_1^I < 0$ and unstable with respect to flocculation where $\Delta\mu_1^I > 0$.

Figure 18 illustrates such a graph where the points (v_1,δ_1) represent simple organic solvents under ambient conditions and the phase boundary has been calculated using typical parameters for n-C7 asphaltenes (v_2 = 1500 cm^3/mol, δ_2 = 22.5 MPa$^{1/2}$). As can be seen, the paraffinic solvents fall in the instability region, while all aromatic solvents lie in the one-phase region, as we would expect. In the final section we will show that this graphical approach can be used to assess the phase stability of the asphaltene component of a live oil.

We have seen in this section that the asphaltenes behave like lyophilic polymer colloids with respect to their phase behavior. The key to developing a functional physical model of their solution behavior was the recognition that, at the conditions for phase separation, the solvent, rather than the asphaltenes, nucleate as a (nearly) pure phase. When this nucleation behavior is assumed, the stability of solutions of asphaltenes in hydrocarbon media can be modelled successfully using a simple modification of Flory's polymer solution theory. It is worthwhile pointing out explicitly that the model developed, like that of Hirschberg *et al.*,[106] treats the asphaltene solution as a simple two component system (solute and solvent). The faithful description of the phase behavior under the conditions studied is remarkable in view of the considerable simplification of the chemical variation within the asphaltene fraction. It is also worth noting that the model developed treats the *threshold* for

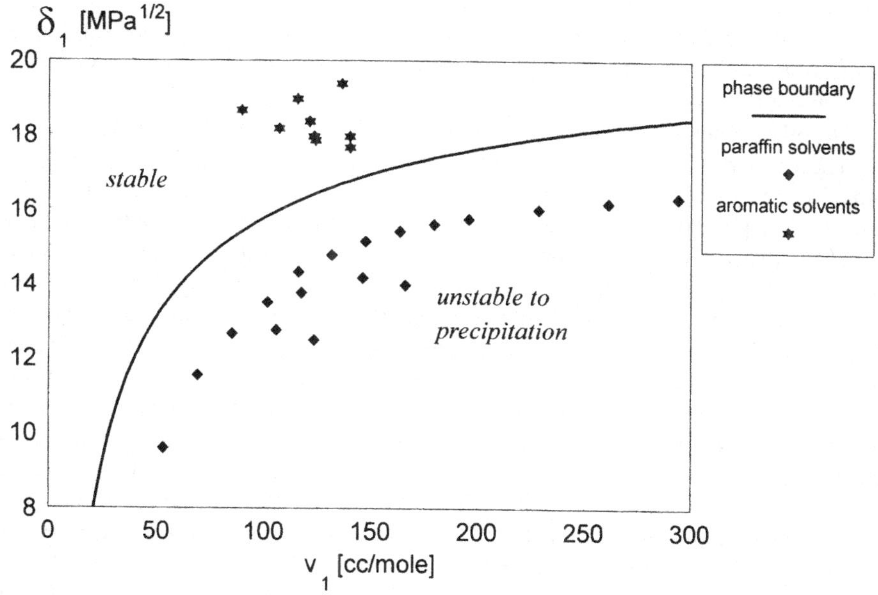

Figure 18. Diagram showing a typical asphaltene phase boundary and the solvent power of paraffinic and aromatic solvents.

asphaltene precipitation. Considering the pronounced compositional polydispersity of the asphaltene fraction we expect that phase-separation at the flocculation threshold will involve only a fraction of the asphaltenic component (that which is least "happy" the hydrocarbon matrix), rather than quantitative asphalt deposition.

Applications of the Model. In order to predict the stability of a live oil at reservoir conditions, we need to construct the phase boundary for asphaltene flocculation and determine the thermodynamic state of the oil as discussed in the preceding section. This requires knowledge of the parameters describing the asphaltene component (v_2, δ_2, and ϕ_2^{II}) and the oil (v_1 and δ_1) at reservoir temperature and pressure.

While it is extremely difficult and costly to recover single-phase oil samples from the level of the reservoir, the live oil can be reconstituted with reasonable accuracy from the stock-tank oil by adding the appropriate low molecular weight gaseous hydrocarbons. A more serious problem is represented by the difficulty of carrying out titration and PVT measurements on the live oil over a range of high temperature and pressure conditions. In view of these considerations, we adopted the following simplifications to characterize the two components of the live oil: The asphaltene parameters v_2, δ_2, and ϕ_2^{II} defining the phase envelope are determined by titration experiments on the stock-tank oil at reservoir temperature; where the reservoir temperature is very high (the case of oil #3) the extrapolation procedure described below can be employed to estimate δ_2 from measurements at lower temperature. The parameters of the oil phase, instead, are calculated from a partial description of the crude oil composition by means of an equation of state (EOS).

The asphaltene parameters δ_2 and ϕ_2^{II} are determined from the titration data using eq.12 as described above. The parameter v_2 can be estimated from SEC molecular weight and X-ray or neutron scattering density data for the n-C7 asphaltene fraction at ambient temperature, although sensitivity analysis suggests that the calculated phase boundary does not depend strongly on v_2. It appears reasonable from our experience,[117] therefore, to employ an approximate value, 1.22 g/cm^3, for the asphaltene density rather than to carry out scattering measurements to characterize the asphaltene fraction of each crude oil. Following the suggestion of Hirschberg *et al.*,[106] and the results of numerical simulations, we have neglected the influence of pressure on the asphaltene parameters.

The oil parameters are estimated using a cubic EOS[125] with data from conventional PVT reports. These reports give the detailed chemical composition, density, and boiling point of each fraction of the live oil up to C_{13}, with the remainder reported as a single, C_{13+} fraction. We have found the boiling point of the C_{13+} fraction to be important for the accuracy of the prediction of the asphaltene phase behavior. In order to better describe the contribution of this fraction to the solvent parameter of the live oil, we have used the value of δ for the oil phase obtained from titration measurements on stock-tank oil to estimate the C_{13+} boiling point. Using this derived boiling point, and tuning the EOS interaction constants to reproduce the live oil bubble pressure and density, v_l and δ_l for the oil phase can be calculated at reservoir temperature and pressure.

An advantage of this approach is that the solvent properties of the live oil medium can be calculated for any temperature and pressure by means of the EOS. In this way it is possible to predict the behavior of the crude oil under different producing conditions. It is not presently practical to define the phase behavior of the live oil in similar detail by experimental means.

The modelling results obtained for oil #6, which is produced from a reservoir at 56°C and *ca.* 398 bar, are presented in Figure 19 (Table 5 reports the PVT data for oils #3 and #6). The predicted onset of asphaltene flocculation for this oil is 210 bar, which compares quite favorably with the field observation that asphaltene deposition occurs in the wellbore at a depth where the pressure is estimated to be 260 bar.

For oil #3, which is produced from an ultra-deep well, it was not practical to carry out the titration experiments at the formation temperature of *ca.* 172°C. Instead, the value of δ_2 was extrapolated to the reservoir temperature from ambient titration results by the application of eq.15, as previously suggested by Hirschberg.[106]

$$\delta_2(T) = \delta_2(T_0)\exp[A_0(T-T_0)] \tag{15}$$

The ambient temperature values of v_2 and ϕ_2^{II}, instead, were used without modification. Figure 20 reports the calculated flocculation phase boundary and solvent properties of

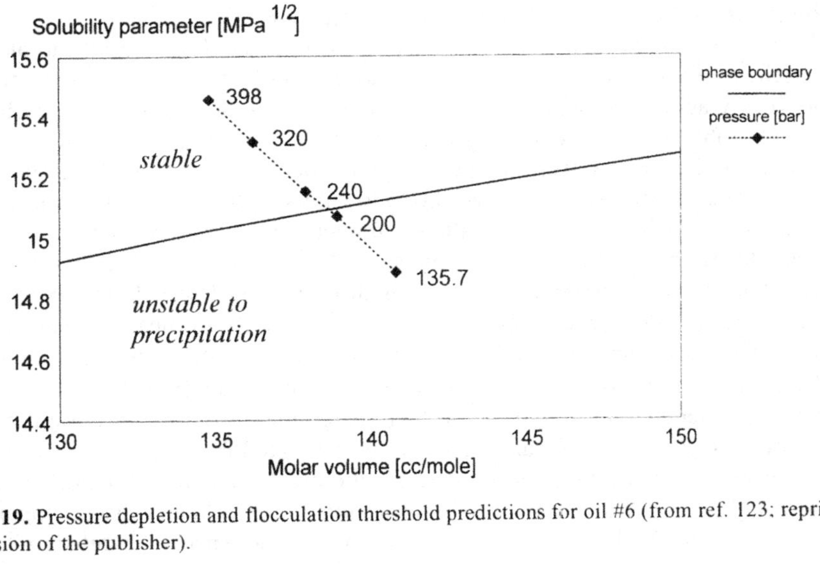

Figure 19. Pressure depletion and flocculation threshold predictions for oil #6 (from ref. 123; reprinted with permission of the publisher).

Table 5. Characterization of oils #3 and #6

	Oil #3	Oil #6
Stock tank API gravity	41.20	35.60
Reservoir pressure (bar)	1060	398
Reservoir temperature (°C)	172.00	56.00
Live oil bubble pressure (bar)	177.80	135.70
n-C_7 asphaltenes M_w	1300	1500

PVT reports

Component	%wt.	%wt.
CO_2	0.00	0.03
N_2	0.00	0.00
methane	0.00	0.33
ethane	0.00	0.86
propane	0.11	1.86
iso-butane	0.27	1.02
n-butane	0.58	2.18
iso-pentane	0.61	1.62
n-pentane	0.84	1.94
hexanes	3.65	3.73
heptanes	5.04	4.57
octanes	7.54	5.15
nonanes	6.23	5.22
decanes	6.43	4.68
undecanes	6.50	3.74
dodecanes	3.79	3.30
C_{13+}	61.32	59.76

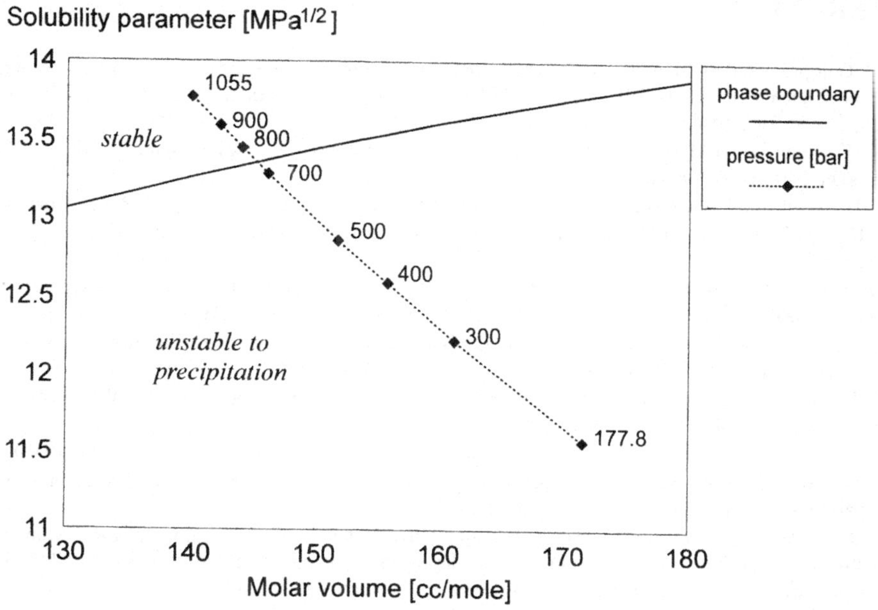

Figure 20. Pressure depletion and flocculation threshold predictions for oil # 3 (from ref. 123; reprinted with permission of the publisher).

the live oil phase over a range of pressures. We see that asphaltene flocculation is predicted to occur at 740 bar, only some 60 - 160 bar below the initial bottom hole flowing pressure of the wells at which the asphaltene deposition problems occurred. Taking into account the neglect of pressure, the extrapolation of the phase behavior from ambient temperature measurements, and the use of laboratory-generated asphaltene samples in these measurements, we consider the results of this calculation to be quite satisfactory.

An extremely noteworthy result of these calculations is the theoretical prediction that the onset of asphaltene flocculation lies far above the bubble point of the live oils (562 bar above the bubble point for oil #3). This result, which is in good accord with the behavior in the field, is in marked contrast with the widely held idea that asphaltene flocculation occurs only very near the bubble point.

The successful application of this model to operative problems involving other crude oils provides additional confidence in its ability to describe to a good approximation the phase behavior of the asphaltene fraction of crude oils under a range of downhole conditions. Given the general nature of the model developed, and its simple application, we expect it to constitute a useful tool for predicting and managing asphaltene deposition in a variety of points in the petroleum cycle.

ACKNOWLEDGMENTS

It is a pleasure to acknowledge the contributions of our colleagues L. Barberis-Canonico, G. Del Piero, L. Faggian, L. Montanari, M. Nali, G. Piro, E. Radaelli, P.A. Sacomani, and F. Stroppa of Eniricerche SpA, and C. Carniani of Agip SpA, to the studies described in this Chapter. We especially wish to recall the contributions of the late E. Mazzolini and L. Bertero of Agip SpA to our work. This research was sponsored by Agip SpA.

REFERENCES

1. M. Galoppini and M. Tambini, Asphaltene deposition monitoring and removal treatments: an experience in ultra deep wells, paper SPE 27622, *1994 Europ. Production Operations Conf.*, Society of Petroleum Engineers: Richardson, TX.
2. K.H. Altgelt and M.M. Boduszynski, "Composition and Analysis of Heavy Petroleum Fractions," Marcel. Dekker, New York (1994).
3. K.H. Altgelt and T.H. Gouw, "Chromatography in Petroleum Analysis," Marcel Dekker, New York (1979).
4. L.R. Snyder and J.J. Kirkland, "Introduction to Modern Liquid Chromatography," Wiley, New York (1979).
5. D.W. Later, M.L. Lee, K.D. Bartle, R.C. Kong and D.L. Vassiloros, Chemical class separation and characterization of organic compounds in synthetic fuels, *Anal. Chem.* 53:1612 (1981).
6. E. Lundanes and T. Greibrockk, Quantitation of boiling fractions of North Sea oil after class separation and gel permeation chromatography, *J. Liq. Chromat.* 8:1035 (1985).
7. M.M. Boduszynski, R.J. Hurtubise and H.S. Silver, Separation of solvent-refined coal into compound-class fractions, *Anal. Chem.* 54:375 (1982).
8. W.D. Dark, Crude oil hydrocarbon group separation quantitation, *J. Liq. Chromat.* 5:1645 (1982).
9. D.W. Later and B.W. Wilson, Standardization of alumina and silica adsorbents used for chemical class separation of polycyclic aromatic compounds, *Anal. Chem.* 57:2979 (1985).
10. C.D. Pearson and S.G. Gharfeh, Automated high-performance liquid chromatography determination of hydrocarbon types in crude oil residue using a flame ionization detector, *Anal. Chem.* 58:307 (1986).
11. M. Bouquet, J.M. Colin, J.P. Durand and R. Boulet, New analytical tools for the upgrading of residual feeds through the FCC process, *ACS Div. Pet. Chem. Preprints* 34:339 (1989).
12. L.W. Corbett and U. Petrossi, Differences in distillation and solvent separated asphalt residue, *I&EC Prod. Res. Develop.* 17:342 (1978).

13. A. Matsunaga, Separation of aromatic and polar compounds in fossil fuel liquids by liquid chromatography, *Anal. Chem.* 55:1375 (1983).
14. E. Lundanes and T. Greibrokk, Group separation of oil residues by supercritical fluid chromatography, *J. Chromatogr.* 349:439 (1985).
15. L. Carbognani and A. Izquierdo, Preparative compound class separation of heavy oil vacuum residue by high-performance liquid chromatography, *Fuel Sci.& Techn. Int'l.* 8:1 (1990).
16. J.G. Speight, "The Chemistry and Technology of Petroleum", Marcel Dekker, New York (1980), Ch.5.
17. J.G. Speight, R.B. Long and T.D. Trowbridge, Factors influencing the separation of asphaltenes from heavy petroleum feedstocks, *Fuel* 63:616 (1984).
18. "Standards for Petroleum and Its Products, Standard No. IP 143/57", Institute of Petroleum, London.
19. D.L. Mitchell and J.G. Speight, The solubility of asphaltenes in hydrocarbon solvents, *Fuel* 52:149 (1973).
20. R.B. Long, The concept of asphaltenes, *ACS Div. Petr. Chem. Preprints* 24:891 (1979).
21. S.E. Moschopedis and J.G. Speight, Investigation of hydrogen bonding by oxygen functions in Athabasca bitumen, *Fuel* 55:187 (1976).
22. J.G. Speight and S.E Moschopedis, Some observations on the molecular "nature" of petroleum asphaltenes, *ACS Div. Pet. Chem. Preprints* 24:910 (1979).
23. J.G. Speight, Solvent effects in the molecular weight of petroleum asphaltenes, *ACS Div. Pet. Chem.Preprints* 26:825 (1981).
24. J.G. Speight, A chemical and physical explanation of incompatibility during refining operations, *Proc. 4th Intl. Conf. on the Stability and Handling of Liquid Fuels, US Dept. Energy,* 169 (1992).
25. R.B. Long and J.G. Speight, Studies in petroleum composition. Development of a compositional "map" for various feedstocks, *Rev. de l'Institut Francaise du Petrole* 44:205 (1989).
26. J.G. Speight, Latest thoughts on the molecular nature of petroleum asphaltenes, *ACS Div. Pet. Chem. Preprints* 34:321 (1989).
27. J.A. Koots and J.G. Speight, Relation of petroleum resins to asphaltenes, *Fuel* 54:179 (1975).
28. J.K. Brown and W.R. Ladner, A study of the hydrogen distribution in coal-like materials by high resolution nuclear magnetic resonance spectroscopy. I-The measurement and interpretation of the spectra, *Fuel* 36:79 (1960).
29. J.K. Brown and W.R. Ladner, A study of the hydrogen distribution in coal-like materials by high resolution nuclear magnetic resonance spectroscopy. II-A comparison with infra-red measurement and the conversion to carbon structure, *Fuel* 36:87 (1960).
30. Y. Maekawa, T. Yoshida and Y. Yoshida, Quantitative ^{13}C nmr spectroscopy of a coal-derived oil and the assignment of chemical shifts, *Fuel* 58:864 (1979).
31. E.M. Dickinson, Structural composition of petroleum fractions using proton and ^{13}C NMR spectroscopy, *Fuel* 59:290 (1980).
32. P.F. Barron, M.R. Bendall, M.J. Armostrong and A.R. Atkins, Application of the DEPT pulse sequence for the fractions of $^{13}CH_n$ sub-spectra of coal-derived liquids, *Fuel* 63:1276 (1984).
33. J.M. Dereppe and C. Moreaux, Measurement of CH_n group abundances in fossil fuel materials using DEPT ^{13}C NMR, *Fuel* 64:1174 (1985).
34. C.E. Snape and M.K. Marsh, Stuctural analysis of heavy fossil fuel fractions using ^{13}C NMR spectral editing, *ACS Div. Pet. Chem. Preprints* 30:20 (1985).
35. C.E. Snape, W.R. Ladner, L. Petrakis and B.C. Gates, The chemical nature of asphaltenes from coal liquefaction processes, *Fuel Proc. Techn.* 8:155 (1984).
36. D.J. Cookson and B.E. Smith, One- and two-dimensional NMR methods for elucidating structural characteristics of aromatic fractions from petroleum and synthetic fuels, *Fuel* 1:11 (1987).
37. C.E. Snape, G.J. Ray and C.D. Price, Two-dimensional NMR analysis of aromatic fractions from a coal liquefaction solvents, *Fuel* 65:877 (1986).
38. B.P. Tissot and D.H. Welte, "Petroleum Formation and Occurrence," Springer-Verlag, New York (1978), Part IV.
39. J.F. McKay, P.J. Amend, P.M. Harnsberger, T.E. Cogswell and D.R. Latham, Composition of petroleum heavy ends. 1. Separation of petroleum > 675°C residues, *Fuel* 60:14 (1981).
40. J.F. McKay, P.J. Amend, P.M. Harnsberger, R.B. Erickson, T.E. Cogswell and D.R. Latham, Composition of petroleum heavy ends. 2. Characterization of compound types in petroleum > 675°C residues, *Fuel* 60:17 (1981).
41. "Geochemistry of Sulfur in Fossil Fuels", edited by W.L. Orr and C.M. White, *ACS Symp. Series 429*, American Chemical Society, Washington, DC (1990).
42. K.D. Rose and M.A. Francisco, A two-step chemistry for highlighting heteroatom species in petroleum materials using ^{13}C NMR spectroscopy, *J. Am. Chem. Soc.* 110:637 (1988).

43. J.M. Ruiz., B.M. Carden, L.J. Lena, E.J. Vincent and J.C. Escalier, Determination of sulfur in asphalts by selective oxidation and photoelectron spectroscopy for chemical analysis, *Anal. Chem.* 54:689 (1982).
44. H.V. Drushel, Sulfur compounds in petroleum-known and unknown, *ACS Div. Pet. Chem. Preprints* 15:C13 (1970).
45. S.R. Keleman, G.N. George and M.L. Gorbaty, Direct determination and quantification of sulphur forms in heavy petroleum and coal. 1. The X-ray photoelectron spectroscopy (XPS) approach, *Fuel* 69:939 (1990).
46. G.S. Waldo, O.C. Mullins, J.E. Penner-Hahn and S.P. Cramer, Determination of the chemical environment of sulphur in petroleum asphaltenes by X-ray absorption spectroscopy, *Fuel* 71:53 (1992).
47. S.E. Moschopedis, R.W. Hawkins and J.G. Speight, Identification of nitrogen functional groups in Athabasca bitumen, *Fuel* 60:397 (1981).
48. J.M. Jacobsen and M.R. Gray, Use of IR spectroscopy and nitrogen titration data in structural group analysis of bitumen, *Fuel* 66:749 (1987).
49. J.F. McKay, J.H. Weber and D.R. Latham, Characterization of nitrogen bases in high-boiling petroleum distillates, *Anal. Chem.* 48:891 (1976).
50. S. Mitra-Kirtley, O.C. Mullins, J. van Elp, S.J. George, J. Chen and S.P. Cramer, Determination of the nitrogen chemical structures in petroleum asphaltenes using XANES spectroscopy, *J. Am. Chem. Soc.* 115:252 (1993).
51. K.D. Rose and M.A. Francisco, Characterization of acidic heteroatoms in heavy petroleum fractions by phase-transfer methylation and NMR spectroscopy, *Energy and Fuels* 1:233 (1987).
52. R.G.S. Ritche, R.S. Roche and W. Steedman, Pyrolysis of Athabasca tar sands: analysis of the condensible products from asphaltene, *Fuel* 58:523 (1979).
53. E.W. Bakar, Mass spectrometric characterization of petroporphyrins, *J. Am. Chem. Soc.* 88:2311 (1966).
54. F.E. Dickson and L. Petrakis, Application of electronmagnetic resonance and electronic spectroscopy to the characterization of vanadium species in petroleum fractions, *Anal. Chem.* 42:978 (1972).
55. M. Popl, V. Dolansky, G. Sebor and M. Stejskal, Hydrocarbons and porphyrins in rock extracts, *Fuel* 57:565 (1978).
56. D.H. Freedman, D.C. Saint Martin and C.J. Boreham, Identification of metalloporhyrins by third-derivative UV/Vis diode array spectroscopy, *Energy and Fuels* 7:194 (1993).
57. R.H. Fish and J.J. Komlenic, Molecular characterization and profile identifications of vanadyl compounds in heavy crude petroleum by liquid chromatography/graphite furace atomic spectrometry, *Anal. Chem.* 56:510 (1984).
58. R.H. Fish, J.J. Komlenic and B.K. Wines, Characterization and comparison of vanadyl and nickel compounds in heavy crude petroleums and asphaltenes by reverse-phase and size exclusion liquid chromatography/graphite furace atomic spectrometry, *Anal. Chem.* 56:2452 (1984).
59. C.D. Pearson and J.B. Green, Vanadium and Nickel complexes in petroleum resid acid, base, and neutral fractions, *Energy and Fuels* 7:338 (1993).
60. J.G. Speight and S.E. Moschopedis, On the polymeric nature of petroleum asphaltenes, *Fuel* 59:440 (1980).
61. D.A. Storm, S.J. DeCaino, M.M. De Tar and V.P. Nero, Upper bound on number average molecular weight of asphaltene, *Fuel* 69:735 (1990).
62. J.F. McKay, P.J. Amend, T.E. Cogswell, P.M. Hamsberg, R.B. Erickson and D.R. Latham, Petroleum asphaltenes-chemistry and composition, *ACS Div. Pet. Chem. Preprints* 22:708 (1977).
63. J. Briant and G. Hotier, Etude de l'étage des asphaltènes dans les melanges d'hydrocarbures: taille des amas moléculaires, *Rev. de l'Institut Francaise du Petrole* 38:83 (1983).
64. J.G. Speight, D.L. Wernick, K.A. Gould, R.E. Overfield, B.M.L. Rao and D.W. Savage, Molecular weight and association of asphaltenes: a critical review, *Rev. de l'Institut Francaise du Petrole* 40:51 (1985).
65. S.E. Moschopedis, J.F. Freyer and J.G. Speight, Investigation of asphaltene molecular weights, *Fuel* 55:227 (1976).
66. M.M. Al-Jarrah and A.N. Al-Dujaili, New findings on the physical nature of asphalts, *Fuel Sci. & Tech. Int'l* 7:69 (1989).
67. M.M. Boduszynski, Composition of heavy petroleum. 2. Molecular characterization, *Energy and Fuels* 2:597 (1988).
68. K.E. Chung, L.L. Anderson and W.H. Wiser, Molecular weight determination by vapor-phase-osmometry, *Fuel* 58:847 (1978).
69. K.H. Altgelt, Asphaltene molecular weight by vapor pressure osmometry, *ACS Div. Petr. Chem. Preprints* 13(3):37 (1968).
70. S. Acevedo, G. Escobar, L.B. Gutierrez and J. D'Aquino, Synthesis and isolation of asphaltenes standard for calibration of G.P.C. columns and determination of asphaltene molecular weight, *Fuel* 71:1077 (1992).

71. R.E. Overfield, E.Y. Sheu, S.K. Sinha and K.S. Liang, SANS study of asphaltene aggregation, *Fuel Sci. & Techn. Int'l.* 7:611 (1989).
72. J.C. Ravey, G. Decouret and D. Espinat, Asphaltene macrostructure by small angle neutron scattering, *Fuel* 67:1560 (1988).
73. E.Y. Sheu, M.M. DeTar, D.A. Storm and S.J. DeCanio, Aggregation and kinetics of asphaltenes in organic solvents, *Fuel* 71:299 (1992).
74. A.F.M. Barton, "CRC Handbook of Solubility Parameters and Other Cohesion Parameters" CRC Press, Boca R aton (1991).
75. C.E. Snape and K.D. Bartle, Definition of fossil fuel-derived asphaltenes in terms of average structural properties, *Fuel* 63:883 (1984)
76. P.R. Waller, A. Williams and K.D. Bartle, The structural nature and solubility of residual fuel oil fractions, *Fuel* 68:520 (1989).
77. C.W. Benson, R.A. Simcox and I.C. Huldal, Tailoring aromatic hydrocarbons for asphaltene removal, *Fourth Symposium on Chemicals in the Oil Industry: Dev. & Appl.*, Ed P.H.Ogden, 215 (1991).
78. G.P. Dayvault and D.E. Patterson, Solvent and acid stimulation increase production in Los Angeles basin waterflood, paper SPE 18816, *1989 SPE Reg. Meeting*, Society of Petroleum Engineers: Richardson, TX.
79. M.L. Samuelson, Alternatives to aromatics for solvency of organic deposits, paper SPE 23816, *1992 SPE Int'l. Symp. on Formation Damage*, Society of Petroleum Engineers: Richardson, TX.
80. G. Broaddus, Well- and formation-damage removal with nonacid fluids, *J.Petroleum Techn.* 685 (June 1988).
81. G. Gonzales and A. Middea, Peptization of asphaltene by various oil soluble amphiphiles, *Colloids and Surfaces* 42:207 (1991).
82. M.G. Trbovich and G.E. King Asphaltene deposit removal: long-lasting treatment with a co-solvent, paper SPE 21038, *1991 SPE Int'l. Symp. on Oilfield Chem.*, Society of Petroleum Engineers: Richardson, TX.
83. M.E. Newberry and K.M. Barker, Formation damage prevention through the control of paraffin and asphaltene deposition, paper SPE 13796, *1991 SPE Production Operations Symp.*, Society of Petroleum Engineers: Richardson, TX.
84. L. Barberis Canonico, A. Del Bianco, G. Piro and F. Stroppa, C. Carniani, E. Mazzolini, A comprehensive approach for the evaluation of chemicals for asphaltene deposit removal. *Recent Advances in Oilfield Chemistry*, Ed. by P. H. Ogden, The Royal Society of Chemistry, p.220 (1994).
85. A. Del Bianco, F. Stroppa and L. Bertero, Tailoring hydrocarbon streams for asphaltene removal, paper SPE 28992, *1995 SPE Int'l. Symp. on Oilfield Chemistry*, Society of Petroleum Engineers: Richardson, TX.
86. F.J. Nellensteyn, The constitution of asphalt, *J. Inst. Petrol. Technol.* 10:311 (1924).
87. F.J. Nellensteyn, The composition of the micelle nucleus of asphalt bitumen and coal tar, and some related problems, *Chem. Weekblad* 36:362 (1939).
88. J.Ph. Pfeiffer and R.N. Saal, Asphaltic bitumen as colloid system, *J. Phys. Chem.* 44:139 (1940)
89. J.M. Swanson, A contribution to the physical chemistry of the asphalts, *J. Phys. Chem.* 46: 141 (1942).
90. T.F. Yen, A macrostructure of petroleum asphalt, *ACS Div. Petr. Chem. Preprints* 35:314 (1990).
91. P.C. Hiemenz, "Principles of Colloid and Surface Chemistry," Ch. 1, 8, and 11, Marcel Dekker, New York, 2nd ed. (1986).
92. J.P. Dickie and T.F. Yen, Macrostructures of the asphaltic fractions by various instrumental methods, *Anal. Chem.* 39:1847 (1967).
93. D. Espinat, and J.C. Ravey, Colloidal structure of asphaltene solutions and heavy oil fractions studied by small angle neutron and x-ray scattering, paper SPE 25187, *1993 SPE Int'l. Symp. on Oilfield Chem.*, Society of Petroleum Engineers: Richardson, TX.
94. K.J. Leontaritis and G.A. Mansoori, Asphaltene flocculation during oil recovery and processing: a thermodynamic-colloidal model, paper SPE 16258, *1987 SPE Int'l. Symp. on Oilfield Chem.*, Society of Petroleum Engineers: Richardson, TX.
95. J.S. Parkand and G.A. Mansoori, Aggregation and deposition of heavy organics in petroleum crudes, *Energy Sources* 10:109 (1988).
96. W.K. Stephenson, Producing asphaltenic crude oils: problems and solutions, *Petrol. Eng. Int'l.* 6:24 (1990).
97. K.J. Leontaritis, Asphaltene deposition: a comprehensive description of problem manifestations, and modeling approaches, paper SPE 18892, *1989 SPE Production Operations Symp.*, Society of Petroleum Engineers: Richardson, TX.
98. K.J. Leontaritis, J.O. Amaefule and R.E. Charles, A systematic approach for the prevention and treatment of formation damage caused by asphaltene deposition, *SPE Production Facilities*, 9:157 (1994).

99. J. Escobedo and G.A. Mansoori, Determination of the onset of asphaltene flocculation (a novel method), paper SPE 28018, Society of Petroleum Engineers: Richardson, TX.
100. J.J. Heithaus, Measurement and significance of asphaltene peptization, *J. Inst. Petrol.* 48:45 (1962).
101. D.L. Katz and K.E. Beu, Nature of asphaltic substances, *Ind. Eng. Chem.* 37:195 (1945).
102. E.Y. Sheu, K.S. Liang, S.K. Sinha and R.E.Overfield, Polydispersity analysis of asphaltene solutions in toluene, *J. Colloid Interface. Sci.* 153:399 (1992).
103. N.F. Carnahan, L. Quintero, D.M. Pfund, H.L. Fulton, R.D. Smith, M. Capel and K. Leontaritis, A small angle x-ray scattering study of the effect of pressure on the aggregation of asphaltene fractions in petroleum fluids under near-critical solvent conditions, *Langmuir* 9:2035 (1993).
104. K.J. Leontaritis, "Asphaltene Deposition: A Thermodynamic-Colloidal Model," Dissertation for the degree of Doctor of Philosophy in Chemical Engineering, University of Illinois, Chicago (1988).
105. H. Tompa, "Polymer Solutions," Butterworths, London (1956).
106. A. Hirschberg, L.N.J. de Jong, B.A. Schipper and J.G.Meijers, Influence of temperature and pressure on asphaltene flocculation, *Soc. Petrol. Eng. J.* 6:283 (1984).
107. J.M. Prausnitz, R.N. Lichtenthaler and E.G. de Azevedo, "Molecular Thermodynamics of Fluid-Phase Equilibria", Prentice-Hall, Englewood Cliffs (1986).
108. N.E. Burke, R.E. Hobbs and S.F. Kashou, Measurement and modeling of asphaltene precipitation, *J. Petroleum Techn.*, 42:1440 (1990) .
109. S.L. Kokal, J. Najman, S.G. Sayegh and A.E. George, Measurement and correlation of asphaltene precipitation from heavy oils by gas injection, *J. Canadian Petrol. Techn.* 31(4):24 (1992).
110. R.L. Scott and M. Magat, The thermodynamics of high-polymer solutions. I. The free energy of mixing of solvents and polymers of heterogeneous distribution, *J. Chem. Phys.* 13:172 (1945).
111. S. Kawanaka, S.J. Park and G.A. Mansoori, Organic deposition from reservoir fluids: a thermodynamic predictive technique, *Soc. Petrol. Eng. Reservoir Eng.* 6:185 (1991).
112. S. Kawanaka, S.J. Park and G.A. Mansoori, The role of asphaltene deposition in EOR gas flooding: a predictive technique, paper SPE/DOE 17376, *1988 SPE/DOE Enhanced Oil Recovery Symp.*, Society of Petroleum Engineers: Richardson, TX.
113. G.A. Mansoori and T.S. Jiang, Asphaltene deposition and its role in Eor miscible gas flooding, *Proceedings, 3rd Europ. Conf. on Enhanced Oil Recovery, Rome,* 16-18 Apr. (1985).
114. F.G. Thomas, D.B. Bennion, D.W. Bennion and B.E. Hunter, Experimental and theoretical studies of solids precipitation from reservoir fluid, *J.Can.Petrol.Techn.*, 31(1):22 (1992).
115. T. H. Chung , Thermodynamic modeling for organic solid precipitation, paper SPE 24851, *67th Ann. Techn. Conf. (1992),* Society of Petroleum Engineers: Richardson, TX.
116. N. Senglet, C. Williams, D.Faure, T. Des Courieres and R. Guilard, Microheterogeneity study of heavy crude petroleum by uv-visible spectroscopy and small angle x-ray scattering, *Fuel* 69:72 (1990).
117. G. Del Piero (Eniricerche), unpublished results.
118. F. Chung, P. Sarathi and R. Jones, Modeling of asphaltene and wax precipitation, Topical Report NIPER-498, UC-122 (1991).
119. K.R.L. Popper, "The Logic of Scientific Discovery," Routledge, New York (1992).
120. K. Kawate, I. Imagawa and M. Nakata, Cloud-point curves of ternary system nitroethane+cyclohexane+polystyrene determined by a novel method, *Polymer J.* 23:233 (1991).
121. G. Hotier and M. Robin, Effects of different diluents on heavy oil products: measurement, interpretation, and a forecast of asphaltene flocculation, *Revue de l'IFP* 38:101 (1983).
122. C. Reichert, B.J. Fuhr and L.L. Klein, Measurement of asphaltene flocculation in bitumen solutions, *J.Can.Petrol.Techn.*, 25(5):33 (1986).
123. R. Cimino, S. Correra, P. Sacomani and C. Carniani, paper SPE28993, *1995 SPE Int'l. Symp. on Oilfield Chemistry,* Society of Petroleum Engineers: Richardson, TX.
124. I. Sanchez, Polymer phase separation, *in:* Encyclopedia of Physical Science and Technology," Meyers, R.A., ed., Academic Press, Orlando (1987).
125. G. Soave, Application of equations of state and the theory of group solutions to phase equilibrium prediction, *Fluid Phase Equilibria* 87: 23 (1993).

Chapter IV

A UNIFIED VIEW OF THE COLLOIDAL NATURE OF ASPHALTENES

Sócrates Acevedo,* María A. Ranaudo, Gastón Escobar, Luis B. Gutiérrez, and Xiomara Gutiérrez†

Centro de Química Orgánica
Escuela de Química, Facultad de Ciencias
Universidad Central de Venezuela
P.O. Box 47102, Caracas, Venezuela

ABSTRACT

Asphaltenes, the high molecular weight component of crude oil is in reality a very complex mixture of compounds which exist in petroleum as a colloidal dispersion. Evidence presented here, together with information in the literature, suggest that these colloidal particles are a mixture of saturated, aromatic, resin, and asphaltene compounds. Presumably saturated and aromatic compounds are adsorbed (or absorbed) on the periphery of the particles due to the interfacial tension effect. As the center of the particle is approached, resins, polar compounds and asphaltenes are encountered. Analysis of the results suggests strongly that the core of the particle is insoluble (solubility < 0.5 g L^{-1}) in all solvents which dissolve the asphaltenes .

The interfacial tension between the crude oil and the particle is considered here of paramount importance in determining the solubility of asphaltenes. The crude oil could be considered as a supersaturated solution with regard to asphaltenes, which, together with the multifunctional chemical nature of the particle explains the tendency of asphaltenes to precipitate, and to be adsorbed by interfaces of quite different nature. In this chapter, the solubility and the adsorption on inorganic solid substrates and at the water-oil interface of asphaltenes are discussed and its consequences, regarding CMC, flocculation of asphaltenes, micelle formation and dispersion, and interfacial (oil-water) tension changes with time are discussed.

INTRODUCTION

The solubility of asphaltenes is a matter of extraordinary importance both from the industrial and academic point of view. Usually, due to the formation of "asphaltene deposits"

---

* Address correspondence to Dr. Acevedo.

† Present Address: INTEVEP, S.A., P.O. Box 76343, Caracas 1070A, Venezuela.

Asphaltenes: Fundamentals and Applications, Edited by E. Y. Sheu and O. C. Mullins
Plenum Press, New York, 1995

that lead to the plugging of tubing and other production facilities, petroleum production must be stopped until the above deposits are mechanically or chemically removed.

This situation has lead to the study of the factors which influence the solubility of asphaltenes in crude oils and several theories have been proposed to explain why asphaltenes precipitate[1,2]. The theory proposed by Hirschberg et al.[2] relies in general on the one proposed for chain polymers by Flory and Huggins[3], and modified to include the solubility parameters δ defined by Hildebrand, Prausnitz and Scott[4] (equation 1)

$$\delta = \left(\frac{\Delta H^{v} - RT}{V} \right)^{1/2} \quad (1)$$

Here ΔH^v and V are the molar enthalpy and molar volume respectively, T is the absolute temperature and R the gas constant. These authors[2] and others[5] have used this model to explain solubility behavior of asphaltenes. In particular, the agreement between predicted and observed behavior near the bubble point is satisfactory.

However, an important limitation of this model is its failure to account for the colloidal nature of asphaltenes which, no doubt, has profound influence in asphaltene solubility.

There is much direct and indirect evidence of the presence of asphaltene in crude oil as a colloidal dispersion. Activities in this direction started at the beginning of this century with the works of Mack[6], Pfeiffer[7], Nellensteyn[8] and others[9, 10]. Among other things these researchers proposed that asphaltenes are colloids dispersed in asphalt and bitumen by resins. In 1945, Kats, on the bases of electron microscope measurements suggested that asphaltene colloids, if present, should have diameters less than 65 Å.[10] Pollack and Yen, using X-ray small angle scattering proposed diameters between 30 and 50 Å for spherical particles of asphaltenes[11]. Leontaris and Mansoori have proposed that flocculation of asphaltenes after the addition of solvents like pentane or hexane, is due to desorption of resins which are adsorbed at the surfaces of asphaltene particles or colloids[12]. In fact similar propositions were made earlier by Van Kerkvoort et al.[13] and Heithaus[14] who defined a flocculation ratio as the minimum amount of "good" solvent (such as toluene) which was required to add to a "non solvent" (heptane) to avoid asphaltene precipitation.

Herzog et al. analyzed the size and shape of asphaltene particles using SAXS and concluded that these are thin disk of 3.4 Å in thickness and radii that follow a power law such as γ^{β} (13 Å$\leq r \leq$ 800 Å)[15]. Andersen and Birdi[16] and Sheu et al.[17] have reported CMC for asphaltenes based on calorimetric and surface tension measurements respectively. Later Andersen and Speight have commented upon the significance of the CMC concept when applied to asphaltenes [18].

As discussed below (cf. results and discussion) the colloidal nature will confere to the asphaltene an excess in chemical potential or instability which could be related to the interfacial tension through the Kelvin equation. Thus, as found earlier by Mitchell and Speight the solubility of asphaltenes is related to the surface tension[19], a fact which has not been generally appreciated.

This instability accounts for the tendency of asphaltenes to precipitate from crude oils under a variety of circumstances and to be adsorbed on interfaces of any kind (see below). For instance, the adsorption of asphaltenes on the reservoir rocks or sands leads to changes in wettability with negative influence of crude oil flow. Studies in this field have shown that asphaltenes adsorb readily on inorganic surfaces[20-22] even in the presence of water[23].

The above studies have been focused towards wettability, which no doubt is of paramount importance in oil recovery.

It should be realized however, that adsorption on surfaces is probably the main source of asphaltene precipitation, a point recently taken up in our laboratories[24] and extended in the present chapter.

Also important is the adsorption of asphaltenes at the oil-water interface which, due to problems associated with emulsion formation, has received considerable attention in the literature[25-33]. Although both the adsorption at the liquid-solid and water-oil interfaces could be attributed to the presence of polar functional groups in the asphaltenes, this is an incomplete and poor argument. Other phenomena, such as the formation of multilayers at both interfaces at very low asphaltene concentration, micelle formation, desorption from surfaces, flocculation, etc., require a more complete and integrated view. It is worth noting, that some recent reports have presented evidence of asphaltene adsorption at the crude oil - gas (methane, natural gas) interface[34,35].

In this chapter we present evidence, based on our work and on data taken from the literature, which suggests that the above phenomena have a common source, that is the inherent instability of asphaltenes which is apparent by their colloidal presence in crude oils.

EXPERIMENTAL AND METHODS

In this section we shall call "asphaltenes" the precipitate obtained according to the technical description below. It is understood however, that this is a SARA mixture of compounds. When stripped of saturates, aromatic and resins only asphaltenes remain.

Materials

Some properties of crude oils used in this work are collected on Table 1.

Precipitation of Asphaltenes

These were obtained from a 1:1 mixture of toluene and crude oil after the addition of 60 volumes of n-heptane as described earlier[36]. When "pure" asphaltenes were required, the precipitate was stripped of resins and other components by soxhlet extraction with n-heptane for 30 hours.

SARA Analysis

The asphaltenes obtained above were placed in a soxhlet, extracted with n-heptane and six liquid fractions were collected during a period of 30 hours. The first two fractions were collected within the first two hours of extraction. Longer periods were employed for the other fractions. The saturated, aromatic, resins and asphaltenes content of these fractions

Table 1. Properties of crude oils employed in this work

Crude oil	°API	asphaltenes %	C %	H %	N %	S %	H/C
Hamaca	—	14	83.4	7.6	1.8	4.5	1.09
Cerro Negro	8	13	81.2	7.7	2.1	5.5	1.16
Ceuta	11	20	86.5	6.8	1.6	—	0.94
Boscan	10	12	81.5	10.3	0.64	5.5	1.52
Furrial	21	10	—	—	—	—	—

(SARA analysis) were determined by an automatic thin layer equipment, provided with a flame ionization detector. Silica rods (0.9 mm diameter), were used and the solvent gradient: hexane, toluene and dichlorometane-methanol (40 %) was employed.

Asphaltenes Fractionation

A simple glass device, constructed with two cylindrical glass tubes of different diameters and lengths was used in this experiment. The device is similar to a cylindrical funnel (diameter: 3 cm.) with a long stem. About 1.3 g of pure asphaltenes (see above) were placed in a Teflon® membrane (pore: 0.2 μm), the membrane was attached to the device and placed in a soxhlet apparatus and the sample was extracted with mixtures of acetone and tetrahydrofurane (THF) during 24 hours for each mixture (seven in total). After each extraction, samples of both the extracted solution (Fi) and the residue asphaltene (Ri) were prepared for analyses.

Isolation of Carboxylic Acids from Asphaltenes and Maltenes

We used the method described in the literature[37] with slight modifications. A chromatography column, provided with a refrigerating case and a solvent recirculating system was used. A saturated solution of KOH in isopropanol was prepared and silica gel was added, together with chloroform. The mixture was stirred for 10 minutes, used to fill the column, which was placed on a boiling flask filled with chloroform. Boiling chloroform was recirculated, the sample (asphaltenes or maltenes) was applied and the neutral and basic compounds were removed by chloroform. Then a 20 % solution of formic acid in chloroform was recycled through the column to remove the acidic fractions. The first fraction (phenols) was discarded. Yields of acidic compounds (carboxylic acids) were 0.09 and 2.38 per cent of crude oil (Cerro Negro) for asphaltenes and maltenes respectively. These mixtures which were characterized by FT-IR (strong hydroxyl and carbonyl bands) were insoluble in aqueous alkaline solutions, though when toluene was added (1:1 relationship) an oil in water emulsion was obtained after shaking. These acids were sparingly soluble in hexane and have solubility in toluene of ca. 75 mg ml^{-1}.

Methylation of Carboxylic Acids

The above acids were methylated with diazomethane in ether using the reported procedure[36]. The resulting ester was characterized by C-13 NMR (methoxy band present).

Adsorption on Solid Surfaces

Adsorbents. The inorganic adsorbent was collected from the incrustations or scale deposited on the production tubing of the Ceuta well. This was powdered in a laboratory mill, soxhlet extracted (chloroform) and sieved to 200 - 270 mesh. Characterization (X-ray) showed the adsorbent to be 97 % SiO_2 with a surface area of 4.8 m^2g^{-1}. The Chloroform was evaporated and the organic residue was analysed by automatic thin-layer chromatography with flame ionization detection.

Adsorption Experiments. The concentration of adsorbed asphaltenes or resins was determined by UV (400 nm) as described elsewhere[20]. Toluene solutions in the range of 0.02 to 10 g L^{-1} were prepared using 10 mL of this placed in a flask containing 200 mg of adsorbent and shaked during 24 hours in a water bath at 26 °C.

Solute solid adsorption isotherms (SSA) on the inorganic solid were determined with the five asphaltenes.

Near changes in the slope of the isotherms, the results were subject to more scatter than in other regions. Averages of three o more measurements were therefore taken near these changes. The other points are averages of at least two determinations.

Mixed Micelles Method

We follow the description given by Tanford[39]. Let $\mu_{i,mic}$ and $\mu^{o}_{i,mic}$ be the chemical and standard potential of amphiphile in the mixed micelle and pure micelle of component i, respectively. Let $\mu_{i,s}$ and $\mu^{o}_{i,s}$ be the chemical and standard potential of component i in solution respectively. Thus, if both, micelle and solution behaves ideally, equations 2 and 3 below will hold:

$$\mu_{i,mic} = \mu^{\circ}_{i,mic} + RT \ln X_{i,mic} \tag{2}$$

$$\mu_{i,s} = \mu^{\circ}_{i,s} + RT \ln X_{i,s} \tag{3}$$

Where $X_{i,mic}$ and $X_{i,s}$ are corresponding mole fractions. Under equilibrium conditions the chemical potentials must be equal and equation 4 is obtained

$$\mu^{\circ}_{i,mic} - \mu^{\circ}_{i,s} = RT \ln\left(\frac{X_{i,s}}{X_{i,mic}}\right) \tag{4}$$

Note that the difference in standard potential is equal to $RT\ln(CMC)^{o}_{i}$, where $(CMC)^{o}_{i}$ is the critical micelle concentration (in mole fraction units) of pure component i. Therefore, after substitution equation 5 is obtained

$$X_{i,s} = X_{i,mic}(CMC)^{\circ}_{i} \tag{5}$$

For a two component system equations 6 and 7could be obtained:

$$X_{f,s} = (CMC)^{\circ}_{2} + \left[(CMC)^{\circ}_{1} - (CMC)^{\circ}_{2}\right]X_{1,mic} \tag{6}$$

Where $X_{f,s}$ is the molar fraction of free amphiphiles in solution:

$$X_{f,s} = X_{1,s} + X_{2,s} \tag{7}$$

Thus a plot of total free amphiphile vs. micelle composition is linear.

Knowledge of both $(CMC)^{o}_{1}$ and $(CMC)^{o}_{2}$ permits the construction of the plots in Figure 7 as follows: for any given micelle composition, $X_{f,s}$ is obtained using equation 6 thus line B could be drawn. Lines E and D are obtained from equation 5. Curve C is obtained calculating the ratios $X_{1,s}/X_{2,s}$ and by normalization ($X_{1,s} + X_{2,s} = 1$) which is equivalent to obtaining the relative molar fractions in solution. Note that $X_{f,s}$ is also the effective CMC.

Figure 1 was plotted assuming that $(CMC)^{o}_{A} = 1\text{x}10^{-5}$ and $(CMC)^{o}_{R} = 100\ (CMC)^{o}_{A}$ where A and R stand for asphaltenes and resins respectively.

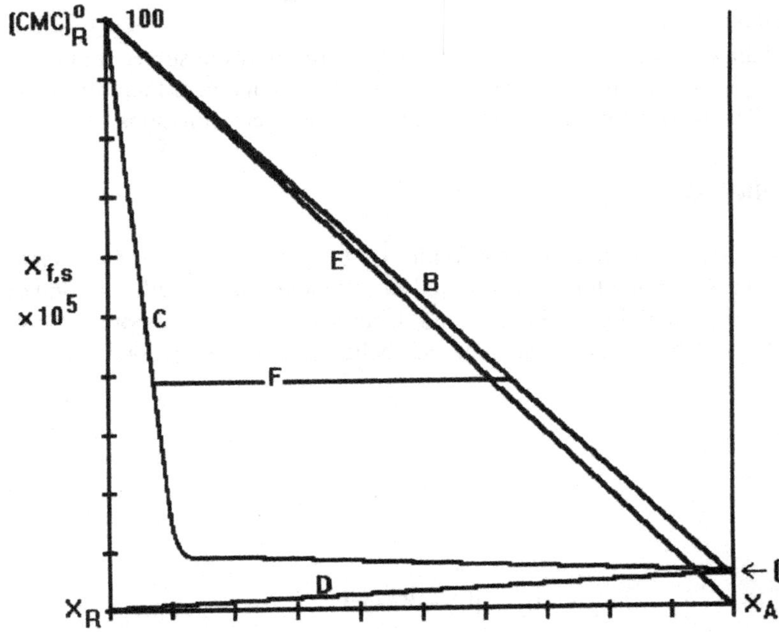

Figure 1. Schematic plots for a hypothetical resins-asphaltene mixed micelle system. B: Effective CMC vs. micelle Composition. C: Effective CMC vs. solution composition. D and E: Plots of equation for asphaltenes (A) and resins respectively. F: Tie line. Regions below curve C and above line B represent solutions and micellar phases respectively. Between them, micelles of compositions given by B are in equilibrium with solutions of composition given by C. These are related by tie lines such as F.

Adsorption at Oil-Water Interface

Interfacial Tension Measurements. Above one dyne cm^{-1} ($mN\ m^{-1}$). The Wilhelmy plate method was employed assuming a zero contact angle (the usual assumption in this case), measurements were performed with a Rovalt Micropore mark II microbalance. The samples were dissolved in 50 ml of toluene to make solutions (0.05 - 0.1 w/v) above presumed CMC values. These were then transferred to flask containing tridistilled water and adjusted to the required pH. The flasks were flushed with nitrogen, stoppered and opened when required. The pH was readjusted and the measurements carried out as usual.

Below One dyne cm^{-1}. A spinning drop apparatus, design at the University of Texas was employed[38]. Since the Cerro Negro crude oil and water have similar densities it was necessary to dilute the sample with xylene (10 % p/p) to have a good displacement of the drop to capillary center. With a syringe, about 0.036 ml of the above solution was placed inside the capillary containing an aqueous Na_2CO_3 solution. After stoppering, the capillary was placed in the measurement chamber, left in stand by for thermal equilibration and then spun to 6600 rpm. Measurements were taken each two minutes during the first hour and at longer periods thereafter. Similar procedures were used with the oleic acid solutions (0.3 - 1 % w/v in a paraffin oil).

Natural Surfactants

The experimental details of the isolation procedure could be found elsewhere[31-33]. The idea of the method is to prepare water in oil emulsions of appropriate samples (crude oil, maltene, solutions of asphaltenes, etc.). The natural surfactants get adsorbed at the oil-water interface and are recovered after washing the excess oil trapped between water drops and breaking the emulsion. These emulsions were prepared at acidic (2.5), neutral (6.5) and alkaline (11.2) pH's.

Magnetic Resonance Spectra

Solid state C-13 spectra were taken in a high resolution MSL-300 Bruker Spectrometer using the high power decoupling and long recycling technique which allows for appropriate nuclei relaxation.

Melting Point Measurements of Resins-Alphaltenes Mixtures

The asphaltenes precipitated from Cerro Negro crude oil (ACN) were extracted with n-heptane (soxhlet) and the extracts were evaporated to dryness and their melting points recorded. With the mixtures containing relatively high quantities of resins, the specimen melted in the usual way. However, as the percentage of asphaltene increases, no accurate value could be normally measured. We circumvent this problem by guessing, from previous m.p. measurements, what the m.p. of the sample could be. The m.p. equipment was then heated slightly below this temperature, the sample was then introduced and a good fusion could be observed.

The first and second extracts (see Table 2) correspond to two successive discharges of the soxhlet (about 0.5 hour). The rest were collected after longer periods of time. The molar fractions of asphaltenes in Table 2 were estimated assuming that the average molecular weight of asphaltenes was 10 times the one for resins.

Electron Microscope Measurements

A drop of Cerro Negro crude oil was frozen (-110°C liquid nitrogen-Freon 22) and fractured in a high vacum chamber surface. Replicas were obtained after shadowing the surface with platinum and covering it with graphite. These were observed using a conventional transmission electron microscope of 100 KV.

A similar experiment was performed with a maltene sample (crude oil without asphaltenes) in which case only very few particles could be observed against the same

Table 2. SARA results for n-heptane extracts of asphaltenes precipitated from Cerro Negro crude oil

Fraction	Extraction time (hours)	Percentages Saturates	Aromatics	Resins	Asphaltenes	Mn
1	0.5	4.6	8.5	66.2	20.7	1220
2	1	—	7.2	68.5	24.3	1230
3	2	—	—	75.2	24.8	1250
4	4	—	—	67.9	32.1	1180
5	24	—	—	62.18	37.8	1306
6	72	—	—	58.9	41.1	1320

bakground as with asphaltenes. Diameters between 25 and 4.5 nm were observed in the crude oil, from which the number (normal) average and weight average were 9.6 and 12.24 nm respectively (polidispersity = 1.28).

RESULTS AND DISCUSSION

SARA Analysis and Fractionation of Cerro Negro Asphaltenes

In Table 2 we have collected the SARA results for the hexane extraction. Two points are worth comment: first, the presence of saturates and aromatics in the precipitate is interesting in view of the poor similarities of these fractions with the asphaltenes (see below). Secondly in spite of the very low solubility of the asphaltenes in n-heptane, it is present in large quantities in all the extracts. These results are consistent with the known view that the asphaltenes could be effectively dispersed by resins.

In Table 3 we present the melting point results of asphaltenes-resins mixture obtained directly from the precipitate after washing it with n-heptane (see experimental). Fortunately, the m.p. (in °C) of the mixtures increases linearly with the percentage of asphaltenes (m.p. = -88.7 + 3.81(%A), c.c. =0.97) thus a fair estimation of the m.p. of pure asphaltene (100% in Table 3) could be made. This value (293°C) compares favorably with the experimental one within a 5 % error.

According to thermodynamics, equation 8 is valid for an ideal two component solution:

$$\ln X_1 = -\frac{\Delta H_f}{RT_o}\frac{\Delta T}{T} \qquad (8)$$

Where:

ΔH_f:	Heat of fusion of solvent or abundant component
T_o:	Melting point of the solvent
T:	Melting point of solution
ΔT:	T_o-T
R:	gas constant

Accordingly, a plot of $\Delta T/T$ vs. $-\ln X_1$ should give a straight line with a zero intercept. Using the results in Table 3 we found the line of equation 9 with a 6% error in the intercept.

Table 3. Asphaltene[a] content, melting points and other properties of asphaltene-resin mixtures

% asphaltenes	m.p. (K)	ΔT	$\Delta T/T$	$-\ln X_1$
77.8	477.5	105.5	0.221	1.35
84.4	513.5	69.5	0.135	1.05
87.6	515.0	68	0.132	0.882
90.1	533.5	49.5	0.093	0.74
93.5	536.5	46.5	0.087	0.52
100.0	583.0	0	0	0

[a]From Cerro Negro (ACN).

$$\ln X_1 = -0.059 - 6.26\frac{\Delta T}{T} \tag{9}$$

From the slope of this line, a value of 7.3 Kcal/mol is found for ΔH_f.

Of course, neither asphaltenes nor resins is an ideal solutions, or a two component thermodynamic system. Therefore, the above results are apparent. However, two conclusions emerge from these experiments: First, resins are not merely adsorbed at the surface of the asphaltene particle, but rather, such a particle could be considered as a resins-asphaltene solution.

Second, equation 9 suggests that the bulk interactions between resins and asphaltenes are non-specific. In fact, these results suggest that interactions between similar pairs (asphaltenes or resins) are quite similar to interactions between asphaltenes and resins. However, as discussed below this is only part of the picture (see, next section).

Mixed Micelles

If the composition and properties of the precipitated asphaltenes reflect those of the micelles in solution then the latter should be considered as mixed micelles. This is a reasonable assumption in view of the large quantities of soluble resins found in the precipitated solid. It should be noted that if mole fractions are used as a composition measure, the one for resins in the precipitates will be greater than for asphaltenes. As discussed in a forthcoming section the flocculation of micelles could be explained in terms of interfacial tension arguments. Any other theory of solubility would have to explain why soluble components are abundant in the precipitates.

Considering the colloidal particles as mixed micelles of resins and asphaltenes is a convenient picture. Using the method described in the previous section, Figure 1 was drawn. In this Figure, line B represent the change in concentration of the free amphiphile (effective CMC) with the composition of micelle. Curve C gives the change of the effective CMC with solution composition. Lines D and E are plots of the equation 5 for asphaltene and resins respectively.

An important corollary of the "phase diagram" of Figure 1 is that for any effective CMC, the micelle will be richer than the solution in the component with lower (CMC)° value, or the asphaltenes in this case. Due to the large expected differences in $(CMC)_i^o$ between resins and asphaltene, more than 90% of this sample will be solubilizated into the micelles in agreement with the fact that they are "more soluble" when in presence of resins. Note that in any crude oil the mole fraction of resins is always larger that the one for asphaltenes and hence the micelles are expected to be richer in resins. This is in agreement with the usual composition of precipitated asphaltenes. Since in any case asphaltenes will be present as mixed micelles, a range of effective CMC, rather than a unique CMC should be expected (see below).

Asphaltenes Fractionation

The results from the solvent fractionation of pure asphaltenes are summarized in Table 4. Here the specimens represented by F_i and R_i are respectively the materials extracted by the solvent mixture and the residue corresponding to that extraction. Also, for the glass device employed, only small quantities of asphaltenes should be extracted. Due to solubility problems C-13 spectra were run with solid samples.

Perhaps the most interesting results in Table 4 are the solubility tests. In particular it is interesting that about 85 % of the sample (R_1) is practically insoluble in toluene, in spite

Table 4. Properties of asphaltenes fractions (Fi) and residue (Ri) obtained by extraction with acetone-THF[a]

	F_1	F_2	F_3	F_4	F_5	F_6	F_7	R_1	R_2	R_3	R_4	R_5	R_6	R_7
% [b]	15	4.6	10.3	21.9	13.4	11.6	6.31	85.0	80.0	69.8	47.9	34.8	23.2	13.0
Mn (GPC)	1800	4500	4600	5300	5700	5700	5500	2900	—	3260	c	c	c	c
H/C	1.21	—	1.33	—	—	—	1.15	1.2	—	—	—	—	—	1.05
Solubility ($g\ l^{-1}$, toluene)	91	72	123	58	3.2	d	d	d	d	d	d	d	d	d
fa[e]	0.47	0.46	—	0.53	0.57	—	0.57	0.51	—	0.50	—	0.56	—	0.57

[a]Solvent sequence: % of acetone 1:60, 2:50, 3:40, 4:30, 5:20, 6:10, 7:0.
[b]Between 1.6 to 1.3 g of asphaltenes in the membrane. Average of three runs.
[c]Due to very low solubility in THF no molecular weight could be measured.
[d]Solubility too low (< 0.5) to be measured (see experimental).
[e]$f_a = C_{Ar}/(C_{AL} + C_{AR})$.

of the fact that the whole sample (i.e., pure asphaltene) has a large solubility in this solvent (100 g L^{-1}). Since in any case crude oils are poorer solvents than toluene, this result is strong evidence in favor of the presence of asphaltenes in crude oils as colloidal particles dispersed by resins and probably by soluble fractions such as F_1.

In general, a decrease in solubility is observed in going from F_1 to F_7, with the exception of F_3. Also, the solubility of R_1 in toluene was recovered when it was remixed with F_1. However, when a similar test was performed with other specimens (for example, R_2 with F_1 plus F_2, etc.) no dissolution or dispersion occur. Thus, although the removal of F_1 appears to be a reversible process, this behavior is exceptional.

The residues R_1 to R_3 were soluble (> 10 g L^{-1}) in THF. However the solubility of R_4 to R_7 in this and other common solvents which dissolve asphaltenes (pyridine, nitrobenzene, chloroform, carbon tetrachloride, etc.) was very low (< 0.5 mg/ml).

When the number average molecular weights (Mn) of Tables 1 and 3 are analyzed, some interesting features emerge. On Table 4, these values gave an average of 1250 ± 50 Daltons, thus no significant differences are detected between them, the first and second fractions contain saturated and/or aromatic compounds. Therefore, these fractions should have Mn values similar to those of resins. Note also that fraction F_1, on Table 4 has a Mn value closer to those of Table 4. On the bases of these Mn values, fraction F_1 appears to be the boundary between resins and asphaltenes.

Both H/C and C-13 results indicates that aromaticity (f_a) increases when going from F_1 to F_7 or from R_1 to R_7 (see Table 4). However, these f_a changes are somewhat low in view of the H/C result for R_7. A sample calculation will show that in passing from an H/C of 1.21 (the one for F_1) to 1.05 for R_7, 8 double bond equivalents will be required, per 100 carbon atoms. Assuming these to be aromatic, f_a values around 63% should be expected. In any case, we believe that these results shows that as the last fractions or residues are approached there is an increase in strong and specific interactions between their molecules, which leads to the formation of insoluble aggregates.

Micelle Structures

From the results discussed above the following micelle structure, depicted schematically in Figure 2 could be proposed: The center or core of the micelle would be occupied by molecules of the type found in R_7 to R_5, which are surrounded by other more soluble

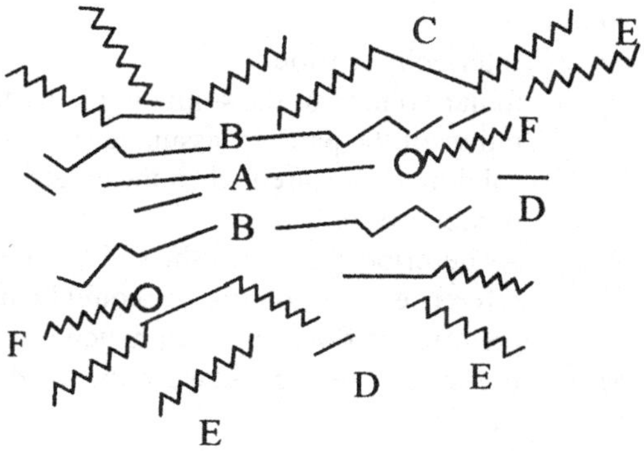

Figure 2. Suggested micelle structure (transverse cut). A and B: type R_5 to R_7 and F_4 to F_2 asphaltene molecules respectevely. C: Resins and F_1 asphaltenes molecules. D: Aromatic compounds. E: Saturated components. F: Amphiphiles.

asphaltene molecules like those found in F_2 to F_4, which would be placed between the center and the periphery. Finally, a peripheral region, made up of resins and small quantities of other components such as asphaltenes (F_1 Type), aromatic and saturated compounds. For reasons to be discussed below, amphiphilic compounds, mainly carboxylic acids, are also included.

Electron microscopy results for ACN as well as those for other asphaltenes[11], shows that the micelles or particles are very small. For ACN we found an average radius of 4.8 nm. Data from the literature points to values smaller than 6 nm[10,11]. The following rough calculation could give an idea of how many molecules to expect in such micelle.
The volume of a spherical particle of radius r is

$$V_p = \frac{4}{3}\pi r^3 \tag{10}$$

for $r = 4.8$ nm, $V_p = 4.63 \times 10^{-9}$ cm^3.

In polar solvents (pyridine) the Mn value of ACN is 2000[36]. Assuming a density $\rho = 1$ for these asphaltenes about 140 molecules per particle could be calculated.

The Kelvin Equation, Supersaturation and Interfacial Tension

A relationship between the chemical potential (μ_2) of a solid phase just being formed due to supersaturation, and the size of the initial crystal is provided by the Kelvin equation (equation 11 below). It is known that this equation could be used for studies of supersaturated solutions and for the calculation of solid-liquid interfacial tension[40].

$$RT\ln\left(\frac{S}{S_o}\right) = \left(\frac{2V\gamma}{r}\right) \tag{11}$$

where

$$\Delta P = \frac{2V\gamma}{r} \tag{12}$$

and

$$\mu_2 - \mu_2^o = RT\ln\left(\frac{S}{S_o}\right) \tag{13}$$

where:

γ: Interfacial tension

V: molar volume of the solute in the particle.

r: radius of the particle, assumed spherical.

S: solubility measured when the solute is in equilibrium between solution and solid particles of radius r.

S_o: As describe above, but the solid is now a flat surface. In the present case, this reference or standard state, could be taken as the solubility of asphaltenes in the presence of flocculated asphaltenes (i.e., the saturated solution).

ΔP: Pressure difference between the inside and outside of the particle (positive, in this case).

R,T: Gas constant and absolute temperature respectively.

μ_2, μ_2^o: Free energy of supersaturated and saturated solutions respectively.

The Kelvin equation measures the degree of supersaturation. Since ΔP is positive, the free energy change is positive and the system would be unstable, thus after the solution becomes supersaturated, small aggregates of solutes or nuclei begin to grow. However, this growing process is hampered by the positive value of ΔP which, when r is in the range of nanometers, could be very significant.

The arguments embodied in equations 11 to 13 could be used conveniently to explain the formation of asphaltenes micelles. When the crude oil is just supersaturated, small particles of asphaltenes, mainly integrated by the less soluble fractions such as R_7 to R_5 begin to form. At this stage, the equivalent r values are very small and these particles would have difficulties in growing, and will remain as such for a time long enough to allow the adsorption of other, more soluble fractions. This adsorption would "shield" the particles from each other providing an additional barrier for flocculation.

At the same time, the above adsorption is promoted by interfacial energy effects. Indeed, the formation of a large quantity of small particles will give rise to a large interfacial area which, unless γ is very small, will eventually lead to flocculation due to increasing total interfacial free energy ΔG. Thus, adsorption will continue, and eventually lead to a micellar structure similar to the one depicted in Figure 2.

Since absorption always leads to interfacial tension reduction, the degree of supersaturation will be lowered. To give a rough idea of the situation, the supersaturation potential energies and interfacial energy (G^s) for the conditions stated in Table 5 were calculated for assumed γ values. G^s, the interfacial contribution to the chemical potential of asphaltenes was calculated using equations 11 and 12 below:

$$G^s = \gamma A \tag{14}$$

$$A = \frac{3V}{r} \tag{15}$$

where A is the area.

These calculations shows that even for small values of γ, significant degrees of supersaturations could be obtained 3 - 40 %.

From a qualitative point of view, this approach offers several advantages. In the first place, considering the crude oil as a supersaturated colloidal dispersion allows one to evaluate the adsorption propensity of asphaltenes at interfaces of any kind, whether they are solid, liquid or gaseous phase. In the sections below we will discuss the adsorption at the solid and crude-water interface. The gas interface adsorption of asphaltenes at the interface of methane microbubbles dispersed in heavy oils has been proposed by Islam as a stabiliza-

Table 5. Calculation of supersaturation and other thermodynamic quantities for assumed values of γ. $V = 2000\ cm^3$, $r = 4.8$ nm, $T = 293$ K

γ (dynes cm^{-1})	$\mu_2 - \mu_2^0$ (cal mol^{-1})	$\gamma \times 10^2$ (dynes cm^{-1})	G^S (cal mol^{-1})
0.1	19.9	1.03	59.8
1	200	1.40	598
5	996	5.54	2990

tion mechanism for these bubbles[34]. The dispersion of these microbubbles appears to be important for recovery of heavy oils.

Second, it is well know that the solubility of asphaltenes changes significantly with pressure[2,5]. It is not the purpose of this work to pursue the predictions of equation 12 when the external pressure on the crude oil is changed. However, it should be noted that the Kelvin equation relates the pressure change directly to solubility, an advantage compared to other theories where the pressure effect is included through molar volumes[2].

Finally, from basic thermodynamics point of view the interfacial tension is related to solubility, another advantage compared to the existing theories where the solubility parameter δ, has to be introduced in a semi empirical manner to account for the enthalpic contribution to solubility. In fact, some time ago Mitchell and Speight[19] found a linear correlation between solubility and δ where this parameter could be related to γ through equation 13 introduced by Hildebrand[41],

$$\delta = \frac{\gamma}{V^{1/3}} \tag{16}$$

where V is the molar volume.

Due to the colloidal nature of asphaltenes in crude oils, it is natural that their solubility should be related to interfacial tension. However, this fact is not properly accounted for in the literature where theories based on the solubility parameter are emphasized.

Adsorption at the Liquid-Solid Interface

The results of the analysis of the deposits are presented in Table 6. The inorganic content varies from 50 to 1 % and in all cases the asphaltenes and resins together account for > 86 % of the organic fraction. Note, however, that the relatively large quantities of saturates and aromatics could be accounted for qualitatively in terms of the interfacial tension arguments discussed in previous sections.

The presence of inorganic materials in these deposits is interesting. In terms of their role during the formation of these deposit. There seems to be no doubt that they act as precipitation nuclei. From the Kelvin equation point of view (see above) these coarse particles ($r > 100\ \mu m$) could be considered as flat particles where asphaltenes *in solution* could deposit. Precipitation in the supersaturated system will then set in since the barrier, due to curvature of the particle is now absent.

In our view, these simple arguments are very important in understanding and solving the very important phenomena of precipitation in production well tubing. From the discussion in previous sections it was clear that the interfacial tension between the micelle and the crude oil should be very small (around 1 dyne cm^{-1} or smaller). Thus, adding surfactants to the crude with the objective of reducing it further and thereby dispersing the asphaltene flocks is, in our view, a very difficult if not impossible task. In any case, the surfactant studies

Table 6. Inorganic content and SARA analysis of the organic content of well deposits found in the Ceuta field

Sample	Inorganic content (%)	Organic analysis (%)			
		Saturates	Aromatics	Resins	Asphaltenes
1	50	10	3	22	65
2	11	6	0	21	73
3	16[a]	5	2	16	76
4	1	5	2	15	78

[a]Sample used as inorganic adsorbent in this work.

should focus on their ability to displace the asphaltenes from the surface they are in contact with, similar to their role in deemulsification of water in oil emulsions.

It should be noted that large quantities of resins are found in the analyzed samples (see Table 6) Thus, the capacity of the resins to keep the colloidal particles dispersed should not be over emphasized.

Adsorption Isotherms

The solid-solute adsorption isotherms (SSA) for pure asphaltenes obtained from different crude oils are shown in Figures 3 to 7. In general, the adsorption behavior has two regimes: at low concentration, it resembles a Langmiur type (L-type) adsorption, at higher concentrations the adsorption increases with the equilibrium concentration in solution (with the exception of ACN Figure 3). These two regimes are separated by a zone where no clear dependence is apparent. The behavior in this zone is consistent with the formation of micelles. At this point we should say that it is very unlikely that asphaltenes would have a critical micelle concentration (CMC) as this term is understood when employing pure surfactants in water. Even for pure surfactants dissolved in organic solvent the existence of such CMC is doubtful. Instead, a progressive formation of aggregates (dimmers, trimmers, etc.) as the concentration of surfactant increases, has been proposed[42]. We strongly believe that essentially, this is the case for asphaltenes, in view of the quite different solubility properties of the fractions present in this mixture (see above). However, from the Figures 3 to 7 it is apparent that there is a zone or range of concentrations where the fluctuations in the plots are significatively larger than in the rest. According to the discussion above on mixed micelles, a range of effective CMC values are to be expected for the asphaltenes. The plots in Figures 3 to 7 reflect this zone. For simplicity, we shall call this zone the micellar zone.

The results from Boscan and Hamaca asphaltenes are interesting since it appears that some desorption from the solid surface occurs in the micellar zone. (see Figures 6 and 7). Presumably, both the micelle and the solid surface are competing for solute in this zone.

It should be mentioned that the desorption behavior in the micellar zone resembles the one presented by surfactants with small amounts of amphiphilic contaminants. One such case is the adsorption at the air-water interface of sodium lauryl sulfate contaminated by small quantities of lauryl alcohol. In this case, the plot γ-log C does not lead to the usual break at the CMC, but to a minimum. This has been attributed to the adsorption of the alcohol together with the sulfate at concentrations below the CMC which leads to γ reductions below the expected value. Above CMC, the alcohol is solubilized into the micelles and the interfacial tension increases again[43].

According to the previous discussion (see above) asphaltene fractions such as R_6 and R_7 are practically insoluble in toluene ($\ll 1\,g\,L^{-1}$). Thus, even for concentration well below

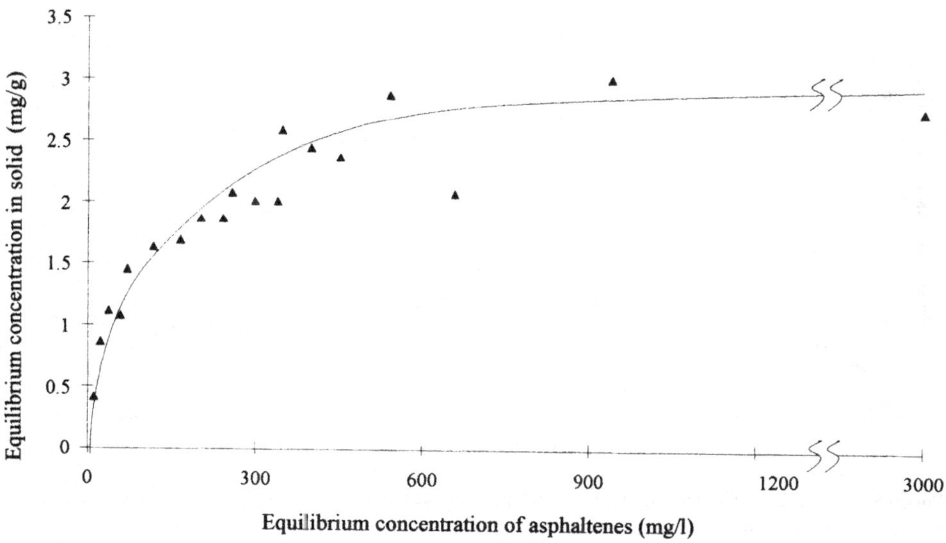

Figure 3. Adsorption of Cerro Negro asphaltenes on the inorganic material. Solvent: toluene, temperature: 26ºC. (Reprinted from ref. 24, with permission.)

the micellar zone any compounds which belong to these fractions would be in dispersed solution as aggregates by more soluble fractions.

Beyond the micellar zone the adsorption increases (except ACN) to a values much larger than those achieved during the L-type behavior. For instance for Ceuta asphaltenes, saturation of the inorganic surfaces occurs for values near 2.7 mg g^{-1}, whereas no saturation for the second regime is apparent at concentrations near 21 mg g-1 (other details for ACN, Ceuta and Furrial asphaltenes could be found elsewhere[24]).

This is consistent with multilayer formation since the adsorption of asphaltenes on themselves is more important than on the inorganic surface[24].

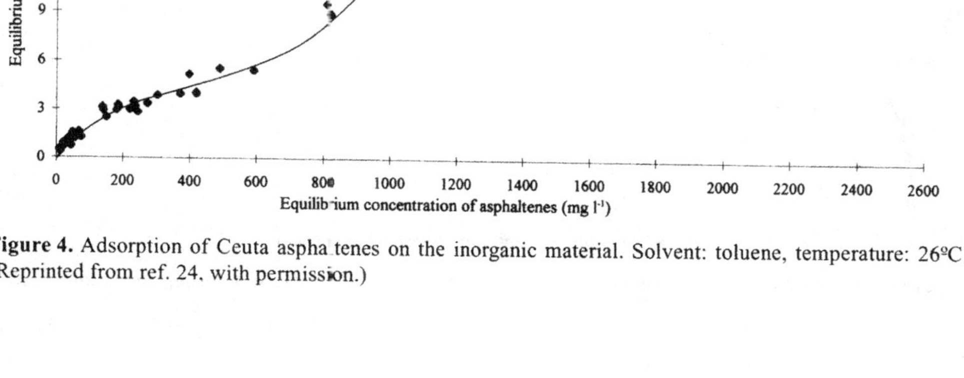

Figure 4. Adsorption of Ceuta asphaltenes on the inorganic material. Solvent: toluene, temperature: 26ºC. (Reprinted from ref. 24, with permission.)

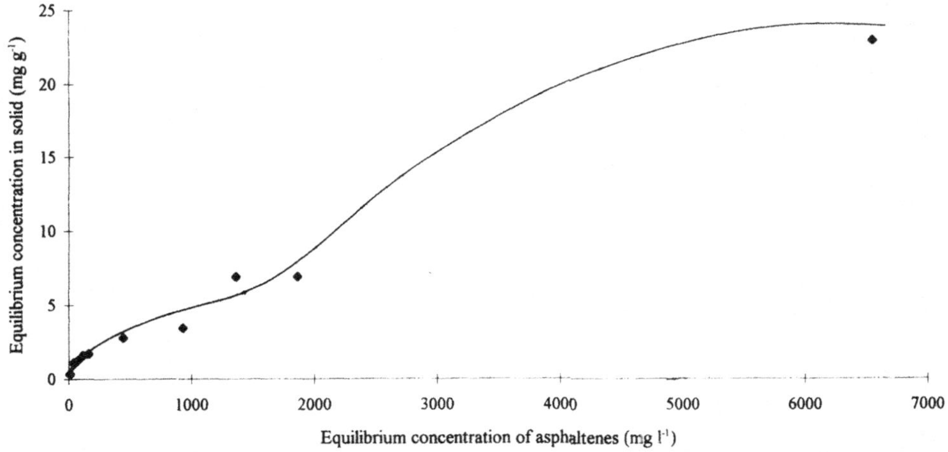

Figure 5. Adsorption of Furrial asphaltenes on the inorganic material. Solvent: toluene, temperature: 26°C. (Reprinted from ref. 24, with permission.)

According to the above results, the adsorption isotherms of asphaltenes on solid surfaces is affected by micelle and multilayer formation. The form of the curve depends strongly on the nature of the asphaltenes, particularly at low concentrations near the micellar zone. In view of the discussion above regarding differences in solubility and structure between the fractions which made up the asphaltenes, this is an expected result. For instance, at high dilution, some of the soluble components of the asphaltenes which disperse the

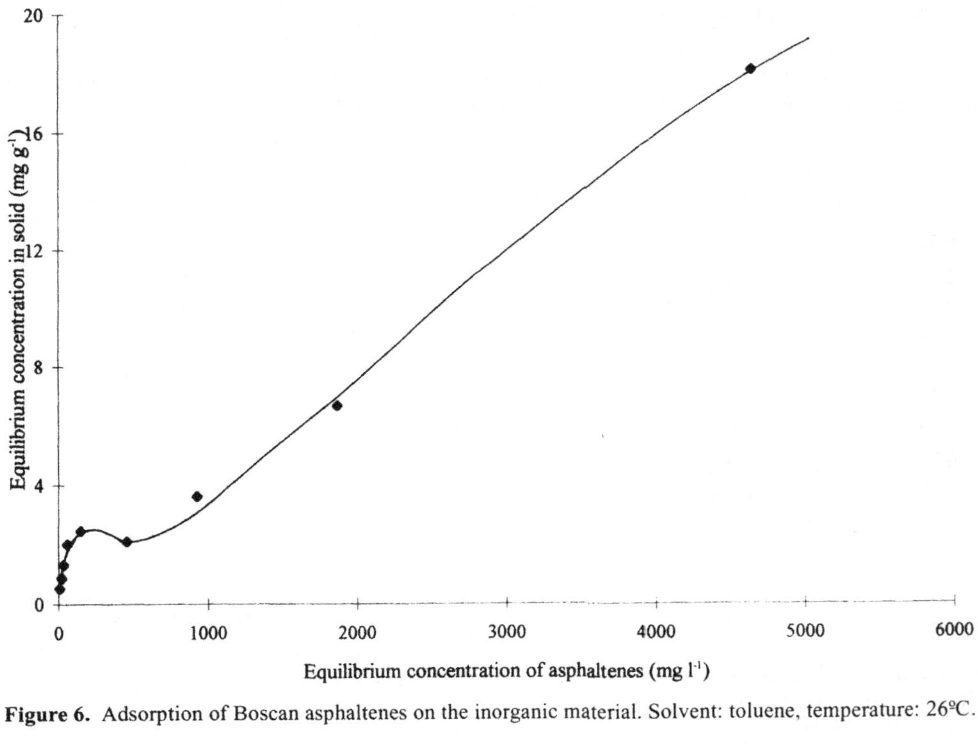

Figure 6. Adsorption of Boscan asphaltenes on the inorganic material. Solvent: toluene, temperature: 26°C.

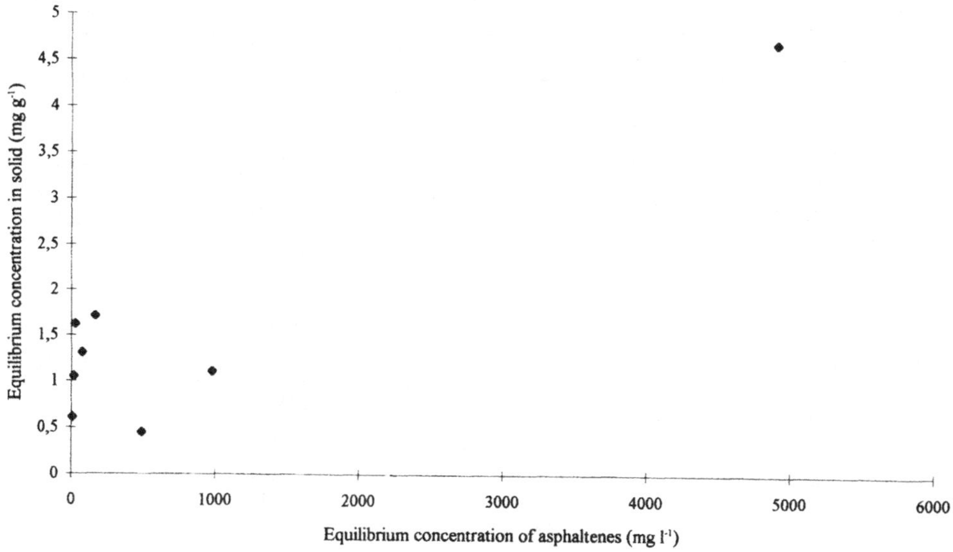

Figure 7. Adsorption of Hamaca asphaltenes on the inorganic material. Solvent: toluene, temperature: 26ºC.

aggregates, will tend to be detached from it, in an extension that depends strongly on specific "inter-asphaltene" forces. This would lead to an increase in adsorption. Also, at high concentration multilayer formation will, of course, depend on these inter-asphaltene forces. Apparently these forces are stronger for Ceuta and Furrial than for ACN (compare Figures 3 to 5). It could be said in part that these differences are due to differences in their aromaticity. For instance solid state C-13 yielded values of 71.4 and 45.9 for Ceuta and ACN asphaltenes respectively (see Figures 8 and 9).

Adsorption at the Oil-Water Interface

The adsorption of asphaltenes and resins (also called natural surfactants) at the oil-water interface has been thoroughly covered in the literature[25-35,44-46]. Here we will outline some conclusions of this work and some results which are relevant to this chapter.

The following are among the most important conclusions and results in this area:

- Asphaltenes and resins both together or separately are adsorbed at the oil-water interface.
- γ values are strongly dependent on pH and salt content of the aqueous phase.
- γ-pH curves of toluene solutions of crude oil in toluene yields bell shaped curves. At low pHs (< 7) adsorption of bases, (ionized to ammonium ions) is dominant, whereas at high pH values (> 8) adsorption of carboxylic acids (ionized to carboxylates) is dominant. Acid-base ion pair are presumably adsorbed in the neutral region ($6 < pH < 8$)[32].
- In dilute solutions the special properties of water (polarity, hydrogen bonding, etc.) controls the adsorption in the sense that only acids and bases are adsorbed during the first minutes of contact between the phases.
- Aging effects are quite significant at all pHs even for dilute solutions. Large changes in γ are observed after days or weeks at acidic and neutral pHs (see Figure 10). In the alkaline region these changes are much faster, generally within one

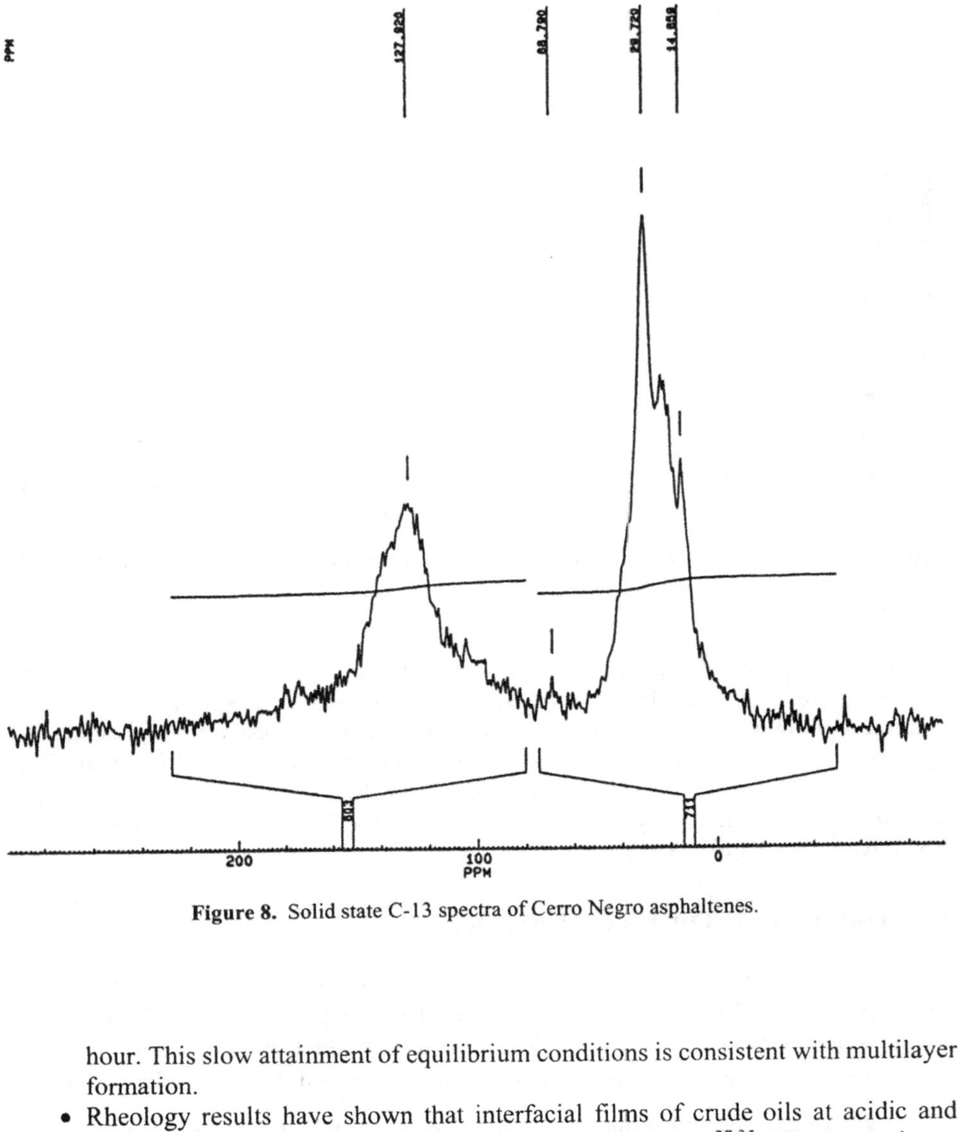

Figure 8. Solid state C-13 spectra of Cerro Negro asphaltenes.

hour. This slow attainment of equilibrium conditions is consistent with multilayer formation.

- Rheology results have shown that interfacial films of crude oils at acidic and neutral pHs are highly viscous and very stiff, solid like films[27,35]. This is consistent with the very high stability of water in crude oil emulsions.
- Drastic changes in interfacial viscosity and elasticity occurs when the aqueous pH changes from neutral to alkaline values (~ 11). The interfacial viscosity and elasticity changes could be two to three orders of magnitude[33].
- Interfacial tensions at alkaline pHs are often ultralow ($\gamma < 0.01$) and nearly constant during the first minutes of contact between phases. After a while γ raises steadily and level off at a value near 1 dyne cm^{-1} (Ref. 30).

As for the case of adsorption on solids, multilayer formation at the oil-water interface is due to the high chemical potential of asphaltenes or high interfacial energy associated with its colloidal nature. Thus, although adsorption of the first "monolayer" is due to specific effects (ionization, hydrogen bonding, etc.), later adsorption at such high dilution (see experimental) is probably due to colloidal instability.

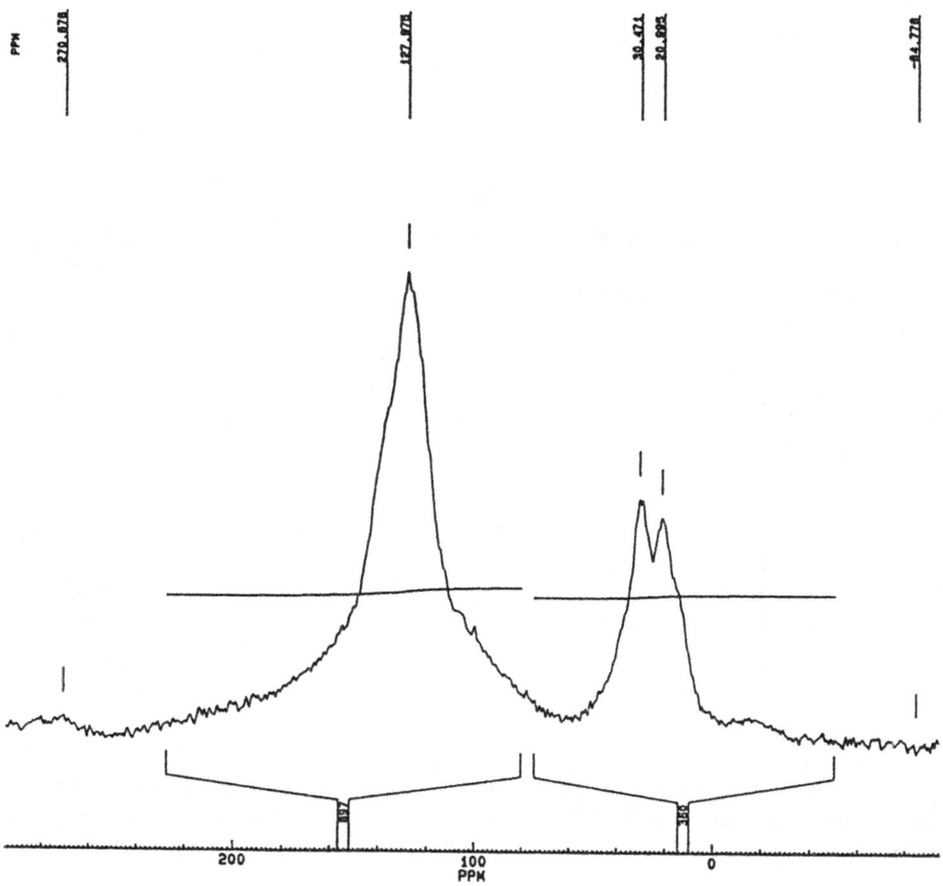

Figure 9. Solid state C-13 spectra of Ceuta asphaltenes.

In Figure 10 we reproduce the time dependence of γ for natural surfactant (NS) obtained previously[32]. These NS are mainly asphaltenes. Note that at pHs below 8, γ changes very slowly so that at least a week is needed to reach equilibrium values. This fact substantiates the multilayer formation, as quoted above. However, for alkaline pHs below 8 γ appears to be constant.

We should mention that the data on Figure 10 were measured by the plate method which is not accurate for values below 1 dine cm^{-1} (see experimental). When a spinning drop is used the ultralow values of γ in this region are reveled[32] (see also Figure 11) together with its time dependence which, for the systems crude oil-aqueous sodium hydroxide, γ remains ultralow for a few minutes, then raises to values near 1 dyne cm^{-1} within 2 hours or less.

The point we want to emphasize is the large differences in the γ relaxation times (time to reach a steady value) between the alkaline and the acidic-neutral region. Although some theories have succeeded in explaining the dependence of γ with time[44-45], they are incomplete in the sense that no account is made for the above differences in time, or for the large changes in interfacial viscosity and elasticity. A proposition consistent with these results is given in the following section.

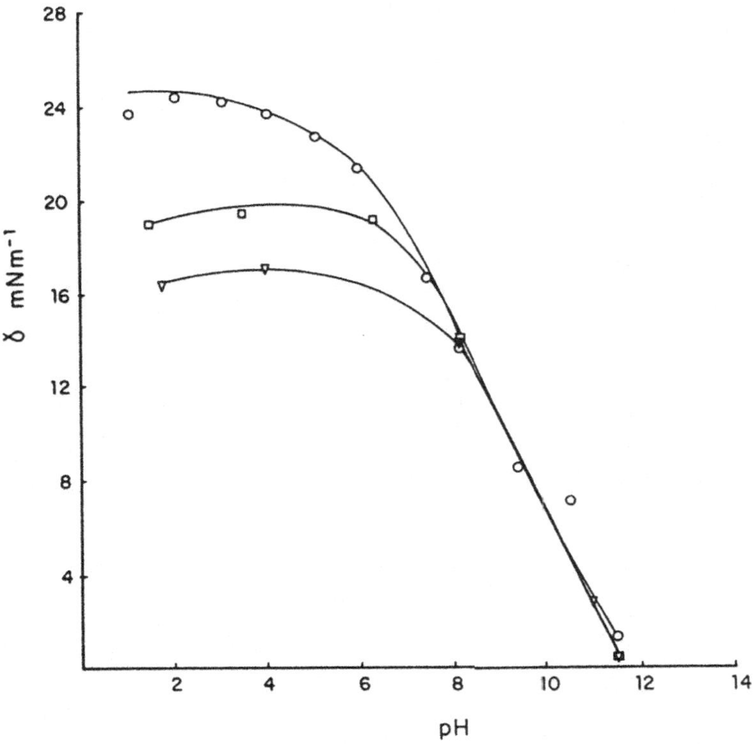

Figure 10. γ-pH curves for NS from Tia Juana crude at different contact times: O 2 h.; □ 3 days; ∇7 days;. Conditions: tridistilled water: sample diluted in toluene (0.1 % w/v). $T = 25°C$. (Reprinted from ref. 32 with permission).

Carboxylic Acids

We suggest that the carboxylic acids, which are responsible for these ultralow γ values, are adsorbed on the asphaltene micelles from the beginning. That is, they are part of the micelle in the crude oil (see Figure 2). This suggestion was substantiated by isolating these carboxylic acids from both asphaltenes and maltenes (see experimental). Actually we found only a small fraction of these acids in the asphaltenes (see Table 7), which could be a consequence of the precipitation method since diluting with 1 volume of toluene and then with 60 volumes of n-hexane will lead to considerable desorption. As discussed below this is of no consequence to the present arguments.

As revealed by GPC, the molecular weight distribution (MWD) of these samples were bimodal (note the values of P in Table 7) and for the asphaltenes case we could manage to measure approximately the values for the lighter components (second entry in Table 7). At this point, these results show that the carboxylic acids are a rather complex mixture of high and low molecular weight values. The MWD of both the heavy components in the ester, and the original asphaltenes were quite similar. C-13 NMR experiments revealed however that aliphatic like acids are the prevailing species in the mixture since the aliphatic to aromatic ratio was >20.

The above results strongly suggest that these low molecular weight aliphatic acids are associated with the asphaltenes and resins in the micelle structure (see Figure 2).

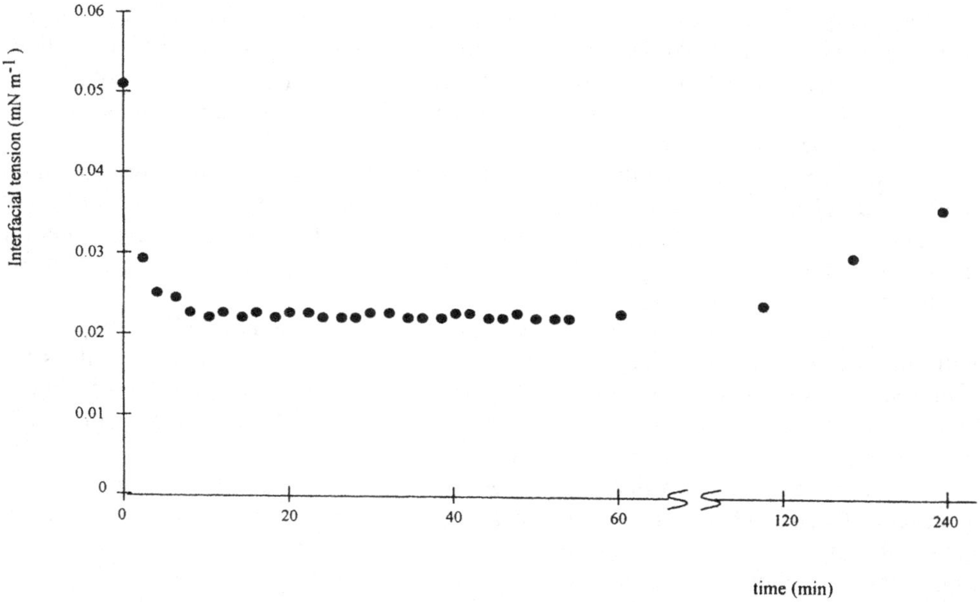

Figure 11. Interfacial tension evolution against time for the system: Cerro Negro bitumen (90 % w/w), aqueous Na_2CO_3 (5 g l^{-1}), $T = 25$°C.

Thus, when the micelles are adsorbed at the oil-water interface, the acids are adsorbed at the same time. Since at the interface these acids have apparent $pK_a > 8$ (see Figure 10) they would have a minor influence on γ at neutral and acidic pH values. In this case amines plays an important role. Amines have high solubiliity in crude oil. It hardly forms micelles in the bulk phase. Moreover, it is structurally inconsistent with the interface. Thus, one anticipates slow diffusion of amines towards the interfaces.

However, in the alkaline region (pH >8), the adsorbed acids will be ionized immediately and γ will drop very fast. Also, the decrease in surface viscosity and elasticity commented above, suggest that the asphaltene micelles will leave the interface. A mechanism

Table 7. Yields and molecular weight[a] values of carboxylic acids extracted from maltenes and asphaltenes[b]

Original sample	Yields (%)	Molecular weight[e] (Daltons) Mn	Mw	P[g]
asphaltenes	0.8 [c] (0.09)[d]	2430	8250	3.4
ester (from asphaltenes)[f]	—	390	500	1.3
maltenes	2.91[c] (2.38)[d]	707	4791	6.7

[a]GPC.
[b]Cerro Negro crude oil.
[c]Respect to sample.
[d]Respect to crude.
[e]Methylated (ester) samples.
[f]See text.
[g]Polidispersity.

for this desorption has been proposed earlier using arguments based on the Young-Dupree equation in which eventually the asphaltenes are redispersed in the crude oil[46]. We believe that in our case this mechanism is inadequatesince, given the high chemical potential of asphaltenes in the crude oil it is difficult to see how a flocculated thick film of asphaltenes could be redispersed in the crude oil.

Rather, we suggest that the micelles, probably with some solubilized oil, will enter the aqueous phase relieving the charge repulsion (the driving force for dispersion) and forming an oil-in-water emulsion near the original interface. This situation will account for the observed reductions in interfacial viscosity and elasticity, and more important, for the observed ultralow values of γ.

Further discussion of these ultralow γ values is beyond the scope of this chapter. We will only mention that according to research in this field when these very low values are achieved it is not possible to distinguish whether one has a w/o or o/w emulsion[47]. In our case this means, as stated above, that the micelles and oil solubilized in water will remain near the original interface leading to the ultralow γ values.

If sufficient alkali is available, more carboxylic acids in the multilayer will be ionized and eventually an emulsion (or microemulsion) will be formed. Of course during the process, the interfacial area will increase several orders of magnitude. This area will be available for asphaltene adsorption thus reducing further its chemical potential in the oil phase.

The above process is another example of synergistic effects in interfacial phenomena, where the carboxylic acids are the surfactants and the asphaltenes the cosurfactants. The adsorption of asphaltene micelles at the interface, promoted in a large extent by its high chemical potential, is an efficient support for the carboxylic acids since it provides a coherent and convenient molecular framework to which the hydrophobic chains of the acids could be attached to provide a barrier for diffusion.

However, if the surface concentration of carboxylates rises unduly, electrostatic repulsion could lead to either detachment or neutralization (by the counter ions) of carboxylates. This will account for the dependence of γ with time mentioned above.

Usually, the rise in γ after it reaches very low values, is observed with strong bases such as sodium hydroxide and thus, no control on pH is warranted. However, when aqueous sodium carbonate was employed γ remained low during all the measurement time (4 hours) (see Figure 11).

CONCLUSIONS

- Resins are not merely adsorbed at the surface of asphaltene colloidal particles. They are forming solutions with them which apparently behaves ideally. It is inferred that the known "enhancement" of asphaltene solubility in the presence of resins is due to mixed-micelles, whose formation could lead to drastic reductions in the chemical potential of "free" asphaltenes.
- The presence of saturates and aromatics, as well as resins and asphaltenes in laboratory and field samples of precipitates suggests that the interfacial tension colloid-crude (γ) should be very small (< 0.1 dynes cm-1). Consequently the expected low interfacial energy explains the relative stability of these colloids in most crude oils. It is suggested that supersaturation is a major source of colloidal instability. Significant reductions in chemical potential are predicted for adsorption of these very small colloidal particles at surfaces of any kind (solid, liquid and gaseous).
- Adsorption of asphaltenes at quartz type solid interfaces has been used to measure what appears to be a micelle zone or concentration region near CMC values. Since

in any case these micelles are mixed-micelles a broad a region of effective CMC values should be expected.

Finally, the evolution of γ (crude oil-water interface) with time is explained in terms of carboxylic acids being adsorbed at the interface together with asphaltene-micelles to which they are attached. This model is consistent with all the experimental evidence reviewed in this chapter.

ACKNOWLEDGMENT

We would like to express our deep acknowledgment to the following colleagues without whom this work could not be done: to Alejandro Amorín, Augusto Marijuan and Almary García for isolation and preparation of samples and to J. Spidel and B. Méndez for NMR spectra. We also thanks CDCH-UCV for financial support, and INTEVEP S.A. for kindly providing instrumental facilities.

REFERENCES

1. Mansoori, G. A. and Jiang, T. S. Proc., Third AGIP SpA Improved Oil Recovery European Meeting, Rome (April 1985). 2.
2. Hirchberg, A., Jong, L.N.J., Schipper, B. A.and Meijers, J.G., SPEJ (June 1984) 283-93.
3. Flory, P. J.: "Principles of Polymer Chemistry"Cornell University Press. Itaca and London, 1953. Chap XII, pp 495.
4. Hildebrand, J. H., Prausnitz, J. M. and Scott, R.L. "Regular solutions", Van Nostrand Reinhold Co. New York 1970, p 27.
5. Burke, N. E., Hobbs, R.E., Kashou, S.F. J. Petr. Technology, (November 1990), 1440.
6. Mack, C., J. Phys Chem., *36*, 2901 (1932).
7. Pfeiffer, J. PH. and Saal, R.N.J., J. Phys. Chem., *44*,139 (1940).
8. Nellensteyn, F.J.: "The Colloidal Structure of Bitumen", The Science of Petroleum, Oxford U. Press, New York, 1938, *4*, 2760.
9. Swuanson, J., J. Phys. Chem., 1942, *46*, 141.
10. Katz, D.L., Ind. Eng. Chem., *37*, 195 (1945).
11. Pollack, S.S. and Yen, T.F., Anal. Chem., *42*, 623 (1970).
12. Leontaritis, K. J., and Mansoori, G.A., SPE (Feb- 4-6), 1987, 149.
13. Van Kerkvoort, W.J., Nievwstac, A.J.J., and Van der Waarden, M. IV Cong. Intern. Chauttage Ind. Paris 1952 (Preprint N° 220).
14. Heithaus, J.J., J. Inst. Petr., *48*, 45 (1962).
15. Hertzog, D., Tchoubar, D., and Espinat, D. Fuel, *67*, 245 (1988).
16. Andersen, S.I., and Birdi, K., J. Colloid and Interface Science, 1991, *142*, 49.
17. Sheu, E. Y., De Tar, M.M., Stormi, D.A., and De Cario, Fuel, 1992, *71*, 299.
18. Andersen, S.I. and Speight, J.G., Fuel, *72*, 1343 (1993).
19. Mitchell, D.L. and Speight, J. E , Fuel, 1973, *52*, 149.
20. González, G., and Middea, A.J., J. Disp. Sci. and Tech., 1987, 8, 525.
21. González, G., and Middea, A., Colloids and Surfaces, 1988, *33*, 217.
22. González, G., and Moreira, M.E.C., Colloids and Surfaces, 1991, *58*, 29.
23. Collins, S.H., and Melrose, J.C.. SPE 11800, 1983, 249.
24. Acevedo, S., Ranaudo, M.A., Escobar, G., Gutiérrez, L.B., and Ortega, P. Fuel, 1995, *74*. To be Published.
25. Reisber, J. and Dosher, T.M. Prod. Monthly 1956, *21*, 43.
26. Strassner, J. E., J. Petr. Technol. 1968, *20*, 303.
27. Eley, D.D., Hey, M.J., and Lee, M. A., Colloids and Surfaces, 1987, *24*, 173.
28. Jones, T.J., Neustadter, E.L., and Whittingham, K.P., J.Can. Petr. 1978, *17*, 100.
29. Jennings, H.Y., Soc. Pet. Eng. J. 1975, *15*, 197.
30. Babu, D.R., Hormof, V. and Neale, G. Can. J. Chem. Eng. 1984, *62*, 156.
31. Acevedo, S., Layrisse, I. and Rivas, H. J. Dysp. Sci. Technol., 1984, *5*, 1.

32. Acevedo, S., Escobar, G., Gutiérrez, L.B., and Rivas, H., Fuel, 1992, *71*, 619.
33. Acevedo, S., Escobar, G., Gutiérrez, L.B., Rivas, H., and Gutiérrez, X. Colloids and Surfaces A., *71*, 65 (1993).
34. Islam, M. R. and Charma, A., "Asphaltene Particles in Fossil Fuel Exploration, Recovery, Refining and Production Processes, Eds M.K. Sharma and T.F. Yen, Plenun Press New York, 1994, p 123.
35. Claridge, E. L., and Prats, M. SPE, paper 29243 (1995).
36. Acevedo, S., Méndez, B., Rojas, A., Layrisse, I., and Rivas, H., Fuel, *64*, 1635 (1985).
37. Ramljak, Z., Solc, A., Arpino, P., Schitler, J.M. and Guiochon, G., Anal. Chem., *49*, 1222, 1977.
38. University of Texas Model 500 Spinning Drop Interfacial Tensiometer Instruction Manual, J. E. Gardner and M.E. Hayes. Univ. Texas, 1981.
39. Tanford, C.: "The hydrophobic effect. Formation of micelles & biological membranes". 2nd Edition. John Wiley & sons, New York 1980, chap. 9
40. Adamson, A.W.: "Physical Chemistry of Surfaces"Third Edition Jhon Wiley & Sons. New York, 1976 Chapter VI, p 334.
41. Hildebrand, J.H. J. Am. Chem. Soc. 1919, *41*, 1067.
42. Kertes, A.S. and Gutman, H. "Surface and Colloid Science"Vol 8 (De. E. Matijevic) Wiley, New York, 1976.
43. Tadros, T.F. and Vicent, "Encyclopedia of Emulsion Technology"(Ed. P. Becher) Vol 1 Marcel Dekker, New York, Chap. 3, p 201-202.
44. Rubins, E. and Radke, C.J. Chem. Eng. Sci., *35*, 1129 (1980).
45. Trujillo, E. M. Soc. Pet. Eng. J. 647 (Aug. 1983).
46. Eley, D.D., Hey, M. J., and Symonds, J.D. Colloids and Surfaces, *32* (1988), 87.
47. Bourrel, M. and Schechter, R.S. Microemulsions and Related Systems. Surfactant Science Series, Vol 30. Marcel Dekker, New York, 1988.

Chapter V

THE EFFECTS OF ASPHALTENES ON THE CHEMICAL AND PHYSICAL CHARACTERISTICS OF ASPHALT

M. S. Lin, K. M. Lunsford, C. J. Glover, R. R. Davison, and J. A. Bullin

Texas A&M University
Department of Chemical Engineering
College Station, Texas 77843-3122

INTRODUCTION

There are about two million miles of asphalt pavement in this country and billions of dollars are spent annually on building and repairing them. Asphalt roads are constructed with layers of graded crushed stone glued together with asphalt. There are several properties that an asphalt must possess to be a good glue. Obviously it must have good adhesion to the stone. It must set up in a reasonable time and it must be sufficiently ductile to resist cracking under stresses from traffic and temperature changes.

Asphalts vary considerably in these properties, but in general they perform satisfactorily when first laid down in new roads. However, oxidation results in progressive hardening characterized by decreasing ductility and increasing viscosities. This hardening is almost entirely due to the formation of asphaltenes, generally defined as paraffin insolubles with heptane, hexane or pentane most commonly used as the solvent.

To discuss asphalt properties, it is convenient to divide the asphalt into several components based on various separation routines. The simplest is the precipitated asphaltenes and unprecipitated maltenes. The maltenes may be variously further separated, but a number of schemes give similar if not identical fractions. Perhaps the most widely used is that of Corbett [1] ASTM D4124 in which the maltenes are separated into saturates, naphthene aromatics and polar aromatics. In other schemes similar fractions are designated saturates, and then oils, cyclics or aromatics corresponding roughly to naphthene aromatics and finally resins corresponding approximately to polar aromatics. When asphalts oxidize, saturates are unaffected, naphthene aromatics are converted to polar aromatics and polar aromatics to asphaltenes [2,3,4]. As a result, naphthene aromatics decrease, asphaltenes increase, while polar aromatics often do not change much.

While asphalt hardening is directly affected by the growth in asphaltenes, the degree of hardening also depends on the size of the asphaltenes and upon the solvent power of the maltenes. Altgelt and Harle [5] demonstrated these effects in various maltenes and in solvents such as cyclohexane, benzene, xylene and others with viscosity rising much more rapidly

Asphaltenes: Fundamentals and Applications, Edited by E. Y. Sheu and O. C. Mullins
Plenum Press, New York, 1995

with increasing asphaltene content, with higher molecular weight asphaltenes and in solvents having lower solubility parameters.

The concept of solubility parameter was developed by Hildebrand [6]. Hansen and others [7,8,9] expanded the idea to include polar and hydrogen binding effects as well as the dispersion force based Hildebrand parameter. In general, mutual solubility is highest when solvents are near each other in the three dimensional parameter space.

Hagen et al. [10] used this concept to study asphalt solubility as it ages. They combined the polar and dispersion force parameters so the solubility could be represented on a two dimensional scale. For virgin asphalt, the region of maximum solubility occurred in the region occupied by such solvents as THF, chloroform and toluene. As the asphalts aged, the area of maximum solubility moved in the direction of an increasing hydrogen bonding parameter. Since the solubility is primarily a function of asphaltene content, this reflects the increasing asphaltene content and the increasing presence of polar and hydrogen bonding groups. Using Field Ionization Spectroscopy (FIMS) Boduszynski et al. [11,4] showed that the average molecular weight of a single asphaltene molecule was in the neighborhood of 1000 and not greatly different from the averages of the other fractions. Molecular weights measured by SEC or osmometry are much higher [12,13,14]. All this supports the widely accepted model of asphaltenes in asphalt as comprised of clumps of molecules of a semi-colloidal nature solubilized by aromatics.

Yen and others [15,16,17] proposed that asphaltenes and resins (heavier polar aromatics) existed as stacks of aromatic discs to which alkyl and naphthenic side chains were attached. These sheets are held together by π–π bonding of condensed aromatic groups and hydrogen bonding groups containing heteroatoms of nitrogen, oxygen and sulfur. These aromatic groups form particles consisting of stacks of molecules. Smaller polar aromatic discs may be included, and being less condensed, help to solubilize the asphaltenes. In poorer solvents and at higher concentrations, the stacks will become larger and eventually precipitate. This will be exacerbated if the polar aromatics are removed from the stack because of relatively higher solubility in the solvent.

The production of asphaltenes during aging results principally from the addition of oxygen to polar aromatics. Much work has been reported on the chemistry of aging. Petersen and co-workers have developed techniques using infrared analysis to measure the concentration of the principal heteroatom chemical groupings in asphalt [18]. The principal groups that form during aging are ketones and sulfoxides, and to a lesser extent carboxylic acids and anhydrides. If one assumes a molecular weight of 1,000, about one oxygen atom might be added per molecule. These polar groups have been shown to greatly enhance the association of the molecules [19,20]. Storm et al. [21] measured intrinsic viscosities for asphaltenes in five vacuum residues noting that they did not vary much despite wide variations in asphaltene heteroatom content. He postulated that these groups must be buried in the particle as implied by the stack model proposed by Yen.

The concept of asphaltenes as displaced particles in a solvent offers much promise in explaining the viscosity building properties of asphaltenes. Asphaltenes would be expected to obey relations developed for the effect of solid suspensions on fluid viscosities.

Many researchers have studied the effect of original asphaltenes, as well as the effect of those produced from air-blowing, on the viscosity of the maltene and solvents. Eilers measured the relative viscosities of solutions of diethyl ether asphaltenes, obtained from several original and air-blown asphalts, in carbon disulfide [22]. The data from all of the asphalts were compared with the Eilers model but did not show good agreement. Eilers's model is an empirical correlation relating the volume fraction to the relative viscosity. The equation implies the assumption that the particle is a perfect hard sphere. Eilers concludes that the deviations between the model and the measured data were associated with the fact that the asphaltenes were non-spherical particles, deformable particles, or polydisperse in size. Reerink and Lizenga [23] measured the viscosity of solutions of n-heptane asphaltenes, obtained from original and air-blown asphalts, in toluene. The solution viscosities were related to the concentration and width of the molecular weight distribution of the asphaltenes as measured by Gel Permeation Chromatography. The data were compared to the Heukelom and Wijga model [24] and did not show good agreement. The Heukelom and Wijga model is a simplification of the Eiler model assuming a linear relation between $1/(\eta_r)^{1/2}$ and the volume fraction. Relative viscosities (η_r) calculated from the model were consistently lower than the measured values.

The Pal-Rhodes Suspension Model

The relationship between suspension viscosity and particle volume fraction for non-interacting spherical particles was first derived by Einstein [25] and given in equation (1),

$$\frac{\eta}{\eta_0} = 1 + 2.5\,\phi_P \tag{1}$$

where ϕ_P, η, and η_0 are the particle volume fraction of the suspension, suspension viscosity, and solvent viscosity, respectively. In most cases, equation (1) is only applicable when ϕ_P is less than 0.05. One of reasons for the deviation of equation (1) at higher concentrations is the presence of interacting hydrodynamic force among particles at large volume fractions. One way to account for this interaction is to apply the mean-field theory. In the mean-field theory, each individual particle is treated as an isolated particle surrounded by the remaining suspension solution with viscosity η, which can be mathematically expressed as equation (2),

$$\frac{\eta + d\eta}{\eta} = 1 + 2.5\,\frac{dV_P^h}{V_T}$$

or

$$\frac{d\eta}{\eta} = 2.5\,\frac{dV_P^h}{V_T} \tag{2}$$

where $d\eta$ is the differential increase in suspension viscosity due to a differential increase in particle hydrodynamic volume, dV_P^h, in a suspension having total volume of V_T. Pal and Rhodes [26] proposed that particles when suspended in solution are hydrated by the solvent and, therefore, have a larger volume than in the dry state. This increase in hydrodynamic volume can be corrected by scaling the dry volume of the particles with a solvation constant K, and equation (2) can be rewritten as follows

$$\frac{d\eta}{\eta} = 2.5\,\frac{K\,d\,V_P^{dry}}{V_T} \tag{3}$$

The total volume of the suspension, V_T, is given in equation (4),

$$V_T = V_S + K\,V_P^{dry} \tag{4}$$

where V_S and V_P^{dry} are the volume of solvent and dry volume of the particle respectively. The particle volume fraction on a dry basis is as follows

$$\phi_P = \frac{V_P^{dry}}{V_T} \tag{5}$$

Differentiating equation (5), equation (6) follows

$$d\phi_P = \frac{V_T\, d\, V_P^{dry} - V_P^{dry}\, K\, d\, V_P^{dry}}{V_T^2} \approx \left(1 - K\phi_P\right)\frac{d\, V_P^{dry}}{V_T}$$

or

$$\frac{d\, V_P^{dry}}{V_T} = \frac{d\phi_P}{\left(1 - K\phi_P\right)} \tag{6}$$

Substituting equation (6) into equation (3) and integrating, yields the Pal-Rhodes model as follows

$$\ln\left[\frac{\eta}{\eta_0}\right] = -2.5\ln\left(1 - K\phi_P\right)$$

or

$$\eta_r = \left(1 - K\phi_P\right)^{-2.5} \tag{7}$$

Sheu et al. [27] relate the n-heptane asphaltene concentration of two Ratawi vacuum residues in toluene to the solution relative viscosity. In this paper, the authors discuss four different solution viscosity models. The solution viscosity data suggest that the asphaltenes associate to form larger particles as a result of solvation by the solvent phase. For the solution relative viscosity, the Pal and Rhodes [26] model showed good agreement with the measured values. No data were reported for maltene/asphaltene systems.

The Pal-Rhodes model assumes that all the particles are solvated by the solvent phase and act like polydisperse hard spheres. Sheu et al. point out that for non-spherical particles the exponent of -2.5 in the Pal and Rhodes model can be a free parameter to account for the particle asymmetry. Storm et al. [21] also use the Pal-Rhodes model to describe the dependence of the relative viscosity of Ratawi vacuum residues on the n-heptane asphaltene. However, in this paper, the concentration of asphaltene was limited to that of original Ratawi vacuum residues. For aged asphalt materials, the concentration of asphaltene may be much higher.

The Relation of Carbonyl Content to Asphaltene Content and Rheological Changes

Oxidative aging, either in the laboratory or in the field, results in an increase in the asphaltene content, viscosity, and the infrared spectrum carbonyl absorbance [20]. Plancher et al. [28] oxidized 4 asphalts in the presence and absence of lime. The data show that the viscosity at 25°C as a function of n-pentane asphaltene content depended on the crude source but was independent of lime treatment. The carbonyl content was also compared with the n-pentane asphaltene content. Lee et al. [29] also reported that the absorbance in the carbonyl region increases linearly with the n-heptane asphaltene fraction of four asphalts after field and laboratory aging. Many researches have reported that viscosity increases with carbonyl content after both field and laboratory aging [29,30,31,32].

Lau et al. [31] showed that for Pressure Oxygen Vessel (POV) aging of five asphalts at temperatures from 60-104.4°C a linear relationship exists between log viscosity and carbonyl content which is dependent only on the crude source and not on the aging conditions of temperature. The Hardening Susceptibility (HS) is defined as

$$HS = \left[\frac{d \ln \eta_0^*}{d\,CA} \right] \tag{8}$$

where η_0^* is the zero shear rate complex dynamic viscosity and CA is carbonyl content relative to unaged materials. Unfortunately in this study, no asphaltene data were reported.

Girdler [12] shows that the produced asphaltenes from oxidative aging are chemically different from those originally present by elemental analysis; the produced asphaltenes have lower nitrogen and sulphur content and higher oxygen content. Because of chemical differences, the effect of the produced asphaltenes on the physical properties of the asphalt, compared to the original asphaltenes, may be different.

Peterson et al. [33] showed that both initial asphaltenes and saturates increased the HS. In two studies [34,35,36] various asphalts were supercritically fractionated and the least soluble fraction (mostly asphaltenes) and most soluble fraction (largely saturates) were discarded. The remaining fractions were blended to make a variety of materials having the usual viscosity range of asphalts. These asphalts were then aged in a POV. Figure 1 shows the HS of several of these blends compared to the original asphalts. The improvement is remarkable. Figure 2 shows the SEC for one of these with the original asphalt. Growth in the large molecular size region is primarily due to asphaltene formation. The difference is quite dramatic. Although we can only speculate at this point, it is likely that very heavy polar aromatics, those that upon reaction are very easily converted to asphaltenes, were removed with the asphaltenes. This would leave behind a maltene that would not produce asphaltenes as readily when oxidized.

It is rather remarkable that a linear relation such as HS exists for all asphalts when carbonyl growth is a relatively minor portion of the infrared spectrum of oxidized asphalt. The research reported here was devised to better understand the relation of carbonyl growth

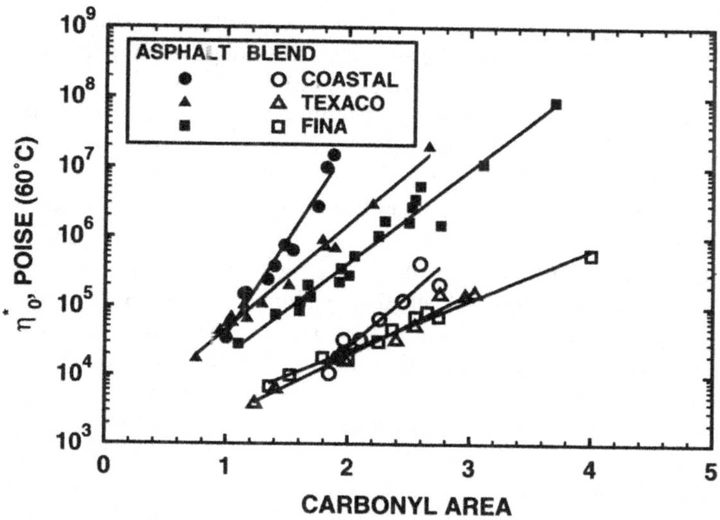

Figure 1. Comparison of HS for various asphalts and their blends.

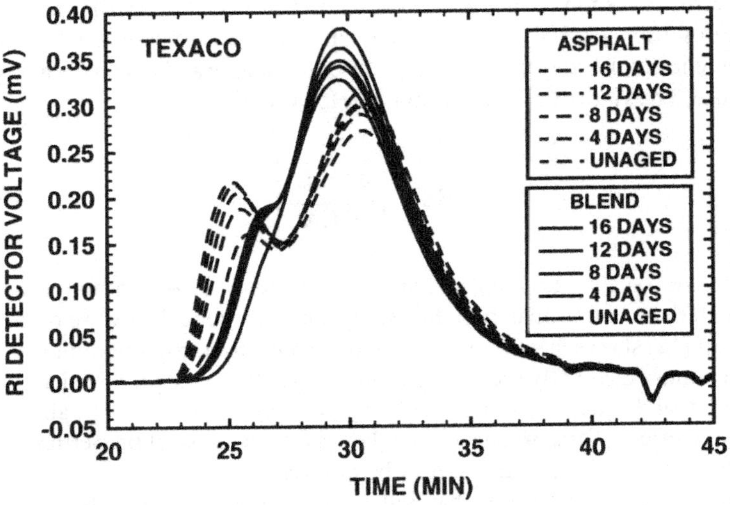

Figure 2. Comparison of SEC for asphalt and blend before and after aging.

to physical and chemical changes, and to the growth in asphaltene content as well as to model viscosity changes in terms of asphaltenes and carbonyl.

EXPERIMENTS AND METHODOLOGY

Seven SHRP whole asphalts (AAA-1, AAB-1, AAC-1, AAD-1, AAF-1, AAG-1, AAK-2) were fractionated by solvent precipitation into asphaltenes and maltenes. Following precipitation, the whole asphalts and maltenes were laboratory aged. For one of the asphalts (AAA-1), unaged asphaltene and maltene were blended to different ratios from the parent unaged whole asphalt.

In the solvent fractionation procedure, 40 grams of whole asphalt was mixed with 4 L of n-hexane yielding a 1 to 100 dilution ratio. The solution was stirred overnight and the asphaltenes were collected by filtering the solution through Whatman no. 41 filter paper. The maltene solution was recovered in a Buchi rotary evaporation apparatus by distilling the n-hexane. This recovery method was used to diminish the effect of the solvent on the maltene properties [37]. Furthermore, the recovered samples were analyzed by gel permeation chromatography (GPC) to confirm complete solvent removal [38].

All seven SHRP whole asphalts and their corresponding maltenes were aged in a POV at 20.7 bar pure oxygen, at temperatures of 60, 71.1, 82.2, 93.3, and 104.4°C for the aging times from 0.25 to 50 days depending on aging temperature. As described by Lau, et. al. [31], 2.4 gram samples were weighed into aluminum trays giving an effective film thickness of 1 mm. These thin films minimized oxygen diffusion effects on the samples.

Blends were made using unaged maltenes, asphaltenes, and whole asphalts from AAA-1. Blend B1 was made from 59 weight percent maltene and 41 weight percent whole asphalt to make a 10 weight percent asphaltene mixture. Physical mixing was used to insure homogeneity. Blend B2 was made by adding 21% asphaltenes to 79% whole asphalt. To insure good mixing, the blend was dissolved with a 85%TCE/15% ethanol solution and then recovered using the procedures described above [38]. The final blend, B3, was made by adding 73.5% whole asphalt to B2. Physical mixing was used.

The changes in chemical, rheological, and compositional properties of the aged whole asphalts and maltenes and unaged blends were measured by Fourier transform infrared spectroscopy (FT-IR), dynamic mechanical analysis (DMA), and solvent fractionation using n-hexane.

Infrared spectra were measured using a Mattson galaxy 5000 FT-IR with the Attenuated Total Reflectance (ATR) method described by Jemison et. al. [32]. Oxidation was monitored by measuring growth in the carbonyl region. The carbonyl area of the aged material is defined as the integrated area from 1650-1820 cm^{-1} relative to the integrated area over the same region for the unaged material. The unaged material has no distinctive carbonyl band and this is defined to be zero carbonyl area.

Two rheological properties, the low frequency limiting viscosity (η_0^*) and the reciprocal of the loss compliance ($1/J''$), were determined from data measured at 60°C with a Carri-Med CSL 500 Control Stress Rheometer using a 2.5 cm composite parallel plate with a 500 μm gap. A 0.1 rad/sec frequency was used to approximate the η_0^* for materials less than 100,000 poise at 60°C. For materials with higher viscosities at 60°C, dynamic rheological measurements were performed at 60, 85, and 90°C. The η_0^* at 60°C was calculated by time-temperature superposition as described by Ferry [39]. For those calculated values of η_0^*, the effective frequency was approximately 1×10^{-5} rad/sec. All of the $1/J''$ values were measured at 60°C and 10 rad/sec in accordance with SHRP specifications [40]. Asphaltenes were measured by solvent precipitation in n-hexane as described by Pearson et. al. [41]. Approximately 0.2 grams of aged material was weighed into a scintillation vial, 20 mL of n-hexane was added, and the solutions were sonicated for 5 minutes. After overnight equilibration, the asphaltenes were separated by filtering the solutions through pre-weighed PTFE membrane, 0.4 micron syringe filters. Following filtration, the filters were dried in an oven at 120°C for 40 minutes. The filters were post-weighed 2 hours after removal from the oven. The weight percentage asphaltene (%AS) is defined as the difference in the filter weight divided by the sample weight.

RESULTS AND DISCUSSION

The change in composition of oxidized asphaltic materials can be clearly detected using infrared spectroscopy. Figure 3 and 4 show the changing infrared spectra of SHRP AAA-1 whole asphalt and maltene, respectively, as they age. Both whole asphalts and maltenes show a definite increase in the carbonyl absorption band from 1650 to 1820 cm^{-1} as aging time increases. Other absorption bands such as C-O single bond stretching (~ 910-1300 cm^{-1}) and sulfoxide (~ 1000 cm^{-1}), are also increasing with time. It is certainly reasonable to expect that there should exist correlations between the increase in infrared absorption of oxygen-containing groups and the oxygen content of aged materials. A very good linear correlation was found between the increase in oxygen content and the increase in carbonyl area for aged maltenes from all of the SHRP asphalts studied. Figure 5 shows that, for both whole asphalts [42] and maltenes, the increase in oxygen content correlates linearly with the increase in carbonyl area. Surprisingly, the %O/CA correlation for SHRP AAA-1 whole asphalt and its maltene is almost the same, while, for SHRP AAG-1, there appears a small parallel shift between the whole asphalt and its maltene. Unfortunately, except for the carbonyl band, most of absorption bands are overlapped with each other and laid on top of a broad absorption band from 910 to 1300, and, therefore, are difficult to quantify precisely. The good linearity between the increase in oxygen content and carbonyl area growth suggests that the formation rate of most oxygen-containing groups are in constant ratio for a given maltene or asphalt over a fairly large aging extent. Therefore,

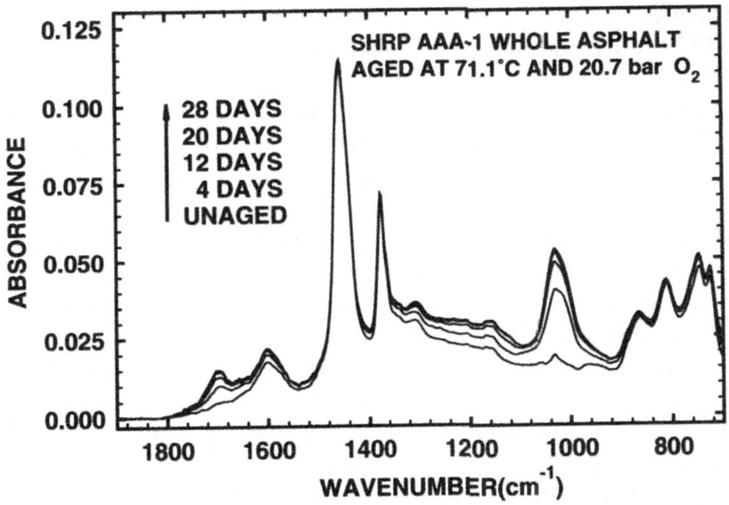

Figure 3. Infrared spectra of SHRP AAA-1 whole asphalt aged at 71.1°C and 20.7 bar O_2.

despite the formation of many other types of functional groups, the increase in carbonyl area can be used as a surrogate for the degree of oxidation for a given aged asphaltic material.

Petersen [20,43] has shown that sulfoxide forms rapidly at the beginning of oxidation and then reaches a "pseudo-steady state" for SHRP AAG-1 Asphalt. This initial rapid sulfoxide formation might explain the nonlinearity of %O/CA correlation in a small initial reaction region observed by Liu et al. [42]. However, some researchers [30] have found a decrease in sulfoxide content with increasing carbonyl in pavement removed from Texas test section. Figures 6 and 7 show the sulfoxide area versus the carbonyl area for SHRP AAA-1 and AAG-1 whole asphalts. Sulfoxide content is characterized by the sulfoxide area, defined as the integrated area of the infrared spectrum from approximately 1100 to 923 cm^{-1} with a

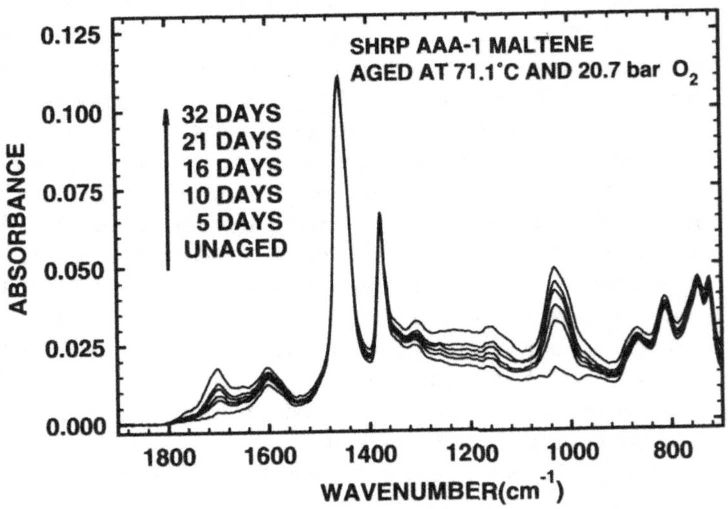

Figure 4. Infrared spectra of SHRP AAA-1 maltene aged at 71.1°C and 20.7 bar O_2.

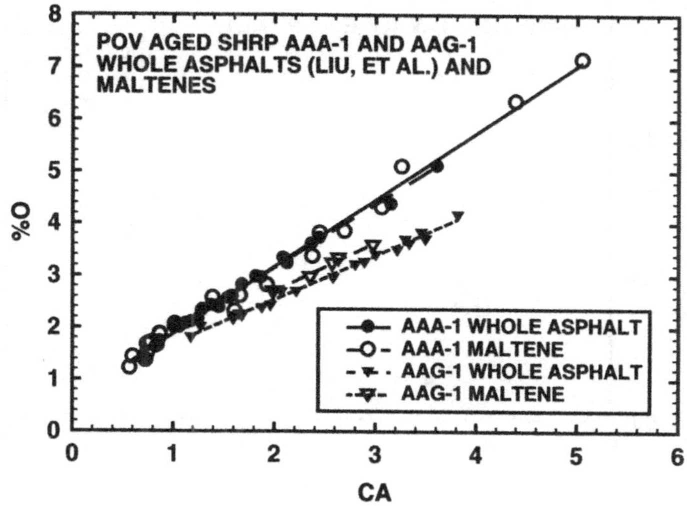

Figure 5. Correlation between %O and CA for SHRP AAA-1 and AAG-1.

selected tangent baseline. After aging, both SHRP AAA-1 and AAG-1 show a rapid increase in sulfoxide area relative to the carbonyl area followed by a decrease after the sulfoxide area reaches a maximum value. Moreover, the maximum sulfoxide content of SHRP AAA-1 is much higher than that of AAG-1. This is probably due to a significantly higher sulfur content originally present in SHRP AAA-1 compared to AAG-1. Figure 8 shows that there exists a fairly good correlation between the maximum sulfoxide area and weight percentage of elemental sulfur originally present in eight SHRP asphalts and their maltenes.

Aging-produced asphaltenes were found to be chemically different from those originally present in asphalts. Figure 9 shows infrared spectra of original asphaltenes from unaged SHRP AAD-1 whole asphalt and aging-produced asphaltenes from aged SHRP AAD-1 maltenes after 9 and 14 days of oxidative aging at 93.3°C and 20.7 bar pure oxygen.

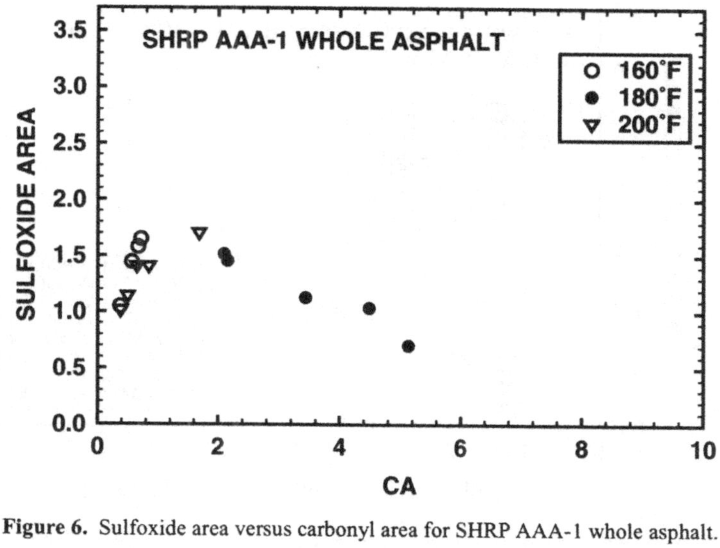

Figure 6. Sulfoxide area versus carbonyl area for SHRP AAA-1 whole asphalt.

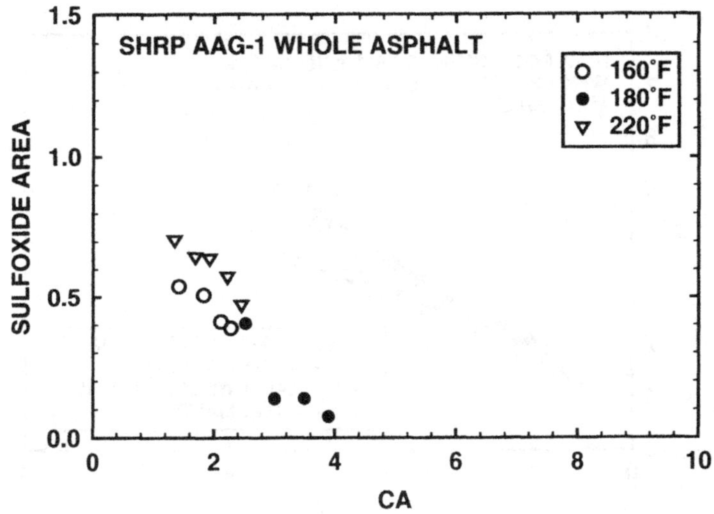

Figure 7. Sulfoxide area versus carbonyl area for SHRP AAG-1 whole asphalt.

Figure 9 clearly shows that those asphaltenes produced by aging maltenes have much higher infrared absorption in the carbonyl band as well as in other oxygen-containing bands. These asphaltene IR spectra strongly indicate that aging-produced asphaltenes are formed due to the very high polarity provided by carbonyl and other oxygen-containing functional groups. Based on these results, a strong correlation between the increase in carbonyl area and the increase in asphaltene formation would be expected. In addition, Figure 9 shows that the carbonyl area of aging-produced asphaltenes grows as the maltene ages. This indicates that those asphaltenes produced at the later aging times required more oxygen-containing groups to give enough polarity to become insoluble in n-hexane than those produced earlier. It is also possible that the existing asphaltenes may continue to oxidize and show increasing carbonyl areas.

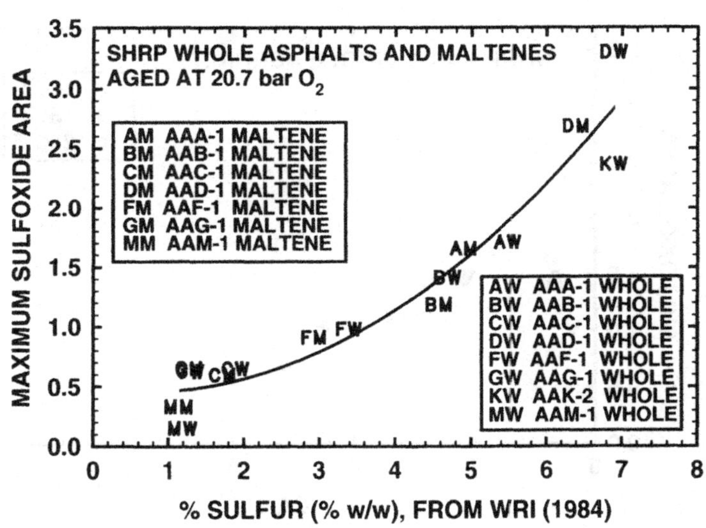

Figure 8. Maximum sulfoxide area versus sulfur content for SHRP asphalts.

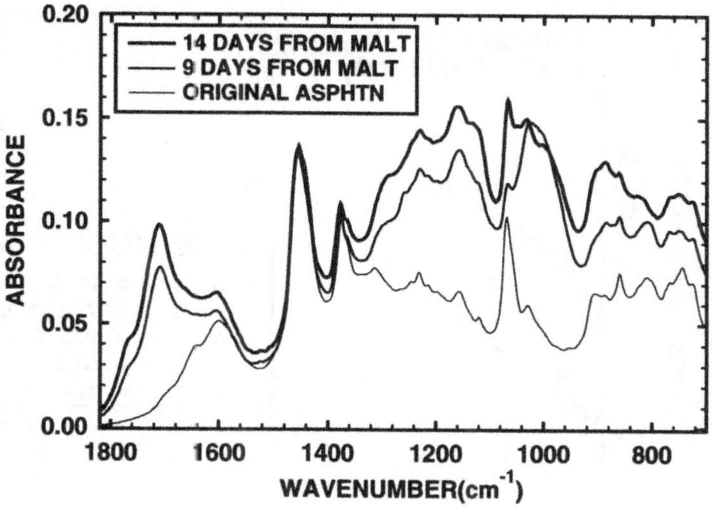

Figure 9. Infrared spectra of aging-produced asphaltenes from SHRP AAD-1 maltene aged at 93.3°C and 20.7 bar O_2.

However, not all of the carbonyl-containing species are found in the asphaltene phase. The maltene fractions obtained from both the 9-day and 14-day aged maltenes also contain carbonyl compounds, as shown in Figure 10, in contrast to the apparent low levels in the unaged maltenes. Furthermore, the maltene separated from 14-day aged maltene shows a larger infrared absorption in the carbonyl band than the maltene recovered from 9-day aged maltene. These results further support the notion that the asphaltenes produced at longer aging time require more carbonyl formation in order to become asphaltenes than those produced earlier. Figure 10 also shows that the maltenes obtained from aged maltenes have fewer aromatic ring structures, as indicated by lower infrared absorption at 1600 cm^{-1}, than

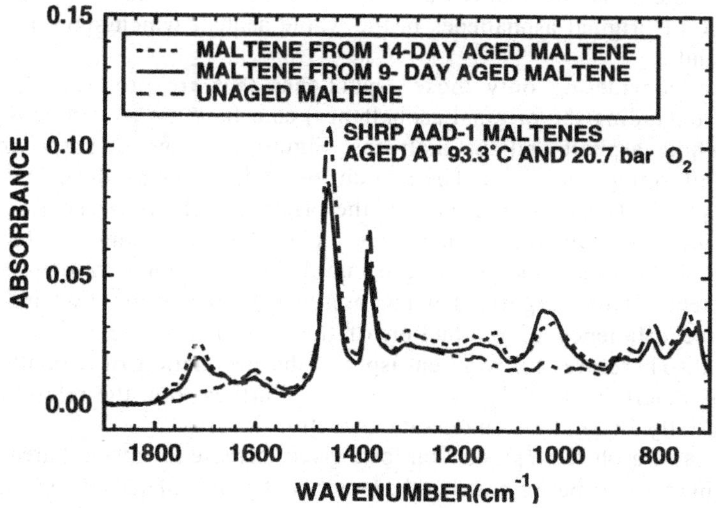

Figure 10. Infrared spectra of maltene from aged maltene at 93.3°C and 20.7 bar O_2.

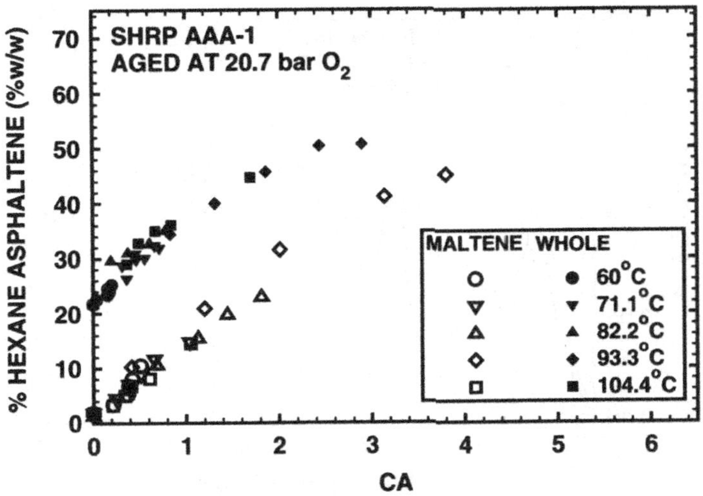

Figure 11. A comparison of %AS and CA for SHRP AAA-1 whole asphalt and maltene.

the unaged maltene. This is because some of the aromatics, especially those heavy, multi-ring aromatics, have been oxidized to become asphaltenes and then precipitated in the asphaltene fraction.

A cross plot of percent asphaltenes versus carbonyl area as AAA-1 whole asphalt and its maltene are aged, for all aging temperatures and times is given in Figure 11. For both the maltene and whole asphalt, the total asphaltene percentage for a given carbonyl area is independent of aging temperature and time. This result suggests that the aging mechanism is independent of temperature for the aging temperature range studied. Both the asphalt and maltene show a linear relationship between carbonyl area and percent asphaltenes for carbonyl areas less than about 2.5. At high carbonyl areas, the asphaltene content appears no longer to increase with carbonyl formation, perhaps, due to the depletion of the original polar aromatic fraction. The whole asphalt shows a higher asphaltene content than the maltene at a given carbonyl area. Again, this is explained by the presence of original asphaltenes in the whole asphalt which show no well-defined carbonyl band.

Next, considering only those asphaltenes that are produced from oxidative aging (i.e., excluding the original asphaltenes) as a function of carbonyl area and for both the whole asphalt and the maltenes, Figure 12 shows a surprising result for AAA-1 at all aging conditions. The amount of asphaltenes produced by aging is not affected appreciably by the presence of the original asphaltene. That is, remarkably, both the whole asphalt and its maltenes alone show the same production of new asphaltenes for a given amount of aging until the asphaltene content approaches its maximum level. This suggests that the presence of the original asphaltenes has no effect on the efficiency of production of the new asphaltene in the aged maltene environment. Therefore, for a given asphalt, the production of asphaltenes per unit oxidation is determined solely by its maltenes. Furthermore, the relationship appears to be essentially linear up to moderate oxidation levels.

The same asphaltene production for a given increase in carbonyl area for the whole asphalt and maltene can be explained by the protective function of maltenes toward asphaltenes as reported by many researchers [44,45,46,47]. Aromatic fractions, particularly polar aromatics, have been found to be easily absorbed on asphaltene surfaces. Authors speculate that the

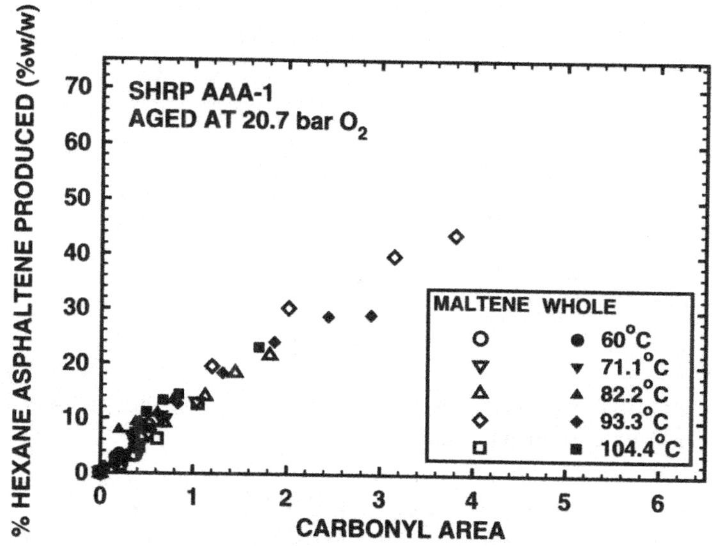

Figure 12. A comparison of produced %AS and CA for SHRP AAA-1 whole asphalt and maltene.

absorption of some maltene molecules on the surface of asphaltene forms a "coating" around asphaltene particle, and therefore reduce the transport of oxygen to the reactive site on asphaltenes. Similarly the clumping of asphaltene molecules to form aggregate structure [3] would tend to mask many possible reactive sites in asphaltenes. Because of this protective action of maltenes, the majority of carbonyl formation occurs in the maltene phase. In addition, new asphaltenes only can be produced by aging maltenes: therefore, AFS should be expected to depend mainly on maltene composition. The produced asphaltene content versus carbonyl area is shown in Figure 13 for all asphalts studied. The lines represent a regression of the data within the linear region. The asphalts have different slopes, indicating that each maltene has a

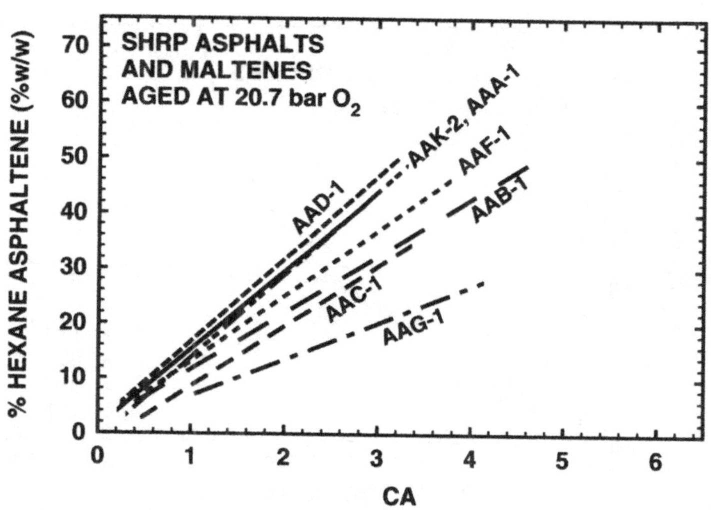

Figure 13. A comparison of AFS for all SHRP asphalts studied.

Table 1. Asphaltene formation susceptibilities of asphalts

SHRP asphalts	AFS$\left(\frac{\%^{w}/_{w}}{CA}\right)$
AAA-1	14.3
AAB-1	10.4
AAC-1	11.0
AAD-1	15.0
AAF-1	11.9
AAG-1	6.7
AAK-2	15.1

different tendency to form asphaltenes during aging. To quantify this, the asphaltene formation susceptibility (AFS), is defined by equation 9

$$AFS = \left[\frac{d\%AS}{dCA}\right] \tag{9}$$

Because this parameter does not change with oxidative aging until high level of oxidation, it is only dependent on the original maltene composition. Table 1 shows the AFS values for all the asphalts studied. Asphalt AAG-1 has a lower AFS than AAB-1; therefore, AAG-1 produces less asphaltene than asphalt AAB-1 for a given change of carbonyl content. Hence, AFS is a measure of productivity of new asphaltenes for a maltene.

As described above, the oxidative aging causes chemical composition changes in the asphalt. The carbonyl formation results in more polar molecules, thus increasing the asphaltene content. This change in chemical composition must ultimately affect the rheological properties.

Figure 14 and Figure 15 give increases in η_0^* on a logarithmic scale versus total %AS increases due to oxidation for AAA-1 and AAC-1 respectively, for the

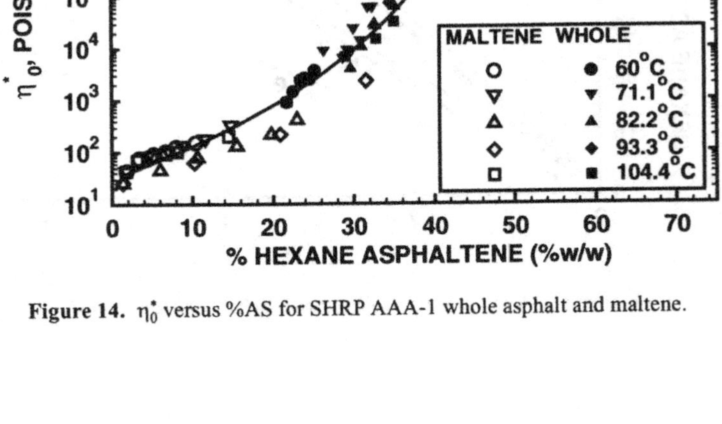

Figure 14. η_0^* versus %AS for SHRP AAA-1 whole asphalt and maltene.

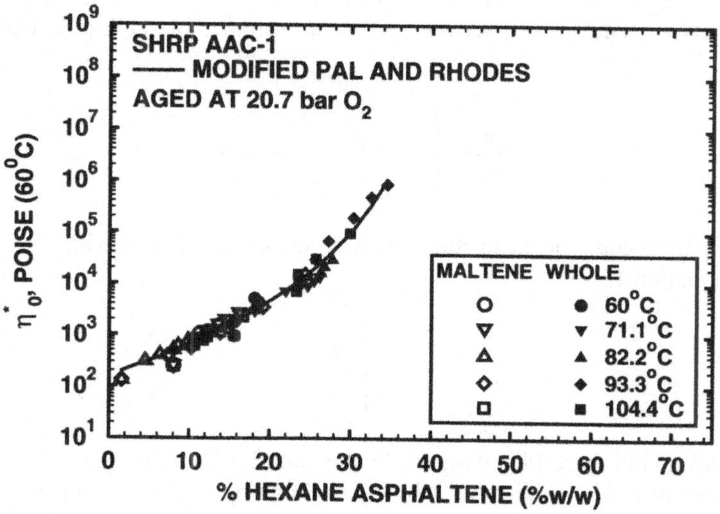

Figure 15. η_0^* versus %AS for SHRP AAC-1 whole asphalt and maltene.

whole asphalts and maltenes at all aging temperatures and times. Again the maltenes are shown with hollow symbols and the solid symbols represent the whole asphalt. Surprisingly, the $\ln(\eta_0^*)$ as a function of total %AS is very similar for both whole asphalt and maltene. Furthermore, the behavior is independent of aging conditions studied. Based on this result, the original asphaltenes and those produced from oxidative aging have almost the same effect, but not the identical, on $\ln(\eta_0^*)$ even though they are chemically very different.

In the previously described Pal-Rhodes model, asphaltenes form colloidal particles that are dispersed by a solvent phase. Furthermore, the asphaltenes were assumed to be solvated polydispersed hard spheres. For such systems, this theory describes the viscosity dependence on asphaltene concentration.

For asphalts, the asphaltene is assumed to be the solvated colloidal particle in the maltene phase. Because the interactions between asphaltenes and maltenes are not solely due to hydrodynamic effects, the assumption that the asphaltene is a solvated hard sphere is not valid in the maltene. Additionally, the shape of asphaltene aggregates may not be spherical. Therefore, the exponent -2.5 in the original Pal-Rhodes model (equation 7) cannot adequately describe the relative viscosity dependence on asphaltene concentration for whole asphalts [26,20]. Instead, the exponent, denoted as ν, is the parameter to be estimated from data. This modified Pal-Rhodes model is given in equation (10)

$$\frac{\eta}{\eta_M} = (1 - K\Phi_{AS})^{-\nu} \tag{10}$$

Where η and η_M are the viscosities of asphalt and unaged maltene and K is the solvation constant. Because the asphaltene weight percent (%AS) was measured and not the volume fraction, the volume fraction is defined by Equation 11

$$\Phi_{AS} = \left(\frac{\rho}{\rho_{AS}}\right) \%AS \tag{11}$$

Φ_{AS} is the volume fraction of the asphaltene in the asphalt. ρ and ρ_{AS} are the densities of asphalt and asphaltene, respectively. Substituting equation 11 into equation 10 gives equation 12.

$$\frac{\eta}{\eta_M} = \left[1 - K\left(\frac{\rho}{\rho_{AS}}\right)\%AS\right]^{-\nu} \tag{12}$$

Comparing equation 12 to equation 10, a solvation constant with respect to %AS is defined in equation 13.

$$K' = \left(\frac{\rho}{\rho_{AS}}\right)K \tag{13}$$

Returning to Figure 14 and Figure 15, the modified Pal-Rhodes model was optimized for SHRP asphalt AAA-1 and AAC-1 by determining optimal (least squares) estimates of K' and ν and the resulting calculated line. This parameter estimation procedure included both the whole asphalt and maltene data. Even though the solvation constant K' and ν are assumed to be constant during the aging process, the agreement of the model and the data is excellent.

To verify that the original asphaltenes and those produced from oxidative aging have a very similar rheological effect, three blends (B1, B2, and B3) from AAA-1 containing different levels of original asphaltenes were made. In Figure 16, these three blends are compared to the optimal modified Pal-Rhodes model based on the least squares regression of the aged asphalt data. The model predicts the blend viscosities reasonably well, confirming that there is a similarity, in terms of rheological response, between the asphaltenes produced by oxidation of maltenes and the asphaltenes originally present in the asphalt.

Figure 17 illustrates the optimal modified Pal-Rhodes models for each asphalt studied. Each asphalt gives unique K' and ν as given in Table 2 The values of ν range from 5.1 to 13.6, much larger than 2.5. Hence the assumptions that other interactions occur between the asphaltenes and maltenes beside hydrodynamic interactions and that the

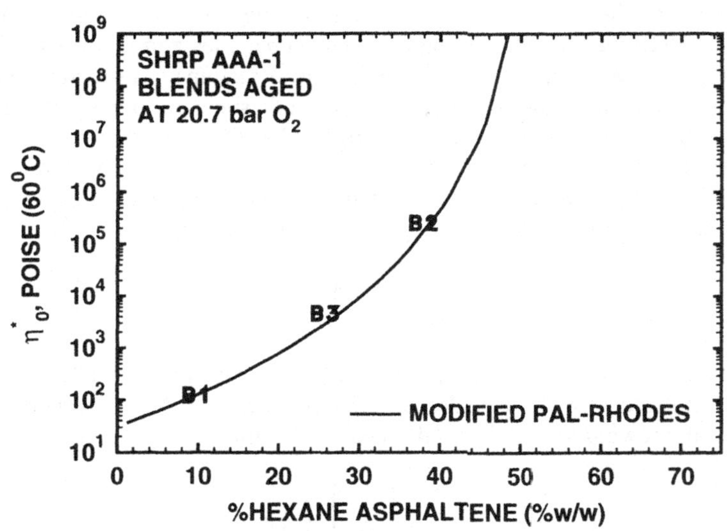

Figure 16. A comparison of η_0^*/%AS function for SHRP AAA-1 whole asphalt, maltene, and blends.

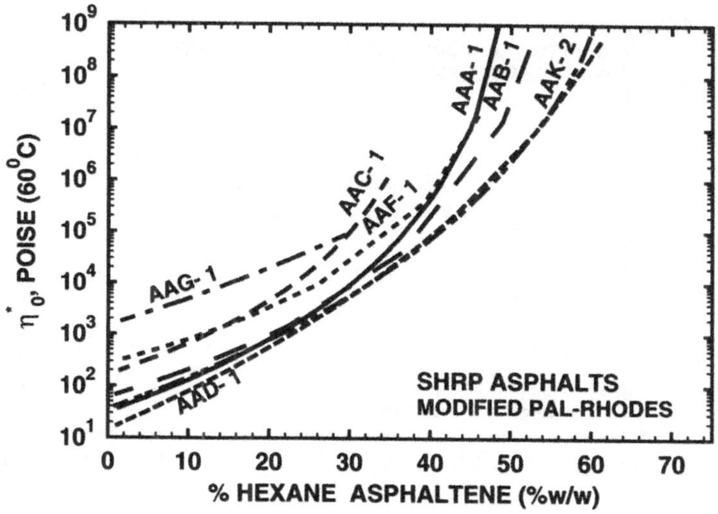

Figure 17. A comparison of η^*_0/%AS for all SHRP asphalts studied.

asphaltenes are non-spherical are validated. Many studies [48,49,50,51] showed that the value of ν is a function of particle shape and shear rate. Figure 18 shows ν as a function of aspect ratio (*a/b*), where a and b are the short and long axis of an ellipsoid, in low shear rate laminar flow conditions. The figure is generated using equations reported by Brodnyan [51]. Figure 18 also shows that the a range of aspect ratio from 0.054 to 0.175 is required to give a range of ν from 5.1 to 13.6 for disk-like particles. These values of aspect ratio for asphaltene particles in maltene media is not unreasonable. Reerink et al. [23] reported a range of aspect ratios from 0.1 to 0.25 measured for various asphaltenes in toluene using rheological measurement at infinite dilute and an ultracentrifugation technique. More interestingly, Ravey [52] used Small Angle Neutron Scattering (SANS) to measure the macrostructure of various asphaltenes in a number of solvents and obtained a range of aspect ratios from 0.043 to 0.1. That the result of this study and two previous studies (Reerink, 1973; Ravey, 1988) give very similar results using three different techniques suggests that the geometry of asphaltene particles in maltenes or organic solvents may be disc-like. The values of *K′* range from 1.14 to 2.36. Low values of *K′* indicate that the maltene is a more powerful solvating agent.

The η^*_0 in logarithm scale versus carbonyl area for the AAA-1 whole asphalt and maltene for all aging conditions is given in Figure 19. In addition, the HS for the whole

Table 2. Parameter ν and solvation constant (*K′*) for asphalts

SHRP asphalts	ν	*K′* (60°C)
AAA-1	6.9	1.88
AAB-1	6.4	1.74
AAC-1	5.2	2.36
AAD-1	13.6	1.18
AAF-1	5.1	1.93
AAG-1	10.0	1.14
AAK-2	9.6	1.37

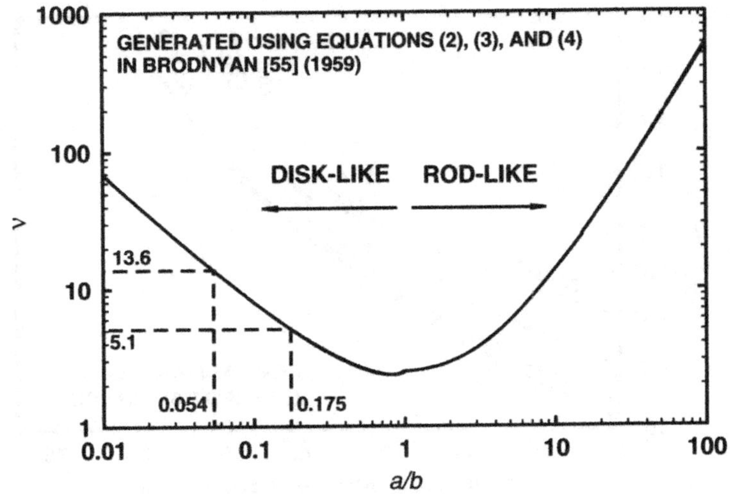

Figure 18. Range of aspect ratio (a/b) and parameter ν of asphaltenes for all asphalts studied.

asphalt and maltene are given in parenthesis. (In Figure 19, there exist two sets of data for maltenes having the same HS but different viscosities. The difference in viscosities may be due to operator effect.) At low levels of aging or low asphaltene concentration, the HS of the maltene is lower than that of the whole asphalt. Only when the maltene reaches the same asphaltene content or viscosity as the whole asphalt, does the HS of the maltene approach that of the whole asphalt. All asphalts studied show similar behavior except for AAG-1 because the HSs of SHRP AAG-1 maltene and asphalt are almost the same.

There appears to be a break point where the HS of the maltene changes abruptly to the value of the whole asphalt. Table 3 shows the %AS, CA and $\ln(\eta_0^*)$ for the aged maltene where the HS of the maltene increases to that of the whole asphalt. For five of the asphalts,

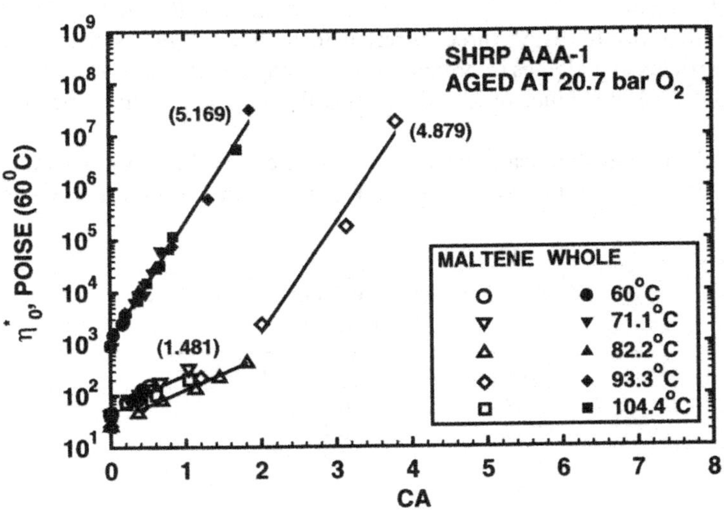

Figure 19. The hardening susceptibilities for SHRP AAA-1 whole asphalt and maltene. The values in parentheses represent hardening susceptibilities.

Table 3. The low frequency limiting viscosities (η_0^*), weight percentage of n-hexane asphaltenes (%AS), and carbonyl areas (CA) of asphalts at the break point

	Maltene at break point			Whole asphalt	
SHRP asphalt	η_0^* (poise, 60°C)	%AS (%w/w)	CA	η_0^* (poise, 60°C)	%AS (%w/w)
AAA-1	1,000	23.00	1.8	915	21.7
AAB-1	1,150	22.0	2.4	1378	22.2
AAC-1	900	12.0	1.5	943	15.6
AAD-1	900	26.0	1.8	1332	26.3
AAF-1	2,000	19.0	1.5	1950	18.0
AAG-1[a]	756	1.8	0.0	1925	10.3
AAK-2	20,000	51.0	4.7	1420	23.3

[a]Listed values for AAG-1 maltene are for unaged maltene because AAG-1 maltene appears to have no break point up to the highest asphaltene content studied.

AAA-1, AAB-1, AAC-1 AAD-1, and AAF-1, the break point is located at the %AS of the whole asphalt. However, the other two asphalts show no logical trend for the location of this break point.

Mathematically, the HS can be expressed in terms of equation 9, 12, and 13. By applying the chain rule to equation 8, the HS is the product of two quantities, the first described by the modified Pal-Rhodes model expressing the content on $\ln(\eta_0^*)$, and the second the AFS (equation 9).

$$HS = \left(\frac{d\ln\eta_0^*}{dCA}\right) = \left(\frac{d\ln\eta_0^*}{d\%AS}\right)\left(\frac{d\%AS}{dCA}\right) = \left(\frac{\nu K'}{1-K'\,\%AS}\right)(AFS) \tag{14}$$

Actually for the whole asphalt, the HS is surprisingly constant for certain ranges of carbonyl and is probably constant during the pavement life. Only in the maltenes studied does the HS change dramatically. This constant HS can be explained by the two competing factors on the right hand side of Equation 14. The first factor is the increasing rate of change of $\ln(\eta_0^*)$ with %AS as described by the modified Pal-Rhodes model (shown in Figure 17), and the second is the decrease in AFS (shown in Figure 12) since the weight percent asphaltene has a limit of 100% while the carbonyl area still grows. Equation 14 also suggests that the HS for a given asphalt and maltene is a weak function of asphaltene concentration only.

Another rheological property ($1/J''$) correlates with the rut depth in pavement mixes [44]. This parameter ($1/J''$) is compared to %AS for SHRP asphalt AAA-1 in Figures 20, and show similar results as that of Figure 14. The same trend is observed for other asphalts studied. Again, the original and produced asphaltenes from oxidative aging have very similar effects on rheological properties.

CONCLUSIONS

Our studies conclude

1. With oxidative aging, n-hexane asphaltene content increases in whole asphalts and maltenes as a result carbonyl formation. The presence of the original asphaltenes has no effect on the formation of the produced asphaltenes per unit oxidation in the whole asphalt until high levels of aging.

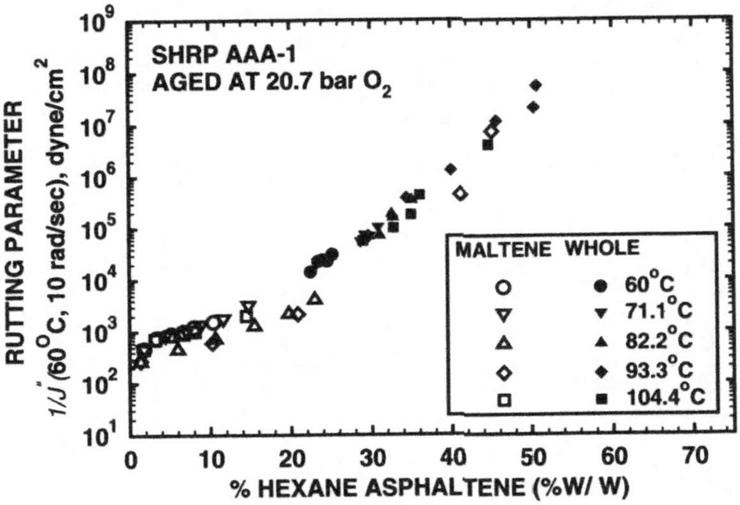

Figure 20. Fatigue parameter versus %AS for SHRP AAA-1 whole asphalt and maltene.

2. The produced asphaltenes have the very similar rheological effect as the original asphaltenes even though they are chemically quite different.
3. A modified Pal-Rhodes model successfully describes the viscosity/asphaltene behavior of both whole asphalts and maltenes.
4. The Hardening Susceptibility is a weak function of asphaltene content for whole asphalts. For maltenes, the HS increases to that of whole asphalt when the %AS content reaches that of the whole asphalt.

ACKNOWLEDGMENT

Support for this work by the Texas Department of Transportation (TxDOT), in cooperation with the Federal Highway Administration, and the Texas Advanced Technology Program (ATP), is gratefully acknowledged. The technical contributions of Ms. Ann Ferry are greatly appreciated.

DISCLAIMER

The contents of the report reflect the views of the authors who are responsible for the facts and the accuracy of the data presented herein. The contents do not necessarily reflect the official views or policies of the Federal Highway Administration or the Texas Department of Transportation. This report does not consititute a standard, specification, or regulation. This report is not intended for construction, bidding, or permit purposes.

REFERENCES

1. Corbett, L. W., Composition of Asphalt Based on Generic Fraction, Using Solvent Deasphalting, Elution-Adsorption Chromatography, and Densimetric Characterization. *Anal. Chem.*, 41: 576 (1969).

2. Corbett, L. W. and Merz, R.E., Asphalt Binder Hardening in the Michigan Test Road After 18 Years of Service. *Trans. Res. Rec.*, 544: 27 (1975)
3. Davison, R.R., Glover, C.J., Burr, B.L., and Bullin, J.A., Size Exclusion Chromatography of Asphalts. In: Chi-San Wu (Editor), *Handbook of Size Exclusion Chromatograhy*, 211-247 (1995).
4. Boduszynski, M.M., Asphaltenes in Petroleum Asphalts: Composition and Formation. In: J.W. Bunger and N.C. Li (Editors), *Chemistry of Asphaltenes*, Advancement in Chemistry Series 195, American Chemical Society, Washington, D.C., pp. 119-135 (1981).
5. Altget, K.H. and Harle, O.L., The Effects of Asphaltenes on Asphalt Viscosity. *Ind. and Eng. Chem. Prod. Res. Dev.*, 14: 240 (1975).
6. Hildebrand, J. and Scott, R., *Solubility of Non-Electrolytes*, 3rd ed., Reinhold, New York (1949).
7. Hansen, C.M. and Skaarup, K., III. Independent Calculation of the parameter components. *J Paint Technol.*, 39:511-514 (1967).
8. Hansen, C.M., The University of the Solubility Parameter. *Ind. Eng. Prod. Res. Dev.*, 8:2-11 (1969)
9. Hansen, C.M. and Beerbower A., Solubility Parameter. In: R.E. Kirk and D.F. Othmer (Editors), *Encyclopedia of Chemical Technology*, 2nd ed., Supplement Volume, John Wiley and Sons, New York, pp. 889-910 (1971).
10. Hagen, A.P., Jones, R., Hofener, R.M., Randolph, B.B., and Johnson, M.P., Characterization of Asphalt by Solubility Parameter. *Proc. AAPT*, 53: 119-137 (1984).
11. Boduszynski, M.M., McKay, J.F., and Latham D.R., Asphaltenes, Where Are You? *Proc. AAPT*, 49: 123-143 (1980).
12. Girdler, R.B., Constitution of Asphaltenes and Related Studies. *Proc. AAPT*, 34: 45-79 (1965).
13. Moschopedis, S.E., Fryer, J.F., and Speight, J.G., Investigation of Hydrogen Bonding by Oxygen Functions in Athabasca Bitumen. *Fuel*, 55: 227-232 (1976).
14. Storm, D.A., DeCanio, S.J., DeTar, S.J., and Nero, V.P.. Upper Bound on Number Average Molecular Weight of Asphaltenes. *Fuel*, 69 735-738 (1990).
15. Yen, T.F., Erdman, J.G., and Pollack, S.S., Investigation of the Structure of Petroleum Asphaltenes by X-Ray Diffraction. *Anal. Chem.*, 33: 1587 (1961).
16. Dickie, J.P. and Yen, T.F., Microstructures of the Asphaltic Frcations by Various Instrumental Methods. *Anal. Chem.*, 39: 1847-1852 (1957).
17. Yen, T.F. and Dickie, J.P., The compactness of the Aromatics Systems in Petroleum Asphaltics. *J. Inst. Petrol. Technol.*, 54: 50-53 (1968).
18. Petersen, J.C., Quantitative Functional Group Analysis of Asphalts Using Differential Infrared Spectrometry and Selective Chemical Reactions-Theory and Application. *Trans. Res. Rec.*, 1096: 1-11 (1986).
19. Petersen, J.C., Relations Between Asphalt Chemical Composition and Performance-Related Properties. Pepr. *Annual Meeting of the Asphalt Emulsion Manufacturers Association*, Las Vegas, Nevada (1982).
20. Petersen, J.C., Chemical Composition of Asphalt as Related to Asphalt Durability; State of the Art. *Trans. Res. Rec.*, 999: 13 (1984).
21. Storm, D.A. and Sheu, E.Y., Rheological Studies of Ratawi Vacuum Residue at 366K. *Fuel*, 72: 233-237 (1993).
22. Eilers, H.J., The Colloidal Structure of Asphalt. *J. Phys. Colloid Chem.*, 53: 1195 (1948).
23. Reerink, H. and Lijzenga, J., Molecular Weight Distributions of Kuwati Asphaltenes as Determined by Ultracentrifugation. Relation with Viscosity of Solutions. *J. Inst. Pet.*, 59: 211 (1973).
24. Heukelom, W. and Wijga, P.W.C, Viscosity of Dispersions as Governed by Concentration and Rate of Shear. *Proc. AAPT*, 40: 418 (1971).
25. Einstein, A., Eine Neue Bestimmung der Moleküldimensionen. *Ann. Phys.*, 19: 289 (1906).
26. Pal R. and Rhodes, E., Viscosity/Concentration Relationships for Emulsions. *J. Rheology*, 33: 1021-1045 (1989).
27. Sheu, E.Y., De Tar, M.M., and Storm, D.A., Rheological Properties of Vacuum Residue Fractions in Organic Solvents. *Fuel*, 70: 1151-1156 (1991).
28. Plancher, H., Green, E.L., and Petersen, J.C., Reduction of Oxidative Hardening of Asphalts by Treatment with Hydrated Lime—A Mechanistic Study. *Proc. Association of Asphalt Paving Technologists*, 45: 1 (1976).
29. Lee, D.Y. and Huang, R.J., Weathering of Asphalts as Characterized by Infrared Multiple Internal Reflectance Spectra. *Applied Spectroscopy*, **27:** 435 (1973).
30. Martin K.L., Davison, R.R., Golver C.J., and J.A. Bullin, Asphalt Aging in Texas Roads and Test Sections. *Trans. Res. Rec.*, 1269: 9-19 (1990).
31. Lau C.K., Lunsford, K.M., Glover, C.J., Davison, R.R., and Bullin, J.A., reaction Rates and Hardening Susceptibilities as Determined from POV Aging of Asphalts. *Trans. Res. Rec.*, 1342: 50-57 (1992).

32. Jemison, H.B., B.L. Burr, Davison, R.R., Bullin, J.A., and Glover, C.J., Application and Use of the ATR, FT-IR Method to Asphalt Aging Studies. *Fuel Sci. Technol. Int'l*, 10: 795-808 (1992).
33. Peterson G.D., "Relationship Between Composition and Performance of Asphalt Recycling Agents," Thesis, Chemical Engineering, Texas A&M University, College Station, Texas (1993).
34. Stegeman, J.R., Davison, R.R., Glover, C.J., and Bullin, J.A., Supercritical Fractionation and Reblending to Produce Improved Asphalts. In *Proceeding of the International Symposium: Chemistry of Bitumens*, Rome, Italy, 1: 366-381 (1991).
35. Stegeman, J.R., Kyle, A.L., Burr, B.L., Jemison, H.B., Davison, R.R., Glover, C.J., and Bullin, J.A., Compositional and Physical Properties of Asphalt Fractions Obtained by Supercritical Extraction. *Fuel Sci. Technol. Int'l*, 10: 767-794 (1992).
36. Jemison, H.B., Davison, R.R., Glover, C.J., and Bullin, J.A., Fractionation of Asphalt Materials by Using Supercritical Cyclohexane and Petane. *Fuel Sci. Technol. Int'l*, 13: 605-638 (1995).
37. Burr, B. L., Davison, R.R, Jemison, H.B., Glover, C.J., and Bullin, J.A., Asphalt Hardening in Extraction Solvents. *Trans. Res. Rec.*, 1323: 70 (1991).
38. Burr, B.L., Davison, R.R., Glover, C.J., and Bullin, J.A., Solvent Removal from Asphalt. *Trans. Res. Rec.*, 1269: 1 (1990).
39. Ferry, J.D. *Viscoelastic Properties of Polymers*. 3rd Ed., John Wiley and Sons, New York (1980).
40. SHRP Binder Specification, Draft No. 7G., SHRP, Federal Highway Administration, Washington D.C. (1992).
41. Pearson, C.D., Huff, G.S., and Gharfeh, S.G., Technique for the Determination of Asphaltenes in Crude Oil Residues. *Anal. Chem.*, 58: 3266 (1986).
42. Liu, Meng, Lunsford, K.M., Davison, R.R., Glover, C.J., and Bullin, J.A., The Kinetics of Carbonyl Formation in Asphalt. *AICHE*, (in press).
43. Petersen, J.C., and Branthaver, J.F., Robertson, R.E., Harnsberger, P.M., Duvall, J.J., and Ensley, K.E., Effects of Physicochemical Factors on Asphalt Oxidation Kinetics. *Trans. Res. Rec.*, 1391: 1 (1993).
44. Traxler, R.N., Relation Between Asphalt Composition and Hardening by Volatization and Oxidation. *Proceeding of AAPT*, **36:** 359 (1960).
45. Thurston, R.R. and Knowles, E.G., Oxygen Absorption Tests on Asphalt Constituents. *Ind. Eng. Chem.*, 28: 88 (1936).
46. Thurston, R.R. and Knowles, E.G., Asphalt and Its Constituents - Oxiadtion at Service Temperatures. *Ind. Eng. Chem.*, 33: 320 (1941).
47. Lee D.G. and Noureldin, N.A., The Effect of Low Temperature Oxidation On the Chemical and Physical Properties of Maltenes and Asphaltenes Derived from Heavy Oil. *Fuel Sci. Technol. Int'l*, 11: 173 (1993).
48. Hinch, E.J. and Leal, L.G., Rheological Properties of A Suspension. *J. Fluid Mech.*, 52: 683-712 (1972).
49. Kuhn, W. and Kuhn, H., Die Abhängigkeit der Viskosität vom Strömungsgefälle Beihochverd_nnten Suspensionen und Lösungen. *Helv. Chim. Acta.*, 28: 97 (1945).
50. Kuhn, W., Huhn, H., and Buchner P., Hydrodynamisches Verhalten von Makromolek_len in Lösung. *Ergeb. Exakt. Naturw.*, 25: 1 (1951).
51. Brodnyan, J.G., The Concentration Dependence of the Newtonian Viscosity of Prolate Ellipsoids. *Trans. Soc. Rheol.*, 3: 61-68 (1959).
52. Ravey J.C., Ducouret, G., and Espinat, D., Asphaltene Macrostructure by Small Angle Neutron Scattering. *Fuel*, 67: 1560-1567 (1988).
53. Burr, B.L., Davison, R.R., Glover, C.J., and Bullin, J.A., Softening of Asphalts in Dilute Solutions at Primary Distillation Conditions. *Trans. Res. Rec.*, 1436: 47-53 (1994).

Chapter VI

ASPHALT EMULSION FOR ENVIRONMENTAL COATING AND ENCAPSULATION

Hsienjen Lian and Teh Fu Yen

Department of Civil Engineering
University of Southern California
Los Angeles, California

INTRODUCTION

In this new era, people are more concerned about their living environment. Environmental problems become a global issue in this world, and many of the environmental problems come from material corrosion such as underground storage tank corrosion, buried pipe line corrosion, etc., which can cause soil and groundwater contamination. Such contaminations directly threaten human health and human life when the underground water is contaminated or when the seepage of volatile organic compounds from underground contaminated soil becomes explosive.

There are two ways to solve these problems: one is remediating the contaminated area; the other is stopping the contamination source by corrosion control. Recently, corrosion studies are threefold: economic, safety and conservation. The primary concern for much of the current research in corrosion is economic losses [1]. In the United States the total combined losses of metallic corrosion is estimated conservatively to be about $70 billion annually, or 4.2% of the Gross National Product (GNP) [2]. About 15% of this total losses could be avoided if currently available corrosion technology were effectively applied.

Asphalt is a dark brown to black cementitious material, solid or semisolid in consistency. The predominating constituents are bitumens which occur in nature or obtained as residue by refining petroleum [3]. Because asphalts possess some special properties such as impermeability to water, pronounced adhesive and cohesive properties, susceptibility to temperature changes and deformation in service, excellent abrasion resistance, chemical resistance to acids, alkalies, air, ground water, corrosive soil conditions, etc, they can be used as pavings, roofings, road joint materials, crack fillers, coatings materials (canal linings, water-proofing cements, pipe dips, sound-deadening products), tiling and flooring coverings materials, electrical insulation products, brake-lining products, etc. In the United States about 70% of all oil asphalts are consumed by the road-paving industry, with some 20% consumed by roofing manufactures and 10% consumed by special usage manufactures.

Asphaltenes: Fundamentals and Applications, Edited by E. Y. Sheu and O. C. Mullins
Plenum Press, New York, 1995

Asphalts can be separated into saturates, aromatics, resins, and asphaltenes fractions by solvents. The polarity of these four fractions increases in the order of saturates, aromatics, resins and asphaltenes. In crude oil, asphaltene micelles are present as discrete or colloidally dispersed particles in the oil phase. Although the asphaltenes themselves are insoluble in gas oil (saturates and aromatics), they can exist as fine or coarse dispersions, depending on the resin content. The resins are part of the oil phase but have a higher polarity than that of gas oil. This property enables the molecules to be easily adsorbed onto the asphaltene micelles and to act as a peptizing agent of the colloid stabilizer by charge neutralization [4].

In its natural state, asphaltene exists in an oil-external (Winsor's terminology) or reversed micelle [5]. The polar groups are oriented toward the center, which can be water, silica (or clay), or metals (V, Ni, Fe, etc.). The driving force of the polar groups assembled toward the center originates from hydrogen-bonding, charge transfer or even salt formation. This oil-external micelle system can be reversed to oil-internal, water-external micelles (usually called Hartley micelles). An aggregate of asphaltene particles with adsorbed resins can form a supermicelle, and oil may be occluded between supermicelles as an intercellular media. Upon further aggregation the supermicelles can coalesce into giant supermicelles, and can even gradually grow into a liquid crystal.

From the above demonstration, it can be noted that micelle structures are predominant in asphalt with a higher asphaltene content. Three different types of asphalt such as sol (micelle, supermicelle, giant supermicelle), sol-gel (supermicelle, giant supermicelle) and gel (liquid crystal) asphalt can be defined. Most of the paving asphalts belong to the sol-gel type of asphalt, and roofing and coating asphalts belong to the gel (air blown) type of asphalt.

Because of the special properties mentioned above and the less expensive material of asphalt, this research focuses on how to make coating and encapsulating asphalt emulsions for buried structures and pipelines to extend their life, and to prevent emission problems with air and for fixation or solidification of hazardous wastes to avoid leaching problems.

EXPERIMENTAL

Asphalt Emulsion (Optimum Condition)

Two asphalt samples, which were supplied by Strategic Highway Research Program (SHRP), were used to make coating and encapsulating asphalt emulsion. The source and refinery process of these two asphalts are listed on Table 1. Thin-Layer Chromatography interfaced with Flame Ionization Detector (TLC-FID) will be used to analyze the change of four components of asphalt through the producing coating and encapsulating asphalt emulsion processes. The selection of solvents for TLC-FID analysis are shown on Table 2 [6]. Three surfactants , hexadecylamine (cationic), aerosol OT (anionic), and span 20 (nonionic), were selected for this research, and their chemical formulas are presented in Figure 1.

Table 1. Sources and refinery processes of the two core asphalt samples

Sample	Source	Refinery process
AAA-1	Lloydminister	distillation
AAD-1	California Coast	distillation

Table 2. Solubility parameters of solvents and solvent mixtures

Solvents	Solubility parameter
Hexane	7.4
Toluene	8.9
Dichloromethane : methanol (95 + 5, v/v)	10.5

Hydrogen peroxide is also selected as an oxidant for this research. A VC-50 low volume, high intensity Ultrasonic Processor was used in the experiments, and the frequency and energy output were 20 KHz and 50 Watts (maximum). The limitation for this instrument is that the concentration of solution (whether organic phase or water) cannot exceed 15% in order to have the ultrasonic tool function well.

First, 4.5 grams of California Coast asphalt were heated to 70 °C, then gently pulled into heated (55°C) water (25.5 grams) which contained 1.5% by weight aerosol OT and 1.1% by weight ferrous sulfate, until well mixed. During this period, light energy output about 8 watts per centimeter from ultrasound enhanced the miscibility of the two phases and reduced the mixing time. After about 10 minutes for mixing, 1% by weight hydrogen peroxide was

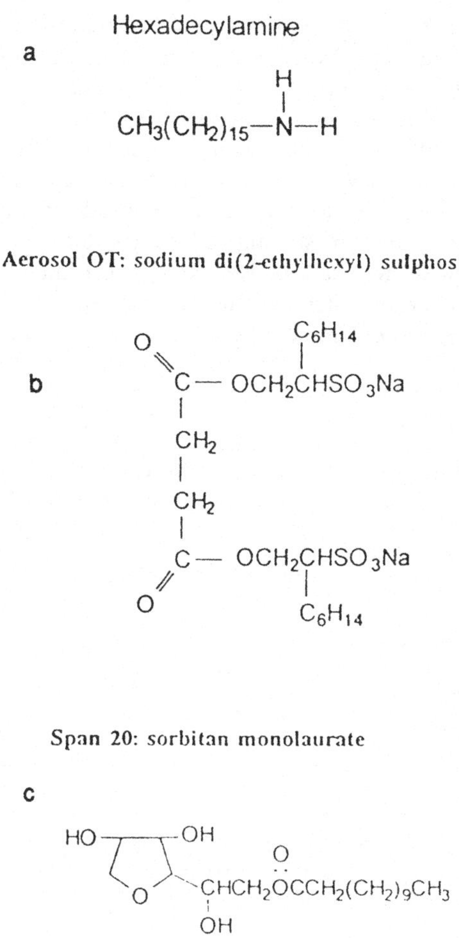

Figure 1. Chemical formula of three surfactants - (a) cationic:hexadecylamine, (b) anionic:aerosol OT, (c) nonionic:span 20.

Table 3. The variation of three components in California Coast asphalt through emulsion polymerization with aerosol OT under ultrasound

Time (minutes)	Gas oil	Resin	Asphaltene
0	35.8	40.8	23.4
10	25.8	42.4	31.8
20	19.7	43.4	36.9
30	9.4	43.1	47.5
40	9	41.1	49.9
50	8.3	40.3	51.4
60	8.8	40	51.2
70	8.3	39.5	52.2
80	8.5	39.9	51.9
90	8.2	39.3	52.5

Note: The coefficients of variation of the above data are in the range of 2.4% - 7.6%.

added once for the 10 minutes of reaction time. At this time, energy output from the microprocessor was 15 watts.

Second, under the same conditions, the experiment was repeated 8 times using 2, 3, 4, 5, 6, 7, 8, and 9% hydrogen peroxide, and adding 1% hydrogen peroxide every 10 minutes throughout all the experiments. After that, dichloromethane was used with the same amount mentioned in the optimum conditions section to obtain a 2% asphalt solution for TLC-FID analysis. The results are listed in Table 3.

With the same procedures, except span 20 or hexadecylamine instead of aerosol OT, the experiments were repeated again. The results are listed in Tables 4 and 5.

During the experiments, the hexadecylamine was partially dissolved in water at 90 °C for 20 minutes by stirring with magnetic bar. For the following experiments on Lloydminister asphalt, we shall focus on aerosol OT and span 20 only.

Lloydminister asphalt was selected to repeat all the experiments mentioned above except hexadecylamine. The results are shown in Tables 6 and 7.

Table 4. The variation of three components in California Coast asphalt through emulsion polymerization with span 20 under ultrasound

Time (minutes)	Gas oil	Resin	Asphaltene
0	35.8	40.8	23.4
10	22.0	43.1	34.9
20	18.4	44.0	37.6
30	9.9	41.2	48.9
40	9.5	40.0	49.5
50	7.9	39.4	52.7
60	9.1	37.7	53.2
70	7.3	37.9	54.8
80	7.1	37.8	55.1
90	6.9	37.8	55.3

Note: The coefficients of variation of the above data are in the range of 2.1% - 7.9%.

Table 5. The variation of three components in California Coast asphalt through emulsion polymerization with hexadecylamine under ultrasound

Time (minutes)	Gas oil	Resin	Asphaltene
0	35.8	40.8	23.4
10	29.8	41.5	28.7
20	28.0	41.9	30.1
30	24.0	42.1	33.9
40	24.1	41.6	34.3
50	23.5	40.7	35.8
60	23.3	41.5	35.2
70	23.9	40.6	35.5
80	23.0	40.9	36.1
90	23.4	40.4	36.2

Note: The coefficients of variation of the above data are in the range of 3.5% - 8.4%.

Coating

Some of the final California Coast and Lloydminister asphalt emulsions, which are listed in Tables 8 and 9, were selected, and two metal wires, copper and aluminum, were used for coating experiments. one tenth of a gram each of copper or aluminum sample was dipped into California Coast or Lloydminister asphalt emulsion for about 20 minutes. Then the sample with coated asphalt was put in a 50°C oven for an hour to remove water. The same procedure was repeated 9 times to ensure a well coated metal surface. After that, the coated metal was placed into acetic acid solution for 72 hours to simulate the EPA toxicity characteristic leaching procedure (TCLP) test. In my encapsulating research, a one-dimensional shaker was used instead of a three-dimensional shaker, but the reaction time was extended to 72 hours, compared with 18 hours for the EPA standard TCLP method. The shaking speed was 20 per minute. The extension of the reaction time was made in order to compensate for the deficit in the mixing process. The final extraction fluid was digested by

Table 6. The variation of three components in Lloydminister asphalt through emulsion polymerization with aerosol OT under ultrasound

Time (minutes)	Gas oil	Resin	Asphaltene
0	52.1	30.4	17.5
10	45.6	32.3	22.1
20	42.0	33.8	24.2
30	37.2	33.0	29.8
40	36.6	32.4	31.0
50	36.1	31.9	32.0
60	36.0	31.7	32.3
70	35.6	31.0	33.4
80	35.5	30.7	33.8
90	35.5	30.5	34.0

Note: The coefficients of variation of the above data are in the range of 1.9% - 7.2%.

Table 7. The variation of three components in Lloydminister asphalt through emulsion polymerization with span 20 under ultrasound

Time (minutes)	Gas oil	Resin	Asphaltene
0	52.1	30.4	17.5
10	41.6	33.5	24.9
20	37.4	34.3	28.3
30	31.5	33.7	34.8
40	31.6	32.0	36.4
50	31.7	31.5	36.8
60	31.2	30.1	38.7
70	31.1	29.8	39.1
80	31.0	29.0	40.2
90	30.9	28.7	40.4

Note: The coefficients of variation of the above data are in the range of 1.9% - 6.8%.

0.1 N nitric acid up to pH 2 or lower than 2 in order to do the analysis of metal ions by the inductively coupled plasma method (EPA 6010A method). All the results are listed in Table 8 and 9.

Encapsulation

From Tables 3 to 7 some of the California Coast and Lloydminister asphalt emulsions are selected for the solidification of hazardous wastes. The selected asphalt samples are listed in Table 10. The highly contaminated soil which contain 200 ppm lead, 650.5 ppm total

Table 8. The copper concentrations of the solution portion after modified TCLP treatment

Composition	Copper (ppm) (EPA 6010A method)	C.V. (%)
Raw copper sample	13.6	0.17
AST (D) (SN 0 min. w/A-OT)	0.13	4.23
AST (D) (SN 10 min. w/A-OT)	0.11	1.55
AST (D) (SN 30 min. w/A-OT)	0.06	4.33
AST (D) (SN 0 min.. w/S-20)	0.09	8.78
AST (D) (SN 30 min. w/S-20)	0.04	0.24
AST (D) (SN 60 min w/S-20)	0.03	1.97
AST (D) (SN 90 min w/S-20)	N.D.	0.35
AST (D) (SN 0 min. w/Hx)	0.19	0.70
AST (D) (SN 90 min. w/Hx)	0.04	2.56
AST (A) (SN 0 min. w/A-OT)	1.10	2.23
AST (A) (SN 10 min. w/A-OT)	0.32	0.70
AST (A) (SN 30 min. w/A-OT)	0.11	0.01
AST (A) (SN 60 min. w/A-OT)	0.04	3.93
AST (A) (SN 0 min. w/S-20)	0.85	1.66
AST (A) (SN 30 min. w/S-20)	0.15	0.24
AST (A) (SN 60 min. w/S-20)	N.D.	0.35

Note: AST(D) - AAD-1 asphalt, AST(A) - AAA-1 asphalt, SN - Sonicated, S-20 - Span 20, A-OT - Aerosol OT. Hx - Hexadecylamine. N.D. - Non-detectable.

Table 9. The aluminum concentrations of the solution portion after modified TCLP treatment

Composition	Aluminum (ppm) (EPA 6010 A method)	C.V. (%)
Raw aluminum sample	10.5	0.56
AST (D) (SN 30 min. w/A-OT)	1.30	1.14
AST (D) (SN 30 min. w/S-20)	1.10	2.14
AST (A) (SN 30 min. w/A-OT)	3.30	1.24
AST (A) (SN 30 min.. w/S-20)	2.90	0.16

Note : AST(D) - AAD-1 asphalt, AST(A) - AAA-1 asphalt, SN - Sonicated, S-20 - Span 20, A-OT - Aerosol OT.

recoverable petroleum hydrocarbon (TRPH) was supplied by the Mobile Oil refinery plant in Torrance, California.

First, 10 grams of contaminated soil were mixed with various amounts of asphalt emulsion, or cement, or both. Then the well mixed sample was placed in an environment with good ventilation for 48 hours. After curing, a modified TCLP was performed to determine the ability of fixation on target compound.

After all the experiments, the extraction fluid was digested with nitric acid to ensure that the pH value of the extraction fluid was lower than 2, in order to analyze for lead concentration by the atomic absorption method (EPA 7420 method). For TRPH in the extraction fluid, 50 ml of dichloromethane were added three times in order to separate organic compounds from the 10 ml extraction fluid. After evaporation of the dichloromethane, the total concentrations of organic compounds can be obtained by a gravimetric method (modified EPA 413.1 method). Freon-113 was used to separate organic compounds

Table 10. The TRPH and lead concentrations of the solution portion after modified TCLP treatment

Composition	TRPH (ppm) (EPA 413.1 method)	Lead (ppm) (EPA 7420 method)
Raw Sample	650.5	220
5% AST	39.1	0.89
10% AST	22.3	0.56
10% CMT	30.6	0.62
10% AST(D) & 10% CMT	17.2	0.45
10% AST(D) (SN 10 min w/S-20)	20.5	0.45
10% AST(D) & 10% CMT(SN 10 min w/S-20)	15.7	0.4
10% AST(D) (SN 20 min w/S-20)	18.3	0.42
10% AST(D) & 10% CMT(SN 20 min w/S-20)	12.9	0.37
10% AST(D) (SN 30 min w/S-20)	16.9	0.37
10% AST(D) & 10% CMT(SN 30 min w/S-20)	10.2	0.34
10% AST(D) (SN 10 min w/A-OT)	21.1	0.47
10% AST(D) & 10% CMT(SN 10 min w/A-OT)	16.4	0.42
10% AST(D) (SN 30 min w/A-OT)	17.5	0.39
10% AST(D) & 10% CMT(SN 30 min w/A-OT)	11.8	0.36
10% AST(A) (SN 30 min w/S-20)	19.3	0.42
10% AST(A) & 10% CMT(SN 30 min w/S-20)	15.9	0.39
10% AST(A) (SN 30 min w/A-OT)	20.7	0.45
10% AST(A) & 10% CMT(SN 30 min w/A-OT)	17.6	0.41

Note: AST(D) - AAD-1 asphalt, AST(A) - AAA-1 asphalt, CMT - Portland cement, SN - Sonicated, S-20 - Span 20, A-OT - Aerosol OT.

from the extraction fluid in EPA 413.1 method. In my experiments, dichloromethane was used instead of freon-113 to determine TRPH in the extraction fluid. All the results, including TRPH and lead concentrations, are listed in Table 10.

DISCUSSIONS

Normally, asphalt emulsions are produced by a traditional colloid mill. The manufacturing plant of this mill is shown in Figure 2 [7]. In the emulsification process, asphalt is heated to temperatures of about 250°F, and the water containing the emulsifying agent (surfactant) is heated to 120-150°F. Both water and asphalt are introduced into a colloid mill which supplies the mechanical energy to break the asphalt down into particles a few microns in diameter. This production of small particles is accomplished by forcing the asphalt to pass through a small clearance (0.25-0.50 mm) between the rotating blades and the stationary drum. Since the shaft is rotated at high speed (1000-6000 rpm), high shearing forces are produced in the material which results in the small diameter particles.

In this research, asphalt emulsions will be made with the same materials except using ultrasound system instead of colloid mill. The scheme of producing asphalt emulsions by

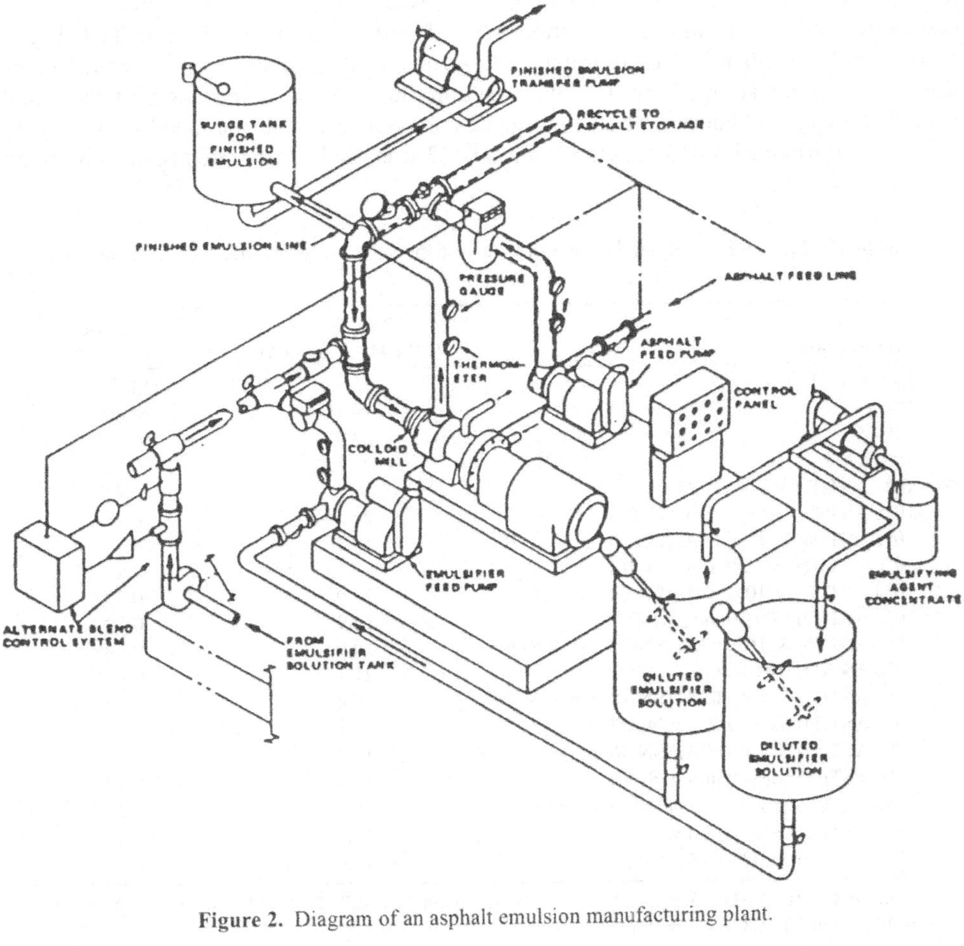

Figure 2. Diagram of an asphalt emulsion manufacturing plant.

using ultrasound is presented in the right part of the system shown in Figure 3. The advantages of the ultrasound system are energy conservation, which is required to heat asphalt to about 150°F and water with surfactant to 125°F, less capital investment and operating costs, and dual effectiveness in producing blown asphalt emulsion compared with the traditional method in which two processes, air-blowing and emulsifying, are required for producing blown asphalt emulsion. In this system, ferrous sulfate is added in the water as the catalyst to enhance the occurrence of hydroxyl free radicals for emulsion polymerization.

Depending upon the electric charge on the asphalt particles, the asphalt emulsion can be classified into cationic, anionic, or nonionic categories. In practice, the first two types are ordinarily used in roadway construction and maintenance. However, nonionic emulsions can be widely used in some special fields such as coating, encapsulating, etc. Emulsion polymerization will occur on asphalt components through hydroxyl free radical oxidation when hydrogen peroxide is exposed to the water containing surfactant, and ferrous sulfate is introduced as a catalyst under ultrasound.

In general, the polymerization process can be divided into three steps, initiation, propagation, and termination. For initiating systems we selected water-soluble oxidants, particularly hydrogen peroxide-ferrous iron of redox systems to make coating asphalts through emulsion polymerization. The reason for choosing hydrogen peroxide-ferrous iron as the initiator was because ferrous ion acts as catalyst to enhance the formation of hydroxyl free radicals under ultrasound. The selection of hydrogen peroxide is due to the easy formation of hydroxyl free radicals by ultrasound, infinite solubility in water, the generation of only nontoxic fume and gas during the reaction, producing water and oxygen as end products. Formation of free radicals by the reaction of this redox couple in aqueous solution was first proposed in 1932 [8], and subsequent work by a number of experimenters has established the validity of this concept [9-11]. The primary reaction between these compounds is as follows [8]:

$$H_2O_2 + Fe^{++} \longrightarrow HO^{\cdot} + OH^{-} + Fe^{+++} \quad (1)$$

If the ferrous iron is initially present in excess over the hydrogen peroxide, then the following reaction will occur:

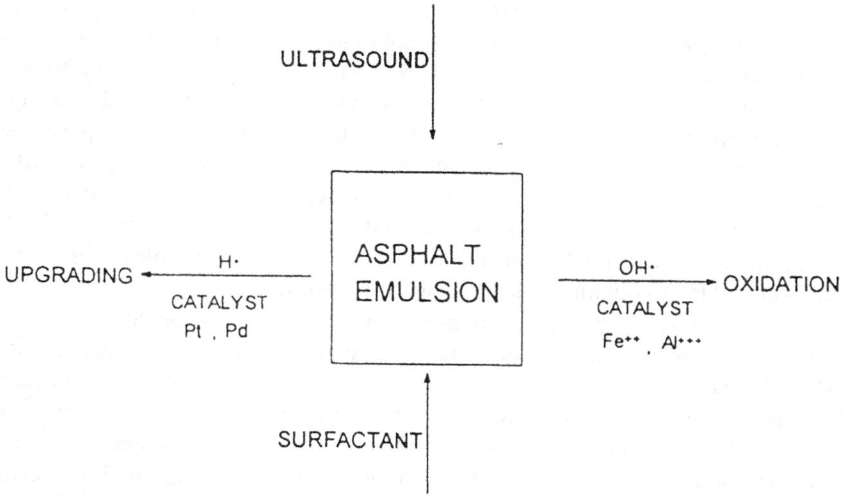

Figure 3. Scheme of ultrasonic system for asphalt hydrogenation or polymerization in emulsion.

$$Fe^{++} + HO^{\cdot} \longrightarrow Fe^{+++} + OH^{-} \quad (2)$$

If the hydrogen peroxide is initially present in large excess, some of it decomposes to oxygen and water. Other reactions occur as follows:

$$H_2O_2 + HO^{\cdot} \longrightarrow H_2O + HO_2^{\cdot} \quad (3)$$

$$Fe^{+++} + HO_2^{\cdot} \longrightarrow Fe^{++} + O_2 + H^{+} \quad (4)$$

In my experiments, the amount of hydrogen peroxide added each time was controlled so as to be equal to or less than the ferrous iron concentration because an excess amount of hydrogen peroxide will form the $HO_2\cdot$ free radical (equation 1) which prevents the polymerization reaction [12]. The free radical polymerization for the hydrogen peroxide - ferrous iron system under sonication is predicted as follows:

$$H_2O \longrightarrow H^{\cdot} + OH^{\cdot} \quad (5)\ \text{Initiation}$$

$$H_2O_2 + Fe^{++} \longrightarrow HO^{\cdot} + OH^{-} + Fe^{+++} \quad (6)\ \text{Initiation}$$

$$RR' + OH^{\cdot} \longrightarrow HOR^{\cdot}R' \quad (7)\ \text{Propagation}$$

$$HOR^{\cdot}R' + HOR^{\cdot}R' \longrightarrow HORR'R'ROH \quad (8)\ \text{Termination}$$

or

$$HOR^{\cdot}R' + OH^{\cdot} \longrightarrow ORR' + H_2O \quad (9)\ \text{Termination}$$

R is a methyl group, and R′ is an aromatic group.

The reaction of free radical polymerization could be equation 5 to 8, or equation 5 to 7, and 9. With surface charge modification under ultrasound membrane mimetic theory, the coating and encapsulating asphalt emulsions were formed through polymerization reaction mentioned as above.

Table 3 to 7 show the same trend, with the asphalt content increasing drastically in the first 30 minutes reaction time, after which the reaction rate becomes slow due to the consumption of surfactants during emulsion polymerization. After 30 minutes of reaction time, 1% surfactants were added to the emulsion system three times for every 20 minutes to stabilize the emulsion system. The total concentration of surfactants added to the emulsion system for the kinetic study is 4.5%, and the total concentration of hydrogen peroxide is 9%. However, even with increasing the surfactant concentration, the aggregation of asphalt particles still occurs during the reaction owing to the conversion of tremendous amount of gas oil to resin or asphaltene which can not withstand the original homogeneity of the asphalt system. At this time the asphalt emulsion is very difficult to form unless high amounts of surfactants are added to obtain stable emulsion properties.

From Tables 3 to 5, the increase of asphaltene content with span 20 is the best, aerosol OT is the second, and hexadecylamine is the last. Asphaltene content increases in the order of 31.9% with span 20 in the system, 29.1% with aerosol OT, and 12.8% for hexadecylamine in California Coast asphalt under ultrasound for 90 minutes. Tables 6 and 7 indicate that the increase of asphaltene content with aerosol OT is 16.5%, and with span 20 is 22.9% in Lloydminister asphalt under ultrasound for 90 minutes. With the same conditions in the same asphalt system, span 20 is the best surfactant for making coating and encapsulating asphalt emulsion, aerosol OT is the second best, and hexadecylamine is third best in this study.

Lin has shown that the polarity of those polar compounds is decreased in the order of the hydroxyl group, carboxylic group, SOO^-group, and amine group [13]. Again, from Figure 1 span 20 possesses three hydroxyl groups which makes itself polar enough to disperse asphalt particle very well in water comparing to aerosol OT with two SOO^- groups, and hexadecylamine is the least effective because it is a long chain amine whose polar groups are much less than aerosol OT and span 20. Additional evidence is that hexadecylamine has been proven not a good peptizing agent in asphalt systems due to poor structure and polarity.

From Table 8, the product which contains high asphaltene content is the better candidate for metal coating. For instance, the copper concentration is 0.06 ppm for California Coast asphalt sonicated with aerosol OT for 30 minutes, compared with 0.13 ppm concentration of lead in solution for the same asphalt sonicated with aerosol OT for 10 minutes. In general, all results for copper concentration are less than 1 ppm, which are lower than 25 ppm State criteria in Title 22. The coefficients of variation for all results are in the range of 0.01-4.23%, except the application of California Coast asphalt emulsion with span 20 without sonication, which has 8.78% of coefficient of variation. The detection limit for copper in ICP instruments is 10 ppb. The California Coast asphalt for coating application is still better than Lloydminister, and the production of asphalt emulsion with span 20 makes a better coating than the products with aerosol OT. For aluminum studies, Table 9 shows that the leaching concentration of aluminum for all applications is less than 5 ppm, and coefficients of variation are less than 3%. Although no criteria for aluminum are listed in Title 22 [14], asphalt emulsions can be used for metal coatings quite well due to low concentrations of copper and aluminum in the leachate.

Many products for asphalt emulsion on metal coatings have been invented in Japan and United States in industry [15-17]. Some of the applications for asphalt emulsion have been used in painting, and for undercoatings of automobiles [18-19]. From Tables 8 and 9, it is clear that our asphalt emulsions made by ultrasound are good coating materials for metals.

Title 22 lists some criteria of inorganic compounds which the limitation of lead concentration is 5 ppm for wastewater. Although the criteria for TRPH in extraction fluid are not shown in Title 22, the limitation of TRPH in TCLP extraction fluid of about 100 ppm has been adopted by industry as the indicator for hazardous wastes treatment. All of the present results, including TRPH and lead concentrations listed in Table 10, are below these two values. The encapsulation of hazardous wastes with asphalt emulsion has been successfully achieved. From Table 10, it is clear that the same concentrations of asphalt can prevent lead or organic compounds from leaching better than portland cement. The increase of asphalt concentration does enhance fixation ability. Also, with the same reaction time for making an encapsulating asphalt emulsion, California Coast asphalt does a better fixation job than Lloydminister asphalt because of its high content of asphaltene in the asphalt system.

Asphalt emulsions produced by the traditional colloid mill method are already widely applied in hazardous waste encapsulation, including wastewater containing heavy metals such as chromium, and arsenic, as well as cyanide, to radioactive materials, wastewater sludge, and ash generated from incineration plants [20-24]. However, the present work is the first application of asphalt emulsions made by ultrasound for hazardous waste encapsulation.

CONCLUSIONS

Asphalt is a composite material, and it can be used in many areas such as paving, roofing, coating, encapsulating, etc., owing to its special properties of high impermeability to water, pronounced adhesive and cohesive properties, low susceptibility to temperature change, excellent abrasion resistance, chemical resistance to acids, alkalies, air, groundwater, corrosive soil conditions, high radioactivity resistance, and good biodegradation resistance.

Asphalt consists of four components: saturates, aromatics, resins, asphaltenes. With the various ratios of four fractions in asphalt, it can be applied in different fields. In colloidal science, asphaltene cannot be dissolved in gas oil (saturates and aromatics). Resin, acting as a peptizing agent, can be adsorbed on asphaltene particles, then enhance asphaltene dispersion in gas oil very well. The polarity of the three components decreases in the order of asphaltene, resin, and gas oil. With the micelle theory, asphalt can be classified into three types, sol, sol-gel, and gel.

The major two advantages for asphalt emulsion on coating and encapsulation are energy conservation and no emission problems. The application of asphalt emulsions in different fields depends on the types of asphalt in the emulsion system. For instance, sol-gel types of asphalt emulsion can be used in road maintenance and hazardous wastes encapsulation, whereas, gel types of asphalt emulsion can be used for coatings that include metals, roofs, bridges, buildings, etc. The gel type of asphalt can also be used for hazardous waste encapsulation. Through emulsion polymerization under ultrasound, the asphaltene content can be highly increased in California Coast and Lloydminister asphalts. The selection of surfactant depends on the surface charge of the target compound. For instance, a cationic asphalt emulsion is better for silica soil with its negatively charged surface.

California Coast and Lloydminister asphalts were polymerized with the addition of surfactants, oxidants, and catalysts in water to make coating and encapsulating asphalt emulsion. The final products used for metal coating and hazardous wastes encapsulation were quite successful. The leaching test shows results which are well under the State lead and copper criteria in Title 22. High asphaltene asphalts yield better results for metal coating and hazardous wastes encapsulation. This method is not only inexpensive, but also innovative in that it combines the air-blowing and emulsifying processes into one unique process.

REFERENCES

1. Uhlig, H.H.; and Revie, R.W., 1985. Corrosion and Corrosion Control: An Introduction to Corrosion Science and Engineering, Third Edition, John Wiley & Sons, Inc., 1-5.
2. Bennett L.H., 1978. Economic Effects of Metallic Corrosion in the United States, A Report to the Congress by the National Bureau of Standards, NBS Spec. Publ. 511:1, Washington, D.C.
3. ASTM D8, 1989. Standards on Petroleum Products and Lubricants, American Society for Testing and Materials, Philadelphia, Pa.
4. Lian, H.J., Lin, J.R., and Yen, T.F., 1994. Peptization studies of asphaltene and solubility parameter spectra, *FUEL*, **73**, 423-428.
5. Yen, T.F., 1988, Asphaltic Materials. *Encyclopedia of Polymer Science and Engineering*, 1-10, Wiley & Sons, New York.
6. Iatroscan Lab. Inc., 1982. Iatroscan TH-10 Instrument Application:Optimising Reproducibility with the Iatroscan TLC/FID analysis, *ANCAL* 5.
7. The Asphalt Institute, 1992. A Basic Asphalt Emulsion Manual:Manual Aeries No. 19, 2nd Edition, College Park, MD.
8. Haber, F.; and Weiss, J., 1932. *Naturwissenschaften* 20:948.
9. Medalia, A.I.; and Kolthoff, I.M., 1949. The reaction between hydrogen peroxide and ferrous iron, *J. Polymer Sci.* 4:377.
10. Weiss, J., 1952. The Reaction between Hydrogen Peroxide and Ferrous Iron, in Advances in Catalysis, Academic Press, New York, 4:343.
11. Baxendale, J.H., 1954. The reaction between hydrogen peroxide and ferrous iron, ibid., 31.
12. Masterton, W.L.; Slowinski, E.J.; and Stanitski, C.L., 1986. Chemical Principles: CBS College Publishing, 356.
13. Lin, J.R., 1992. Dissertation:Asphalt Waste Minimization, chapter 5. Department of Civil and Environmental Engineering, University of Southern California, Los Angeles.

14. California Code of Regulations, 1993. Title 22. Social Security: Division 4. Environmental Health, vol. 29A, published by Barclays Law Publishers, S. San Francisco.
15. Roadway International Corp., 1976. Coating Composition for Protecting Metal against Corrosion, Neth. Appl. NL 7605185, 10 pp.
16. Sekiyu, K.K., 1985. Two-Component Coating Compositions, shell, Jpn. KoKai Tokkyo Koho JP 60008373 A2, 5 pp.
17. Masaru, M., 1985. Asphalt Coating Compositions, Jpn. Kokai Toyyko Koho JP 60133067 A2, 2 pp.
18. Roediger,J.C.; Tompkins, E.E; and Schoenbam, G.H.,1969. Undercoating Automobiles with an Asphalt Emulsion, U.S. US 3427172, 3 pp.
19. Ladislav, N., 1978. Anticorrosion Paint, Czech. CS 171859, 2 pp.
20. Fumiaki, F.; Tasuku, S.; Kohei, K.; Hiroshi, A.; and Arata, O., 1976. Solidification of Sludges Containing Heavy Metals and Arsenic, Nichireki Chemical Industry Co., Japan, *Japan Kokai JP 51069483* 15.
21. Tsuneo, M., 1979. Fixation Treatment of Electrostatically Collected Dust from an Incineration Plant, JP 54057468,Showa 3.
22. Tomizo, O.; Katsumi, F.; Noriyuki, S.; and Nobuaki, S., 1980. Reduction Treatment of Chromium (VI) Discharged from Chromium Slag, *Hokkaidoritsu Kogyo Shikenjo Hokoku* 279:155-163.
23. Akira, I.; Mitsuhiko, N.; and Hideo, K., 1985. Solidification of detergent-containing radioactive waste solution, Jpn. Tokkyo Koho JP 60046680, B4, Showa, 5.
24. Conner, J.R., 1986. Method for rendering hazardous wastes less permeable and more resistant to leaching, Chem-Technics Inc., *USA, U.S. US 4623469 A* 7.

Chapter VII

POTENTIAL OF ULTRASONIC GENERATORS FOR USE IN OIL WELLS AND HEAVY CRUDE OIL/BITUMEN TRANSPORTATION FACILITIES

M. R. Islam

Department of Geological Engineering
South Dakota School of Mines and Technology
501 East St. Joesph Street
Rapid City, South Dakota 57701

INTRODUCTION

Crude oils produced in many parts of the world contain asphaltenes. Asphaltenes are known to deposit in the vicinity of production wells during miscible floods, after acid stimulation, during thermodynamic changes, and in the pipe lines during transportation or during pressure changes in a mature field with an asphaltic crude. Asphaltene deposition leads to production loss and involves expensive corrective measures. This study proposes a new technique for cleaning asphaltenes with an ultrasonic device.

The principle behind this new technique is ultrasonic irradiation. The ultrasonic treatment has been proven to reduce viscosity to facilitate flow in pipe lines. The preliminary experiments conducted by the author show that the same ultrasonic technique can be used to clean asphaltene clogged sections of oil wells and reservoirs. Little has been reported, however, on the optimization of amplitude, power generation and time required for cleaning and viscosity reduction, and the effect of ultrasonication on adsorption properties of the reservoir rocks and fluids.

Ultrasonic generators can be used in oil wells and pipe-lines that have asphaltene deposition problems. Emphasis should be placed on using the technique with simultaneous pumping operations. The same device can be used in countering asphaltene based fouling mechanisms and processes generally used in asphaltene problems. These devices can be used for a variety of applications of ultrasonics, i.e., enhancement of flow in clayey sands to facilitate soil and ground water remediation, and increased ground water production.

Asphaltenes are complex organic materials that are considered to be arranged in stacked, multi-ring structures. They contain nitrogen, oxygen and sulfur atoms in addition to carbon and hydrogen atoms within the repeating unit[1-4]. Asphaltenes have a wide range of structures and also vary in physical properties from reservoir to reservoir. The actual structure of asphaltenes depends on the source of the oil as well as asphaltic materials.

Asphaltenes: Fundamentals and Applications, Edited by E. Y. Sheu and O. C. Mullins
Plenum Press, New York, 1995

Asphaltenes are not truly soluble in most crude oils. They can be characterized as multipolymers, often referred to as polycondensates with a great variety of repeating blocks[5]. These particulates are maintained in suspension by materials called resins. These smaller suspending molecules are soluble and act in what has been described as "micellar type" arrangements in order to keep the asphaltenes in suspension. When stabilizing forces are altered, the asphaltenes coalesce under certain pressure, temperature, or compositional conditions. This leads to the formation of larger groups, called flocs, that precipitate and have been reported to have a density of 1.2 g/ml[6]. Asphaltene content of crude oils can vary from 0 to 60%[7]. On the other hand, resin-to asphaltene volume ratios vary from 1:1 to 20:1 in oils that are stable and to less than 1:1 in oils that are characterized by rapid precipitation of asphaltenes[8]. Even though it is generally understood that asphaltene content of a crude oil increases with decreasing API gravity, asphaltene precipitation has been reported in light oils as well.

Asphalt, Asphaltenes and Resins

Asphalt is defined by ASTM as a dark to black cementious material in which the predominating constituents are bituminous materials that occur in nature or are obtained in petroleum processing. Although asphaltene contents vary in each asphalt, there is no asphalt which does not contain aphaltene. Also, the amount of asphaltene present in the system controls the physical properties of asphalts. Asphaltenes occur in native petroleum and are enriched in asphalts when the light fractions of petroleum are removed during the refining process. Asphalt is a colloid composed of several fractions, the major ones being asphaltenes. Because the asphalt system is not a true solution, it can be fractionated into maltenes (saturates, aromatics and resins), and asphaltenes by solvent fraction method[9]. The colloidal make-up of the asphalt depends on the chemical nature and percentage of these fractions and their relationship to each other. Asphaltenes are the dispersed phase in the asphalt whereas the maltenes are the continuous phase. The asphaltenes are thought to furnish hardness while the maltenes are believed to provide the adhesive and ductile properties of asphalt[10]. The maltenes present have an influence on the viscosity or flow properties of the asphalt. Similarly, asphaltenes also impact the viscosity of crude oil[11].

In crude oil, asphaltene micelles are present as discrete or dispersed particles in the oily phase. Although the asphaltenes themselves are insoluble in gas-oil (saturates and aromatics), they can exist as fine or coarse dispersions, depending on the resin content. The resins are part of the oily medium but have a polarity higher than gas-oil. This property enables the molecules to be easily adsorbed onto the asphaltene micelles and to act as a peptizing agent of the colloid stabilizer by charge neutralization[12,13]. An aggregate of asphaltene particles with adsorbed resins form a super micelle, and oil can be accumulated in the micelle as intercellular medium. The super micelles aggregate further to form giant super micelle that can grow gradually to form a liquid crystal. Lian et al.[14] in their research concluded that the different compositions of asphaltenes in asphalts have different miscibility in solvents because asphaltenes do not dissolve well in solvents. They also showed resins to be the best peptizing agent in the asphalt colloidal system due to their molecular weight and high aromatic portion and hydroxyl group.

Asphaltenes and resins are heterocompounds and form the most polar fraction of crude oil. The stronger tendency of resins to associate with asphaltenes reduces the aggregation of asphaltenes[15] that in turn determines, to a large extent, their dispersion in crude oil[16]. The most common model for asphaltene/resin interaction is the colloidal model[17]. Asphaltene micelles are assumed to be kept in solution by a layer of resins.

Asphaltenes are a constituent of the heavier fractions of crude oils. The remaining species are resins and aromatics. Structural studies indicate that the asphaltene and resin molecules have a similar molecular weight range of 500 to 1,000. Considering the stabilized nature of asphaltenes as colloidal particles peptized by resins, any actions of chemical, electrical, or mechanical nature that depeptize these particles will lead to flocculation and precipitation.

Structurally, asphaltenes contain flat sheets of condensed aromatic systems that may be interconnected by sulfide, ether, aliphatic chains or naphthenic ring linkages. Gaps and holes appear as defect centers in the aromatic systems with heterocyclic atoms coordinated to transition metals such as vanadium and nickel, most likely caused by free radicals[13]. Due to their complex structure and large size, asphaltene particles are characterized as colloids, determined by X-ray diffraction and other methods[19-21].

ASPHALTENE PRECIPITATION

Gas injection, re-injection, and miscible flooding of oil reservoirs are considered to be economically viable techniques. Unfortunately, these technique can also initiate asphaltene precipitation[22]. For instance, the role of CO_2 in destabilizing asphaltene-crude oil equillibria is well documented[1,23-25]. Carbon dioxide lowers the pH of the system and, therefore, de-stabilizes asphaltene-crude oil equillibria. Consequently, some degree of asphaltene precipitation is always present in a CO_2 flood operation[12,25]. Similarly, flooding of a rich gas during miscible flooding de-stabilizes the asphaltene-crude oil mixture by lowering the carbon-to-hydrogen ratio in the solution[26]. Usually stripping gas from the oil has been found to improve the solubility of asphaltenes[27]. The straight chain hydrocarbons have less affinity for the asphaltene ring structures than oils that have a higher carbon-to-hydrogen ratio. The negative effect of rich gas is at a maximum near the bubble point[22,27-29]. This effect is alleviated after the bubble point is reached[15]. This peculiar behavior near the bubble point is more pronounced for a rich gas than for carbon dioxide[22]. Islam[12] recently reviewed mechanisms involved in asphaltene precipitation during enhanced oil recovery.

The mechanism of asphaltene precipitation is very complex and despite the presence of a wealth of research on the topic, controversy as to the nature of asphaltene solution in crude oil persists. There are two differing schools of thought as to the nature of asphaltenes in solutions. The first one considers asphaltenes to be dissolved in oil in a true liquid state. In this case, asphaltene precipitation is considered to depend on thermodynamic conditions of temperature, pressure and composition. This particular school of thought recognizes asphaltene precipitation as a thermodynamically reversible process[30-31]. The second school of thought considers asphaltenes to be solid particles which are suspended colloidally in the crude oil and are stabilized by large resin molecules. Consequently, the deposition process is considered to be irreversible. This particular approach uses methods used to describe colloids to describe the phenomena of asphaltene precipitation [32,33,34].

Asphaltene precipitation may occur when the equilibrium of the stabilizing forces has been perturbed. The most important compositional change occurs during miscible production. However, pressure changes even during primary depletion can cause asphaltene precipitation. Also, during acid stimulation or cyclic CO_2 injection, similar destabilizing may occur. Following are the major de-stabilizing forces which may cause asphaltene precipitation [12,35]: a) CO_2 injection, b) miscible flooding, c) pH shift, d) mixing of crude streams, e) incomplete organic chemicals (e.g. isopropyl alcohol, methyl alcohol, acetone, and some other solvents), f) well stimulation, such as acidizing, surfactant treatment, wellbore heating; g) shear, h) neutralization of particles due to "streaming potential", i) temperature drop, and j) charged, bare metal surfaces.

Dickakian and Seay[36] have shown that asphaltene precipitation and subsequent carbonization are the major mechanisms behind the crude oil heat exchanger fouling. Such fouling in crude oil heat exchangers is caused by precipitation of asphaltenes also costs refiners in many ways. The effect of asphaltenes on thermal fouling test can be seen readily

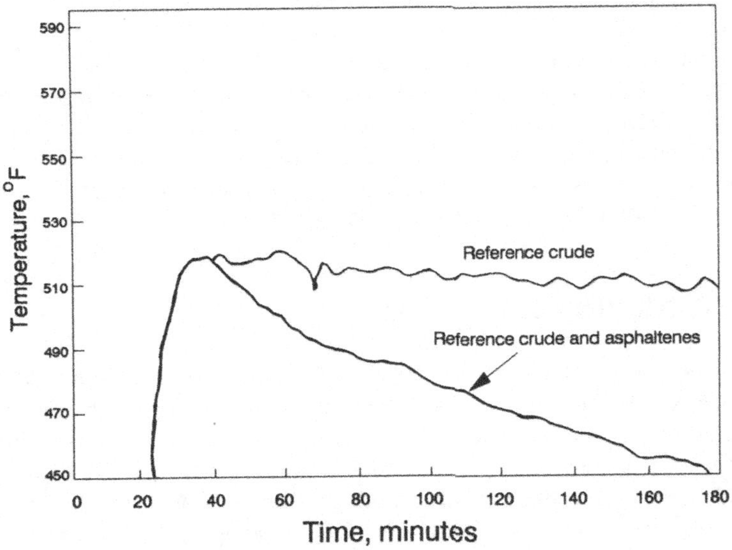

Figure 1. Effect of asphaltenes on thermal fouling of a heat exchanger (from Ref. 36).

by comparing characteristics of a typical crude oil with those of a crude oil containing asphaltenes. Figure 1 shows temperature variation in a reference crude and in the same crude with 1.3% asphaltenes. Analyses by Dickakian and Seay[36] showed initially asphaltenes precipitated and adhered to the metal surface and were then carbonized on the surface into infusible coke. The time sequence of this transition is shown in Fig. 2. The following reasons were identified as the primary causes of heat exchanger fouling:

- incompatibility leading to asphaltene precipitation
- adherence of precipitated asphaltenes to the hot metal surface
- carbonization of asphaltenes into hot infusible coke.

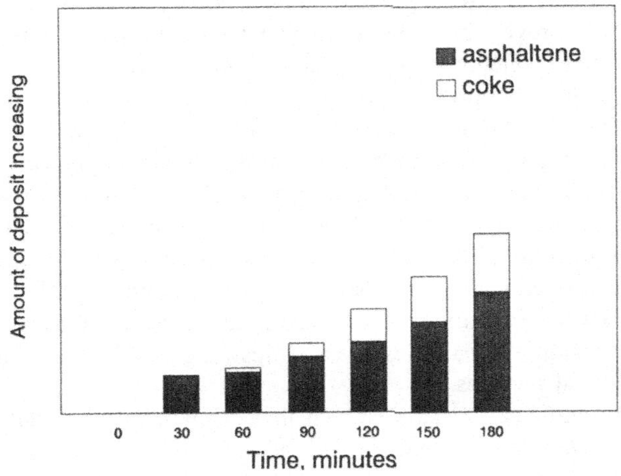

Figure 2. Composition of fouling deposits as a function of time (from Ref. 36).

In transportation of heavy crude, the problem is mainly to decrease the overall viscosity. The presence of asphaltene makes the viscosity of the crude oil so high that the transportation becomes a real problem. However, asphaltene precipitation during transportation is not considered to be a problem if the asphaltene contents are removed prior to transportation of the crude oil.

Field Conditions for Asphaltene Precipitation

In 1965, Haskett and Tartera[37] used the argument that since the addition of low-molecular weight paraffin hydrocarbons causes the precipitation of asphaltenes, associated resins, and naphthenes, it could be expected that the release of these components, when the pressure of the crude oil in the tubing falls below the bubble point, reduces the tendency for asphaltenes to become depeptized. Consequently, the precipitation of asphaltenes would occur generally in the single-phase crude oil. This single-phase crude oil, which undergoes no compositional change, is subject to changes in pressure and temperature, turbulence and possible electrical effects near the well-bore. As one of the first field examples of asphaltene deposition during primary depletion, Haskett and Tartera[37] presented results observed in the giant oil field of Hassi Messaoud, Algeria. This reservoir is situated at a depth of 3610 m and the initial reservoir pressure was 471 atm. The bubble point for this 42° API crude varied from 147 to 199 atm. The gas-oil ratio varied from 183-248 std m^3/m^3. It was observed that:

- the asphaltene deposits became less severe with increasing depth. This was true regardless of the depth at which the deposit was found. This indicated that the most favorable condition for deposition was just before the bubble point was reached. This corresponded to the lowest pressure, the lowest temperature and the longest flowing time for the given tubing diameter, flow rate, and pressure.
- the deposits of asphaltene were re-peptized when the well-head pressure was once again reduced. Otherwise, asphaltene deposits were extended all the way to the surface. Locations of these deposits correlated to positions where the bubble points were reached in the tubing.
- an extension was found even when the only change invoked was closing the well for measuring the asphalt profile. This was attributed to the re-adsorption of the liberated gas as the oil level rose in the tubing. This created an excess of hydrocarbons in the oil phase, which caused the immediate precipitation of asphaltenes and associated elements.
- one well, which produced at a well-head pressure considerably above the bubble point, showed asphaltene deposition up to the surface, except on the downstream side of the choke where the pressure was below the bubble point.

From the above observation, it was recommended that the Hassi Messaoud wells be put on production at low well head pressure so that two-phase flow would occur sufficiently deep in the tubing so as to minimize asphaltene deposition. Mechanical choke deep into the tubing was tried for a few cases with success.

Asphaltene problems in some Venezuelan medium oil reservoirs were reported by Lichaa[33]. He presented asphaltene problems in the Mata-Acama field of Venezuela. This sandstone reservoir is located at a depth of 3773 m and at a temperature of 135° C. The crude contained 25% of light fractions while the rest was C_7+. Even though this crude contained only 0.4 to 9.8% asphaltene, the wells showed considerable asphaltene deposition problems during natural depletion or acid stimulation. Interestingly, no asphaltene precipitation was observed for a nearby Boscan heavy oil reservoir, even though this crude contained an asphaltene concentration higher than 17.2 percent. This formation is at a depth of 2789 m and a temperature of 82° C. This Boscan crude was found to prevent electro-deposition of asphaltene from the Mata-Acema crude. Lichaa[33] reported the use of Boscan crude as a stabilizing agent for emulsions injected as displacing fluids in enhanced oil recovery processes. Laboratory investigations on the crude of these fields revealed the following:

- the resins which serve as peptizing agents for the asphaltenes are easily oxidized. This process increases the crude viscosity and provides favorable conditions for precipitation of the depeptized asphaltenes.
- asphaltenes are found to migrate and deposit at the cathode upon application of an electric field. Consequently, the particles migrate back and deposit at the other electrode, if the polarity of electrodes is reversed.
- effective inhibition of asphaltene deposits in the formation appeared to be possible by injecting sufficient amounts of peptizing agents to the formation.
- electrical effects play an important role in asphaltene deposition. Controlling the electro-deposition of asphaltene particles could lead to the development of a preventive technique for asphaltene deposition.

According to Lichaa[33], a crude oil with high asphaltene concentration may not necessarily show deposition problems due to the lack of peptizing agents. Inasmuch as asphaltene stability is very sensitive to the dissolution of condensate in the reservoir crude, asphaltene precipitation may occur near the well-bore during retrograde condensation. Once again, such a problem can be reduced by constraining the producing gas oil ratio.

TRANSPORTATION OF ASPHALTIC CRUDE OIL

The most convenient and usually the most economical method for transporting crude oil over land is through pipelines. The power required to transport crude oil by the method is proportional to the viscosity of the crude oil because the flow is considered to be laminar. The pressure gradients with heavy asphaltic crude oil being very high, the major obstacle to increasing bitumen production is considered to be the difficulty of moving the viscous bitumen to the market. The currently used method involves the use of natural gas condensate to reduce bitumen viscosity and density for pipeline transportation. However, as bitumen production outgrows condensate supplies, new techniques are needed for decreasing crude oil viscosity. To eliminate problems in pipe line transportation of crude oil, viscosity reduction has been the main option. The oil-in-water emulsion technology has been one notable development in this regard. In this process, the extra heavy crude or bitumen is suspended in water in the form of micro spheres, leading to a reduction in viscosity. Several researchers have addressed this process[38-41]. The impact of bitumen content on viscosity is shown in Figure 3. It can be seen that the viscosity increases exponentially with increasing hydrocarbon content. This graph shows the basis for selection of commercial emulsions. The type of surfactant used during emulsification has a profound impact on diameter of the dispersed phase that, in turn, dictates the emulsion viscosity value. Figure 4 shows the effect of emulsion droplet size on viscosity at different shears. The emulsion viscosity increases rapidly with decreasing mean droplet diameter with an increasing tendency to pseudoplasticity. More recently, Nunez and Rivas[42] reported further developments of the emulsion transport technology. They used a 1.2 km long pipeloop to evaluate some of the emulsion properties in a full scale. They found that at nominal bitumen-to-water ratio, the fluid mainly presents a pseudoplastic behavior at the time of its manufacturing. However, shear degradation takes place along the length of the pipeline and the fluid gradually reaches Newtonian asymptote. This behavior was attributed to the formation of particle aggregates that immobilized fractions of the continuous phase. These structures could be eliminated only with

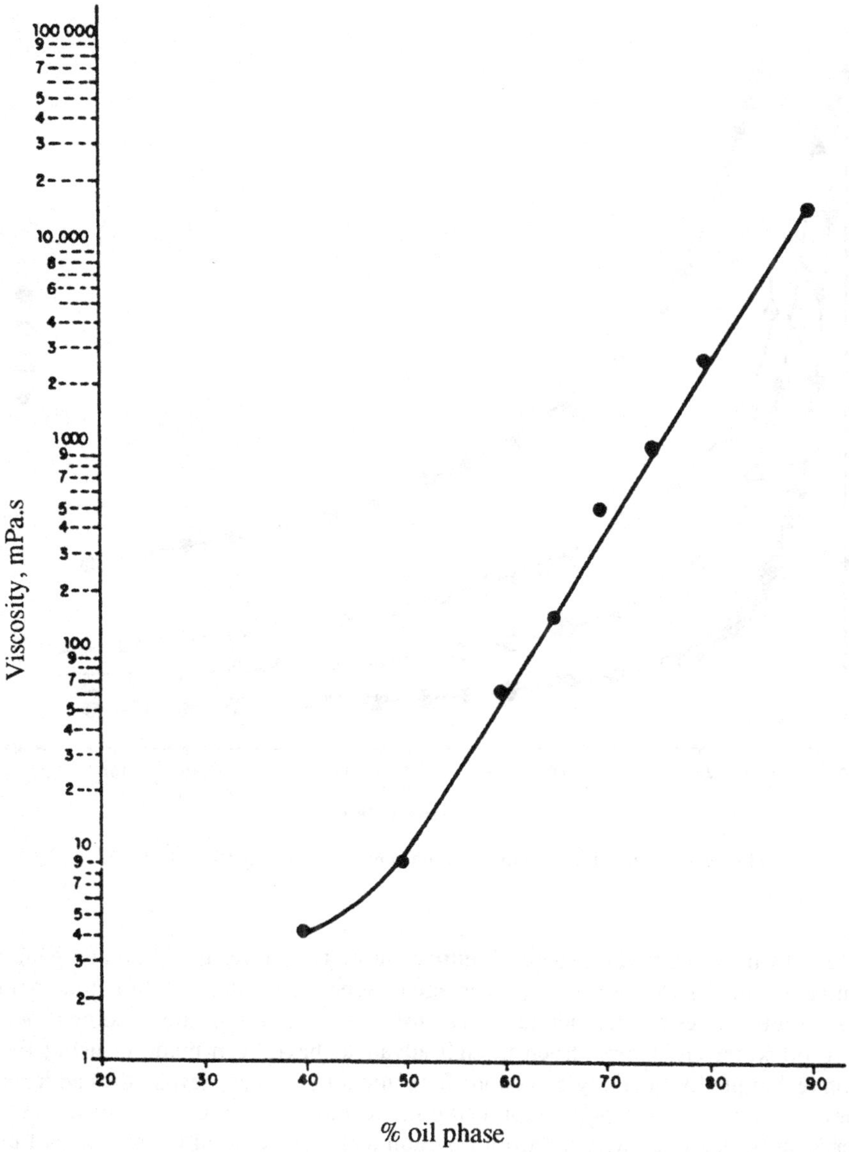

Figure 3. Viscosity of an emulsion (oil is the dispersed phase) at 25°C (from Ref. 40).

additional shear to the emulsion. Based on this study, they[42] recommended several modifications to the technology. However, the prospect of ultrasonic treatment was not studied, even though apparently the use of ultrasound can provide additional shear to the system in order to control the rise in viscosity due to aggregation in the asphaltene phase. Some of the latest developments in this area appear to be in developing core annular flow[43].

The emulsion transportation process requires a two-stage emulsification process at the upstream end and a de-emulsification process at the downstream end of the pipeline. The de-emulsification process is as challenging, as temperature, water pH, water salinity, and surfactant concentration play an important role. The bitumen recovered from the de-emulsification process essentially remains as viscous as the original bitumen and needs to be

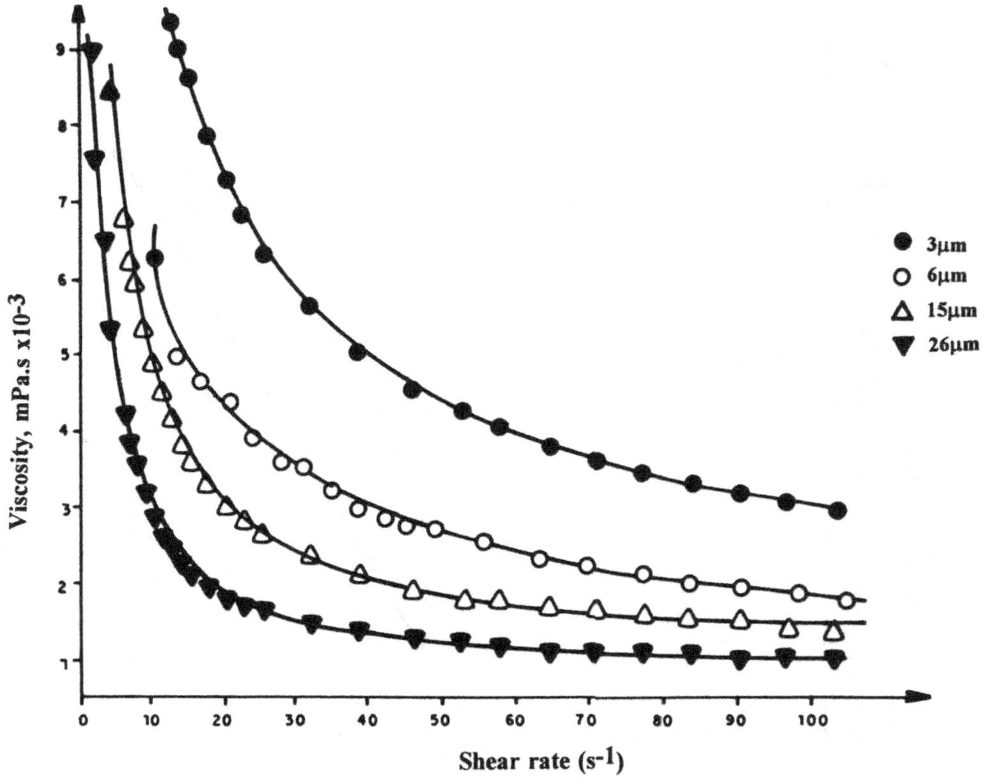

Figure 4. Effect of droplet diameter on emulsion viscosity (from Ref. 40).

subjected to various upgrading stages. Figure 5 shows typical results of breaking studies of an emulsion. This figure shows the percentage of separated water as a function of time for various salinity values and temperatures. The use of froth flotation shows some promises in breaking oil/water emulsions. Even though advances have been made in using the froth flotation technique for breaking emulsions for some time[44], very few studies addressed the problem of heavy oil-water separation using the technique[45]. Figure 6 shows that a high recovery can be achieved with the froth flotation technique only at the expense of product quality.

UPGRADING ASPHALTIC CRUDE OIL

One traditional viscosity reduction process involving a low severity thermal cracking, known as visbreaking, is superior to oil-in-water transportation in the sense that it not only reduces viscosity but also upgrades the bitumen to some extent[46]. This process reduces the upgrading efforts required in a downstream refinery. However, this process poses a coke formation problem that results from thermal cracking of high viscosity crudes.

The Jet Cracking process that operates under low severity conditions to reduce coke formation has been under development[47]. In this process the bitumen is subjected to high mechanical shear conditions at very short residence times while minimizing coke formation. In essence, bitumen, preheated to about 400°C is passed through a nozzle at very high

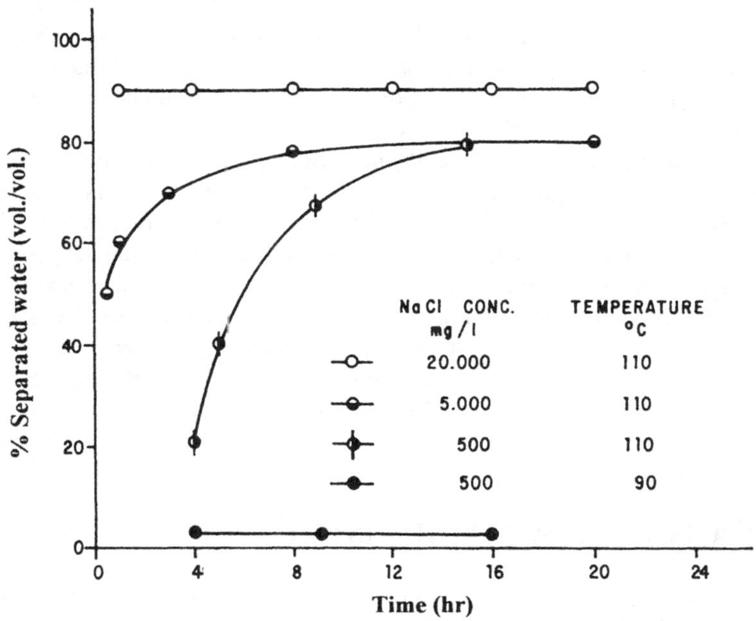

Figure 5. Separation of emulsions stabilized by nonionic surfactants (from Ref. 40).

velocities. As a result of the high shear experienced by the bitumen during its passage through the nozzle, very fine droplets (20-30 microns average diameter) are formed. As the bitumen droplets enter the reactor chamber, additional heat is applied through heating elements placed around the reactor wall. The heaters are controlled in such a way so as to maintain isothermal conditions inside the reactor. At the exit of the reactor, the effluents are cooled down and

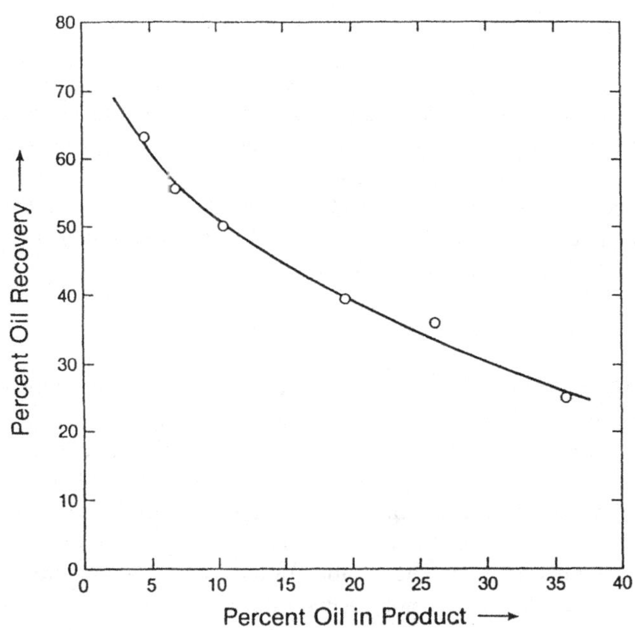

Figure 6. De-emulsification through froth flotation (from Ref. 45).

separated into gas and liquid fractions in a separator. The main idea behind this process is that when subjected to high mechanical shear conditions, asphaltene aggregates will break up thus resulting in viscosity reduction. Islam and Chakma[48] showed that the jet cracking process has the ability to decrease viscosity of a crude (with an asphaltene content of 15.3%) from an initial value of 70,000 mPa.s down to 200 mPa.s.

Chakma and Berruti[49] found that the ultrasonic treatment provides the shear needed to breakup the asphaltene aggregates. They found the viscosity of the mixture to be a strong function of asphaltene content, and the asphaltenes with higher molecular weight resulted in higher viscosity of the mixture than those with lower molecular weights.

Asphaltene deposition in the well tubing and pipe lines cause a number of problems in wireline operations and production facilities. These problems lead to extensive delays in wireline operations and changes to running procedures are required to overcome these problems. Extensive studies were carried out to study the efficiency of asphaltene dissolvers to eliminate the time-consuming physical removal of deposited asphaltenes from a production system.

INTRODUCTION TO SONOCHEMISTRY

A series of alternate compressions and rarefactions is associated with any sound wave propagating through an elastic medium. This creates an acoustic pressure in the medium, P_a, which is related to time, t, according to the following relationship[50]:

$$P_a = P_{\max} \sin 2\pi f t \tag{1}$$

where f is the frequency and $P_{\max}$ is the maximum pressure amplitude. Acoustic intensity, then, is given by

$$I = \frac{(P_{\max})^2}{2\rho c} \tag{2}$$

where ρ is the density of the fluid and c is the speed of sound in that fluid. This definition of acoustic intensity, however, has no practical use because the intensity is measured in terms of the energy fed into the reacting system. It is customary to either monitor a standard chemical reaction or monitor the temperature rise in a known volume of water during sonication compared to electrical heating of the same. It is well-known that the intensity of the ultrasound will depend on the distance, d, traveled in a medium. This attenuation due to viscous forces of the medium can be represented as:

$$I = I_o \exp(-2\alpha d) \tag{3}$$

where α is the absorption coefficient of the medium and I_o is the intensity at the point of origin. This relationship, along with the fact that (α / f^2) for a medium is constant for a given temperature, indicates that the attenuation is larger at higher frequencies. In this process of attenuation of intensity, sonication causes enhanced movement of molecules and can be responsible for increased molecular mixing leading to enhanced chemical reactions. On the mechanical side, the same effect can cause disruption of the liquid/liquid phase interfaces leading to intense emulsification. The combination of the chemical and mechanical effects is, however, more complex and is not yet well defined. In general, sonochemical effects become increasingly intense as the intensity of sonication is increased. However, recent investigations show that many reactions have an optimum intensity above which the

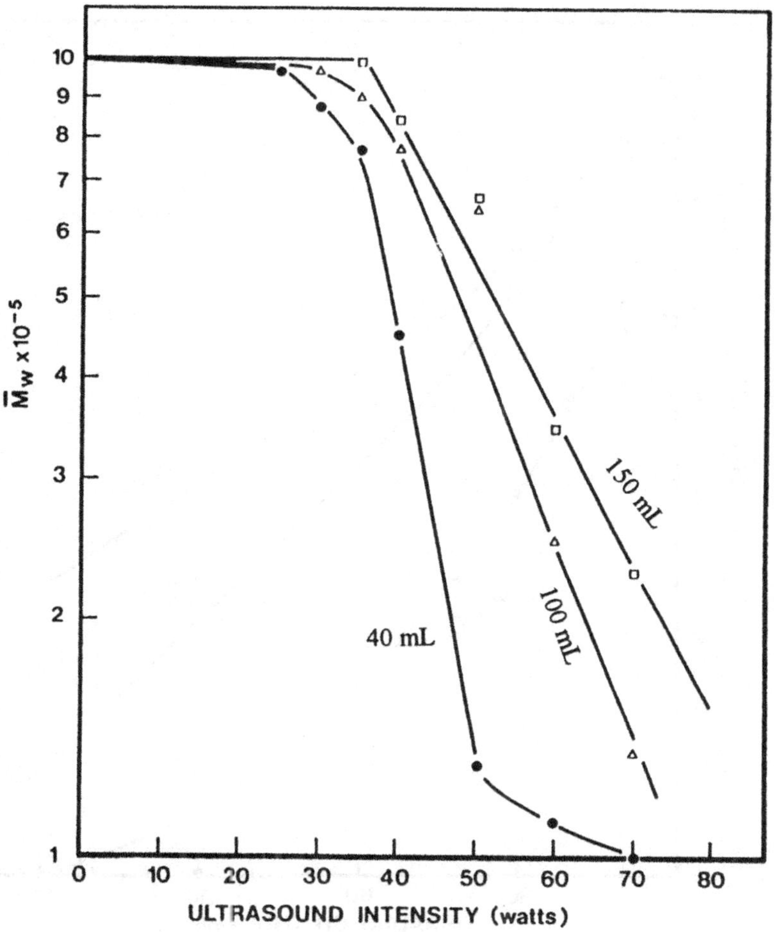

Figure 7. Effect of ultrasound intensity on the degradation of 0.5% hydroxypropyl cellulose solution (from Ref. 54).

efficiency fails[51]. More importantly, there have been several recent studies that reported that the products of a reaction can depend on the intensity and/or source of ultrasound used[52,53]. Ultrasonic intensity plays an important role in degradation of high molecular weight solutions. Even though, no such study is available for asphaltenes, such study has been carried out for other long-chain molecules. One such example is given in Figure 7[54]. In this study, Malhotra observed that after the onset of cavitation the decrease in molecular weight as a function of ultrasound intensity was also volume dependent. This behavior is expected to persist for all long-chain molecules affected by ultrasonic irradiation. As further evidence of ultrasonic degradation of long-chain molecules, Malhotra[54] presented Figure 8. It is clear from these data that the degradation of HPC (hydroxypropyl cellulose) molecules is volume dependent. However, because other governing factors, such as the ratio of the probe diameter to reactor diameter and the immersion depth of the probe in solution were also varied among different cases, this result is not of quantitative importance.

In the sonication process, ultrasound frequency plays an important role. Most previous works involved in sonochemistry have used frequencies of around 20-50 kHz, dictated by the allowed frequency range used by commercial equipment. Recent studies,

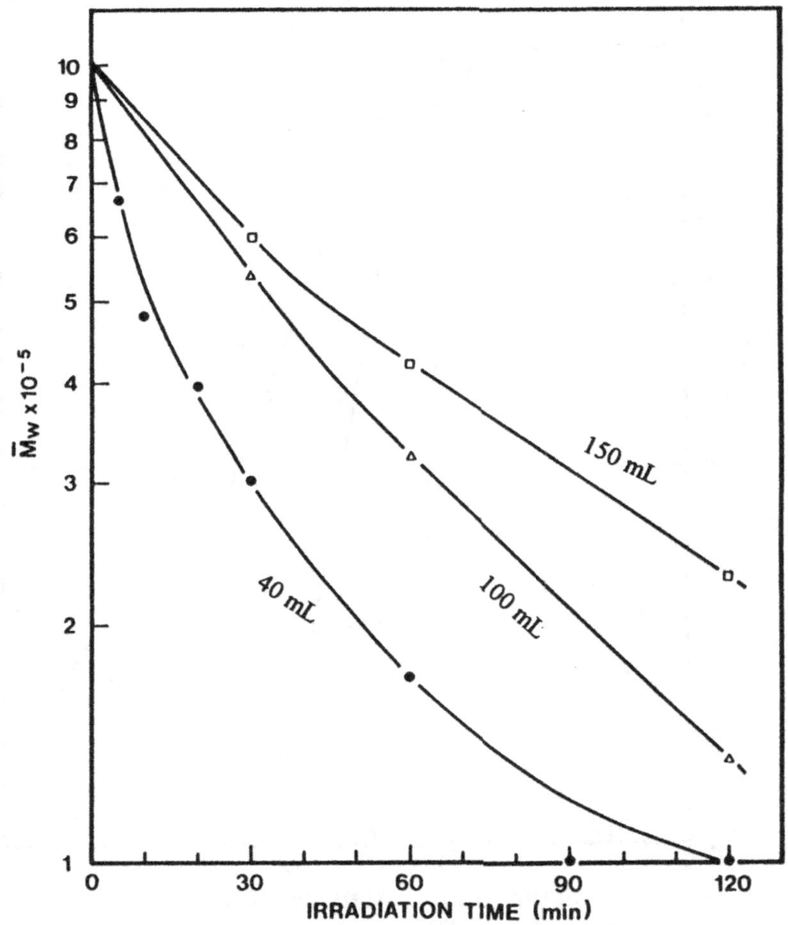

Figure 8. Effect of molecular weight as a function of irradiation time for the 70-W ultrasonic degradation of 0.5% hydroxypropyl cellulose solution (from Ref. 54).

however, have indicated that sonochemical effects are lessened at higher frequencies(50). This can be explained by the fact that bubbles have less time to grow and collapse at high frequencies. This should lead to lower final temperatures and pressures due to sonication.

ULTRASONIC TREATMENT IN HEAVY OIL UPGRADING

Conventional refining processes use thermal cracking or chemical conversion (at high temperature) for breaking asphaltenes into other products (hydrogenation, solvent refining, degradation by acid, etc.)(55). Typically, this is carried out under high temperature conditions. The rate of reaction is often increased by using high pressure in conjunction with high temperatures. This upgrading process can greatly benefit from an ultrasonic process which can be used at ambient temperature and pressure conditions.

In an ultrasound process, a sound wave passing through a viscous liquid such as water creates a vibration pattern that sets the liquid in motion. The ultrasonic vibration patterns form water molecule layers that stretch, compress, bend, and relax. Interacting layers generate tiny vacuum spaces called cavitations within the liquid. Imploding cavitations scrub part surfaces and pull away foreign matter. In commercial or laboratory ultrasonic processors ultrasonic sound waves are modulated by a sinusoidal frequency using a preset analog wave form. However, going to digital wave forms permits adjusting the cavitation response to the task. Power burst duration determines cleaning intensity. Regulating the time between power bursts allows sufficient trapped gas release.

Several studies have been reported on the use of ultrasound in upgrading heavy oil and bitumen[55,56]. Lee et al.[56] demonstrated that the ultrasound technology can be effectively used in converting the heavy fractions of the crude oil and asphaltenes into oil and resin fractions in an alkaline aqueous solution under ambient temperature and pressure conditions. They identified thermal scission of bonds of the heavy oil fractions and simultaneous formation of radicals from the water molecules as the main reason for conversion. These two phenomena were reported to cause cracking of heavy fractions and act as a source of hydrogen in hydrogenation of the crude oil, respectively. Due to ultrasonic irradiation following events were suggested to occur

- inversion of reversed micelles;
- creation of self-induced surfactant molecules;
- exchange between cavitation process and amphiphiles;
- transport and migration of radicals in micelles and intermicellar medium.

It is well known that ultrasonic radiation generates very high temperature and pressure at localized centers in an aqueous solution. However, these microscopic "hot spots" do not necessarily heat up the total volume involved in the process. This makes the ultrasonic process an excellent process in terms of energy consumption for extraction of heavy crude and bitumen.

Three phenomena are invoked by ultrasonic irradiation. They are: cavitation, rapid variations in sonic pressures leading to alternating compression and rarefaction in the medium, and microstreaming in which a large amount of vibrational energy is stored in a small amount without heating up the overall system. Cavitation, is by far the most widely understood phenomenon among the three cited above. Microbubbles are formed during cavitation leading to chemical changes in the medium. Centers of these microbubbles exhibit very large pressure and temperature changes (as high as 5100 atm and 7200°C)[57]. This very property of ultrasonic irradiation makes the ultrasound process very desirable in heavy oil upgrading, because it makes it possible to break down large molecules without raising the temperature of the overall system. This concept of ultrasonic upgrading was tested by Lee et al.[56] who postulated the following advantages of ultrasonic irradiation:

- ultrasound can greatly enhance interfacial reactions between acid-bearing functional groups in the heavy oil and alkaline reagents in the aqueous phase generating surface active agents;
- an emulsion can be formed in the presence of ultrasound leading to increased interfacial area, hence, accelerating chemical reactions;
- free radical reactions and polymerization reactions can occur in the emulsion phase, the microenvironment of which can enhance the reaction rate.

While the works of Yen and associates[55,56] demonstrate that ultrasonic treatment can be used as an attractive upgrading method, data presented have been preliminary in nature. Much of the work has been exploratory and needs a great deal of research before optimizing relevant parameters, such as ultrasound frequency, kinetics of chemical additives, energy consumption, and most importantly economic considerations.

Chakma and Berruti[49] studied the effect of ultrasound on the viscosity of Athabasca bitumen and found that the viscosity reduction of up to 15% can be achieved by the application of ultrasound. They concluded that ultrasound causes the asphaltene aggregates to break up to some extent, resulting in viscosity reduction. Temperatures were monitored during the ultrasound treatment, but no correlation was found between viscosity reduction and temperature increase. For instance, as the ultrasound treatment was continued for a long time, the temperature of the product was the highest, but the viscosity started to increase. Even though temperature values were not reported, it can be inferred that the main reason for viscosity reduction was the change in structure of asphaltene molecules.

Figure 9 shows the percent viscosity reduction as a function of treatment time. This figure shows that the ultrasonic treatment indeed results in viscosity reduction. This figure also indicates that the there is an optimal time of ultrasonic treatment, beyond which the viscosity of bitumen actually tends to increase slightly. Interestingly, the reduction of viscosity was less intense when a mixture of toluene and bitumen was used. Chakma and Berruti attributed this behavior to the solubility of asphaltene molecules, leading to less resistance to flow as compared to that of asphaltenes in pure bitumen. A significant reduction in viscosity was, however, observed when an aqueous NaOH solution was used as the solvent. Figure 10 shows viscosity variations of a 50% bitumen and 50% caustic mixture after a treatment of 60 minutes as a function of caustic concentration. Note that the viscosity reduction is drastic as compared to the pure bitumen case. This reduction was attributed to the ability of caustic to form stable emulsions. As the bitumen/solvent ratio is increased to 2, the reduction in viscosity is more intense. This is reflected in Figure 11. Note that the viscosity is decreased below 10 mPa.s. even with the addition of a small amount of NaOH to the mixture. A very fine and stable emulsion formation was reported for this case. This emulsion can account for a such a reduced viscosity. Chakma and Berruti[49] reported the existence of an optimal solvent/bitumen ratio for different NaOH concentration values. Figure 12 shows this trend. The optimal concentration value did not appear to vary for different concentration values.

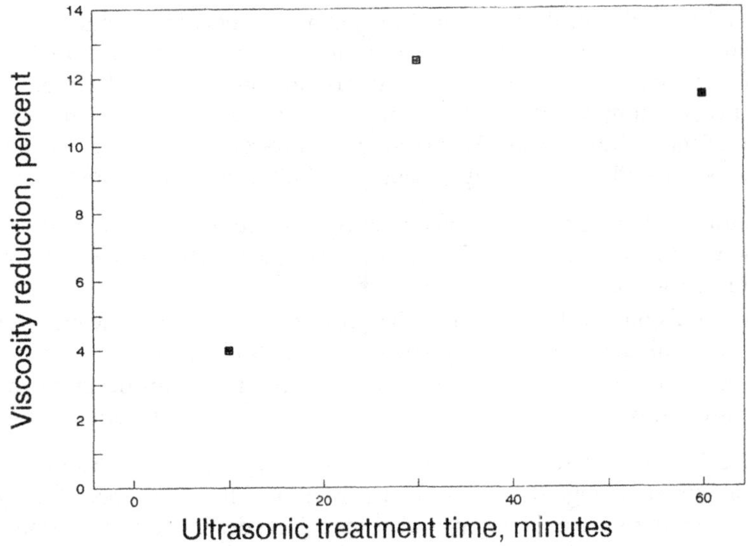

Figure 9. Effect of ultrasonic irradiation on Athabasca bitumen (from Ref. 49).

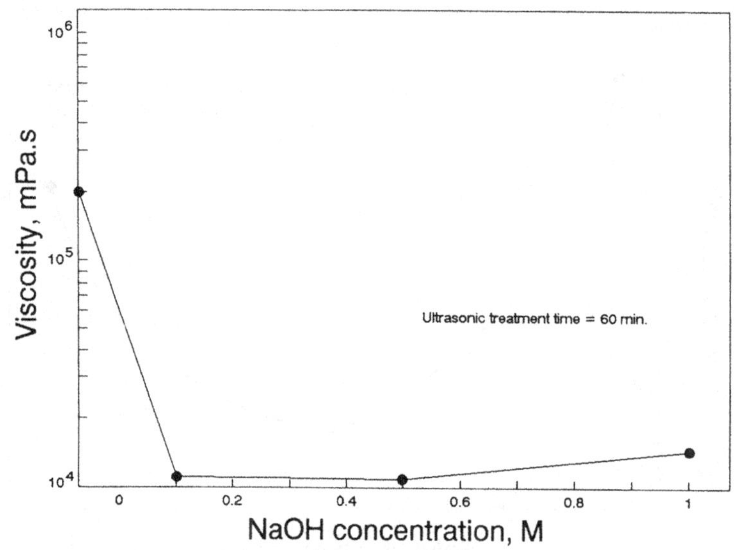

Figure 10. Viscosity of 50% bitumen/50% NaOH solution mixture exposed to ultrasonic irradiation (from Ref. 49).

Recently, Smordal and Islam[58] demonstrated that the combination of ultrasonic treatment with jet reactor can be attractive. It appears that by using ultrasound, shear can be provided to create microbubbles in the feedstock leading to enhanced mixing which would limit the pressure drop across the nozzle to a lower value. Conventional jet reactors often need very high pressure drop in order to ensure effective upgrading with a jet reactor[56]. No experimental work has been presented in support of this numerical observation of Smordal and Islam[58].

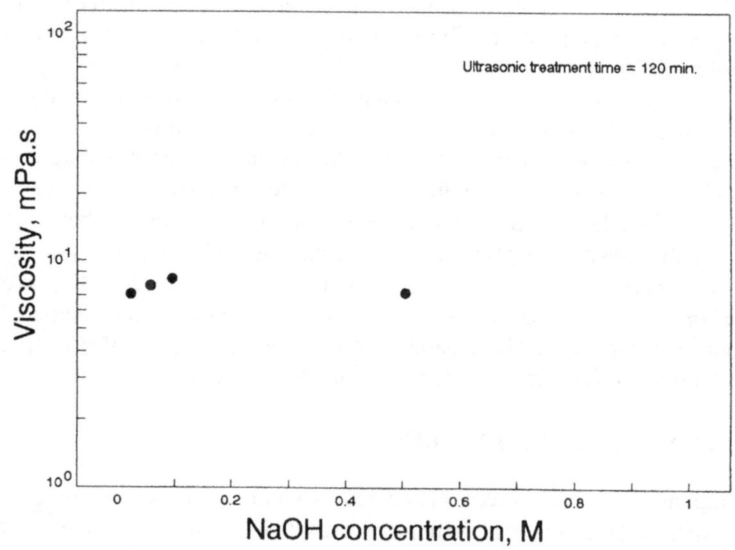

Figure 11. Viscosity of 66.7% bitumen/33.3% NaOH solution mixture (from Ref. 49).

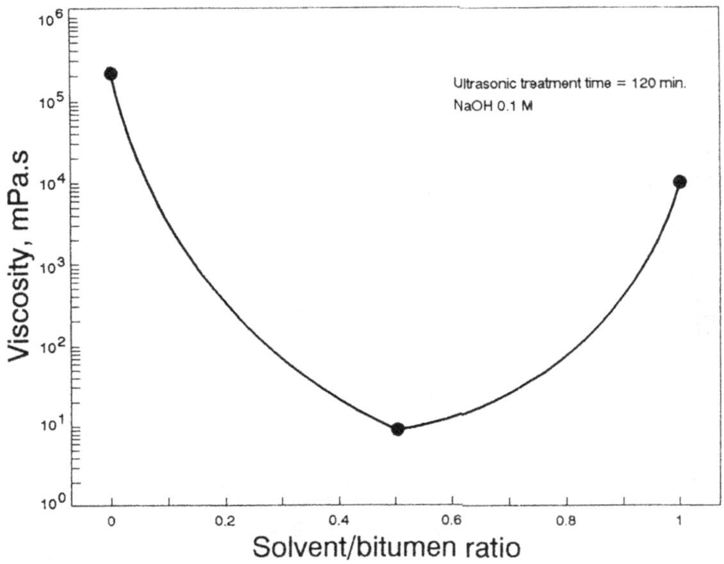

Figure 12. Viscosity of bitumen/NaOH (0.1M) mixtures (from Ref. 49).

WELLBORE CLEANUP WITH ULTRASOUND

Ultrasonic or ultra sound, which involves vibratory waves above the detection limits of human ears (about 16 kHz or 16,000 cps), is already successful in the pharmaceutical, aerospace, food, medical, and mineral processing industries. Ultrasonic energy has been demonstrated to be successful in reducing sulfur and ash significantly to make ultra fine coal, a marketable product. However, the use of the ultrasonication process in the petroleum industry is new and much more research needs to be done before any commercial application can be accomplished.

Gollapudi et al.[(59)] reported the use of an ultrasonic generator in order to decrease or eliminate asphaltene precipitation. The following is a description of the process developed by Gollapudi et al.[(59)].

A commercial ultrasonic generator at a frequency of 10 kHz with a total power output of 250 watt was used. The High Intensity Ultrasonic Processor transmits the high frequency electrical energy to the piezoelectric transducer within the converter, where it is changed to mechanical vibrations. The vibrations from the converter are intensified by the probe, creating pressure waves in the solution. This action forms millions of microscopic bubbles (cavities) which expand during the negative pressure excursion, and implode violently during the positive excursion. It is this phenomenon, referred to as cavitation, which produces the powerful shearing action at the horn tip, and causes the molecules in the solution to become intensely agitated. The ultrasonic power generator has the capability of providing power at different levels, operating under pulse or continuous operations. However, this study used only the continuous operation.

Preliminary Experimental Results

Preliminary experiments were conducted by Gollapudi et al.[(59)] in order to observe the effectiveness of ultrasonic treatment for removal of asphaltic materials from the wellbore. In their experiments, Gollapudi et al. used foundation coating (containing mainly asphalts) in order to carry out the experiments. In the first set of experiments, the asphalt was irradiated in a container

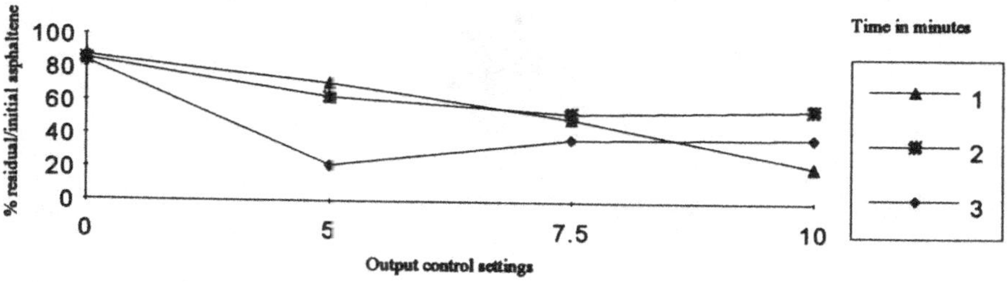

Figure 13. Residual asphaltene after ultrasonic irradiation in water.

using a standard tip of a high intensity ultrasonic processor for one, two, and three minutes, separately. Following this, the residual asphalt was measured. Different runs were conducted with water and kerosene as the medium. A different set of experiments was conducted in a laboratory flow cell containing unconsolidated sands. In these porous-media experiments, single-phase flow was observed with water and kerosene in different settings while asphalt was used as the immobile phase. Flow rates in these experiments were used to correlate the performance of the ultrasonic treatment. The final set of experiments was conducted in flow cells with an inch long asphalt/sand mixture placed at the top of a sand column in a flow cell, and flow rates were measured at atmospheric pressure. The sand column was treated with ultrasound and the flow rates after the treatment were measured and compared. During the ultrasonic treatment the standard tip of the ultrasonicator was placed in such a manner that the sandy tar portion of the sand pack was in good contact. Emphasis was given to prove that ultrasonic treatment will improve flow rates in the affected section of a reservoir. The asphalt was exposed to ultrasonic treatment at different output powers. The selection of the time and output control was on the basis of cleaning performance of the exposed ultrasonic conditions.

The experiments proved that the ultrasonic treatment can be used to clean asphaltene in the media of water or kerosene. However, the cleaning performance is much higher in the presence of kerosene. During the ultrasonic treatment, the sample heated up due to ultrasonic power dissipation. This temperature rise alone, however, would not account for any appreciable change in viscosity of the asphalt. Previous studies indicate that a much higher temperature rise is needed for any measurable viscosity change for bitumen and other asphaltic materials[49].

In the case of an aqueous medium, treatment time had more impact than did the output control setting. Smallest amounts of residuals were left when the asphalt was treated in water at a output control settings of 5 (110 watts) and 7.5 (180 watts) for one and two minutes, respectively (Fig. 13). The three minutes treatment shows a steady decline in the residuals left with increased or decreased output control settings. But the amount of residual left behind as residual is more or less the same as that of one minute treatment at an output control setting of 5. The performance of ultrasonic cleaning of asphalt is shown in Fig. 14.

The ultrasonic treatment in kerosene shows that there is a strong correlation between time of treatment and output control settings. Note that higher the treated time the lower the residual left behind. Output control settings seem to play insignificant roles at larger output power settings (Figs. 15 and 16). The colloidal make-up of the asphalt depends on the chemical nature and percentage of the fractions, maltenes (saturates, aromatics and resins) and their relationship to each other. The asphaltenes are the dispersed phase in the asphalt whereas the maltenes are the continuous phase. The asphaltenes are thought to furnish hardness while the maltenes are believed to provide the adhesive and ductile properties of asphalt[(10)]. The maltenes present have an influence on the viscosity or flow properties of the asphalt. The ultrasonic treatment appears to be disrupting maltenes to facilitate asphaltene cleaning and viscosity reduction.

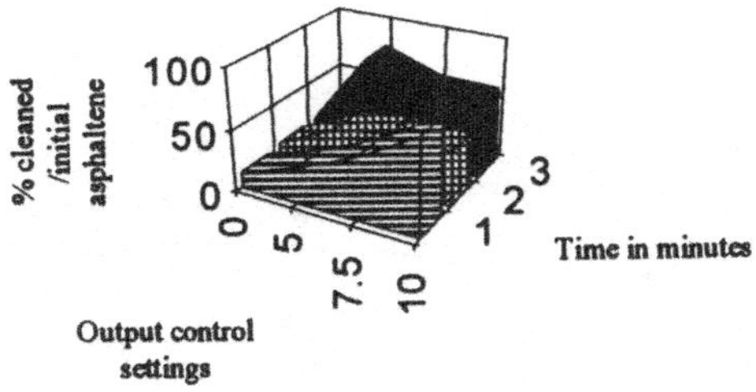

Figure 14. Performance of ultrasonic cleaning in water (from Ref. 59).

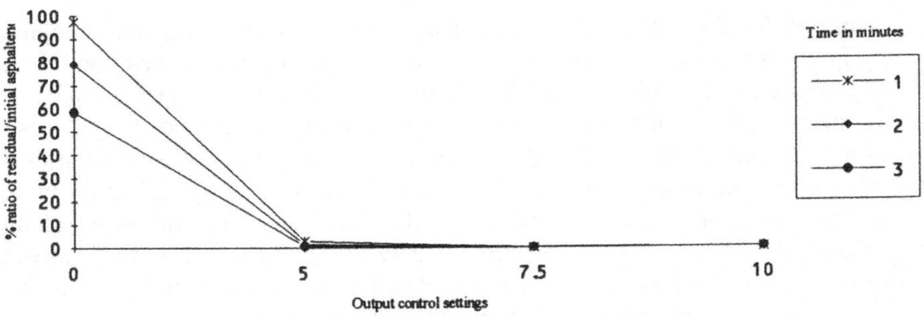

Figure 15. Residual asphaltene after ultrasonic irradiation in kerosene (from Ref. 59).

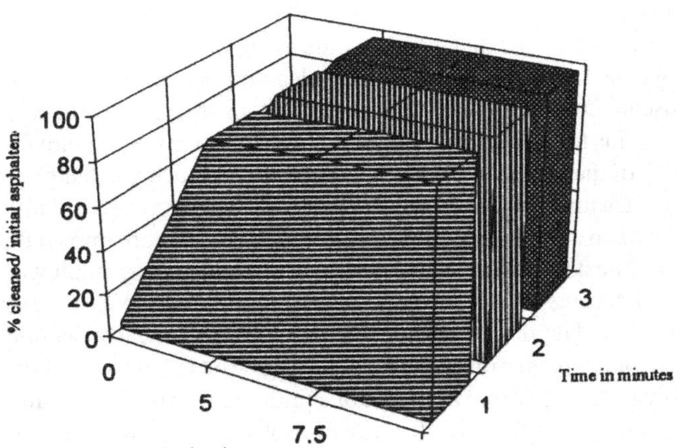

Figure 16. Performance of ultrasonic cleaning in kerosene (from Ref. 59).

Asphaltenes and resins are heterocompounds and form the most polar fraction of crude oil. The strong tendency of resins to associate with asphaltenes reduces the aggregation of asphaltenes[(15)]. The most common model for asphaltene/resin interaction is the colloidal model[(17)]. Asphaltene micelles are assumed to be kept in solution by a layer of resins. Here, it is also anticipated that the difficulties associated with electrodeposition of the asphaltenes are overcome by using the ultrasonic treatment.

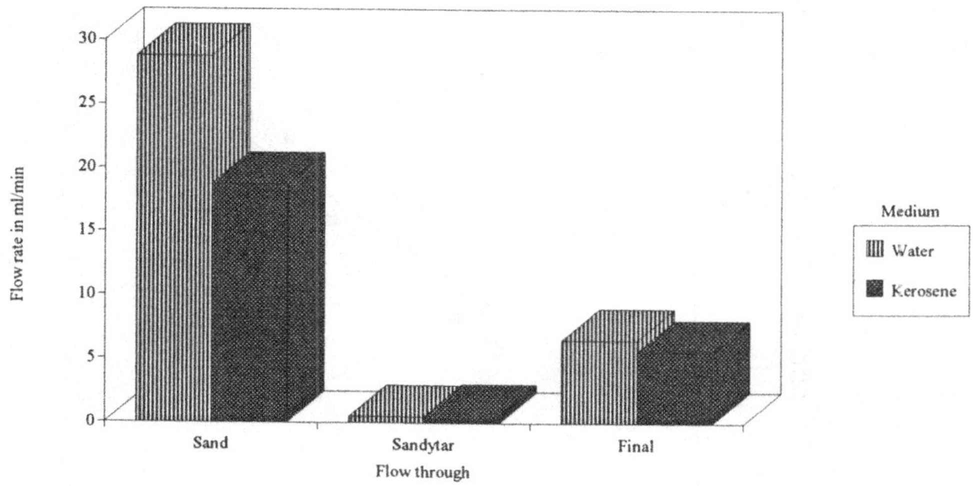

Figure 17. Effect of ultrasonic treatment on flow rates (from Ref. 59).

The sand packs used in the second series of experiments have permeabilities in the ranges of 28 to 30 Darcy. The ultrasonic treatment cleaned 0.5 inch of the sand/asphaltene column, and the separated asphaltene was found to have traveled half way through the sand column. The ultrasonic treatment appeared to be completely, more or less, cleaning the one half inch of the sand/asphaltene by improving permeability around 494 times in aqueous medium and 275 times in kerosene medium. However, the bottom sand/asphaltene part appeared to be cleaned to some extent only, and the top section (one to one and a half inch) of the sand column that is located right below the sand/asphalt column appeared to be contaminated with asphaltenes. Assuming that the bottom half inch of the initially placed sand/asphalt column was not cleaned at all, and the cleaned asphaltene traversed down from the sand/asphalt section to invade the upper portion of one half inch of the sand column; the permeabilities of the infected section were found to be reduced by 80 times and 50 times, in aqueous and kerosene media, respectively. Temperature was monitored throughout and could not account for even a small fraction of the viscosity reduction.

A considerable improvement in flow rates of over 500% is noticed (Fig. 17) after the ultrasonic treatment for two minutes at an output control setting of 7.5 (180 watts). However, the improvements are not substantial when compared to the initial flow rates in sand packed flow cells. This may be due to the reason that the separated asphaltenes get re-deposited at lower levels of flow cells and reduce flow rates. Chakma et al.[47] found the viscosity of a crude oil to be a strong function of asphaltene content, and the asphaltenes with higher molecular weight resulted in higher viscosity of the mixture than those with lower molecular weights. Figure 18 shows the viscosity variation of Athabasca bitumen as a function of asphaltene concentration. Figure 19 shows how temperature alone can account for viscosity reduction, especially if asphaltene is precipitated out. In either media, i.e. aqueous or kerosene, ultrasonic treatment appears to be reducing viscosity to facilitate removal of asphaltene for cleaning and viscosity reduction.

EFFECT OF PRESSURE ON ULTRASONIC TREATMENT

Before ultrasonic treatment can be proposed for removal of asphaltic materials, the effect of pressure on asphaltic materials should be studied. Several publications have

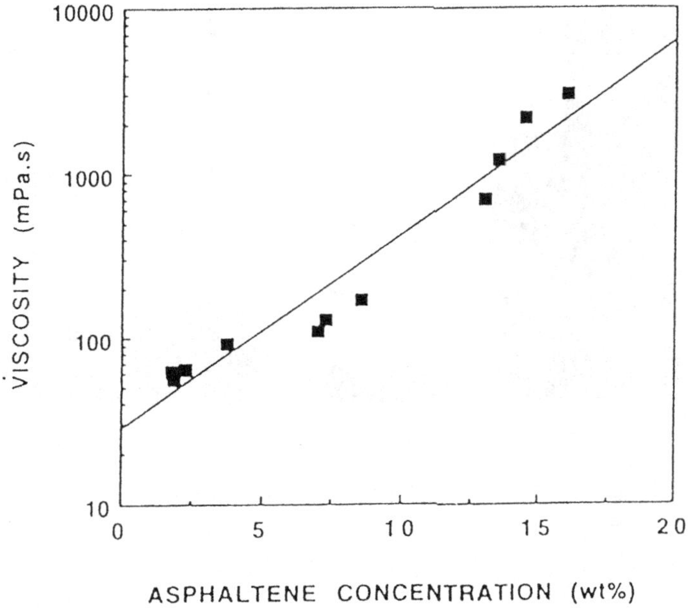

Figure 18. Effect of asphaltene concentration on the viscosity of Athabasca bitumen (from Ref. 46).

reported the study of the effect of pressure on reactions under sonication. Unfortunately, no study has been reported on the effect of confinement pressure in ultrasonic treatment of asphaltic crudes.

In general, it is found that the rate of pressure exerted on a reacting solution is increased. Schmid and Rommel[60] investigated the ultrasonic degradation of polystyrene in toluene solutions at various pressures. In their study, they found that the rate of degradation

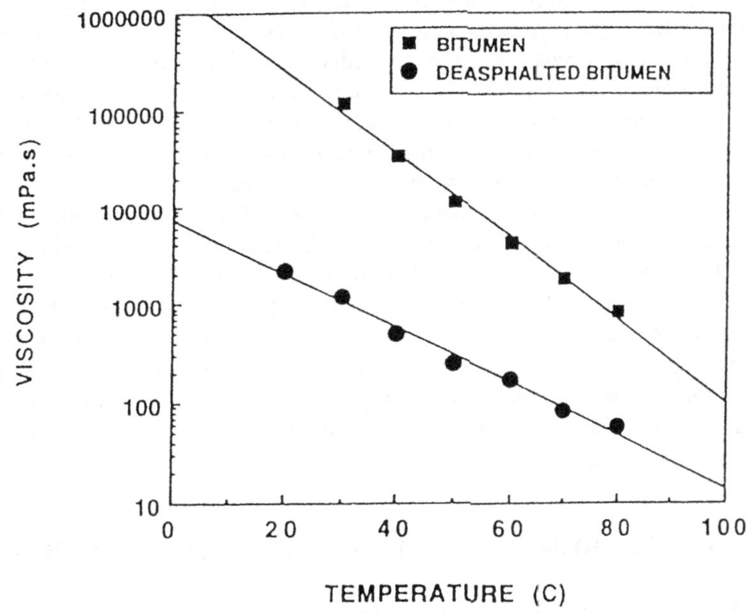

Figure 19. Effect of temperature on Athabasca bitumen (from Ref. 46).

decreased with increasing pressure leading to increased final chain lengths. More precise investigations indicated that the reaction rate vs. pressure curve goes through a maximum[61]. Ono[61] observed that the degradation first increased with pressure, then passed through a broad maximum and decreased slowly with further increase in pressure.

It is well known that an increase in confining pressure tends to suppress cavitation. However, if gas pressure is used to vary pressure in a system, increased concentration of gas is likely to counteract the suppression of cavitation effect. Even if these two effects are isolated, the rate constants for degradation passes through a maximum as the pressure is increased. However, unlike the gas pressure case, the rate constant decreases rapidly to zero, after reaching a maximum value as pressure is increased. Figure 20 shows intrinsic viscosities of long-chain molecules as a function of time and mercury column pressure[62]. This figure indicates that ultrasonic treatment of asphaltic crude cannot be effective under high

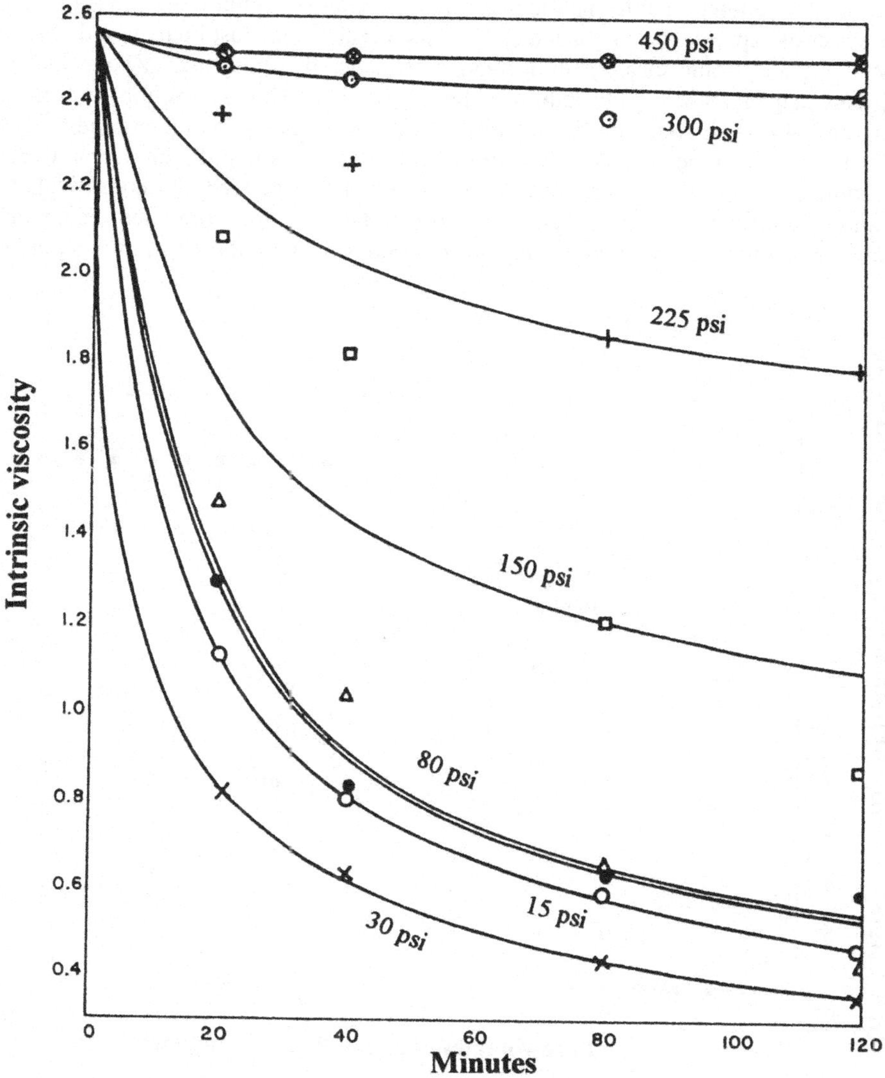

Figure 20. Intrinsic viscosities as a function of time and pressure (from Ref. 62).

confining pressures. However, because most of asphaltene precipitation takes place at low pressures, the effect of a high pressure on effectiveness of ultrasonic cleaning should not be a concern.

EFFECT OF FORMATION PERMEABILITY ON ULTRASONIC TREATMENT

The effect of sand consolidation on ultrasonic cleaning of asphaltic crude has not been studied in detail. However, it is conceivable that the wellbore cleanup will not be affected by the nature of rocks surrounding the wellbore. However, as far as cleaning the vicinity of the wellbore is concerned, low permeability will affect cleanup of asphaltic precipitation adversely. The recent study of Zhong and Islam[63] indicates that asphaltene does not propagate well in tight formations. Figure 21 shows the nature of asphaltene propagation as a function of formation permeability for an asphaltene concentration of 0.1%. In the series of experiments conducted by Gollapudi et al.[59] unconsolidated sands were used for studying ultrasonic cleanup of formation damage of asphaltenes. Zhong and Islam complemented this study with additional tests. Figure 22 shows how the flow rates are affected by ultrasonic treatment for different permeability cases. All other conditions were similar to the ones reported by Gollapudi et al.[59]. It appears that the ultrasonic treatment has a more profound impact on remediation of asphaltene damage in a consolidated core than an unconsolidated core. This effect is expected because, for more intense damages for the low permeability cases, there is more room to improve with the ultrasonic treatment.

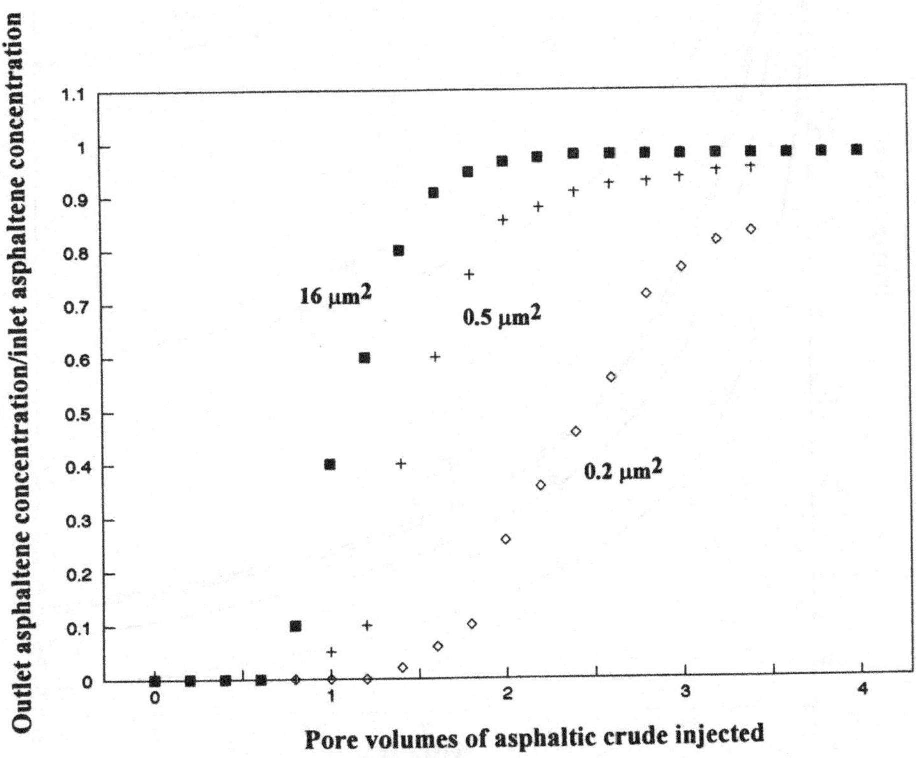

Figure 21. Asphaltene propagation in different porous media.

DESIGN OF AN ULTRASONIC DEVICE FOR PETROLEUM APPLICATIONS

While designing the required ultrasonic generators, the parameters should be taken into consideration:

a. accessibility to any part of the well for wire line operations,
b. ability of the generator to work simultaneously with pumping operation, if possible,
c. production of the required power and amplitude range for special use in oil wells and pipe-lines,
d. attachable to any part of the pipe line to reduce viscosity, and
e. usage in primary crude oil heat exchangers.

Designs should be developed for three separate devices for use in oil wells, to attach any part of crude oil/bitumen transporting pipe line or facility, and for use in primary crude oil heat exchangers. The developed design of the ultrasonicator is also anticipated to control asphaltene based fouling mechanism in heat exchangers, with minimal alterations followed by scaled design.

The following factors should be carefully considered in order to design a dev:ce for ultrasonic cleaning.

Optimization of Parameters

Efforts should be concentrated mainly to simulate actual pipe line and oil well conditions. All the samples should be ultrasonically treated at continuous and pulse modes

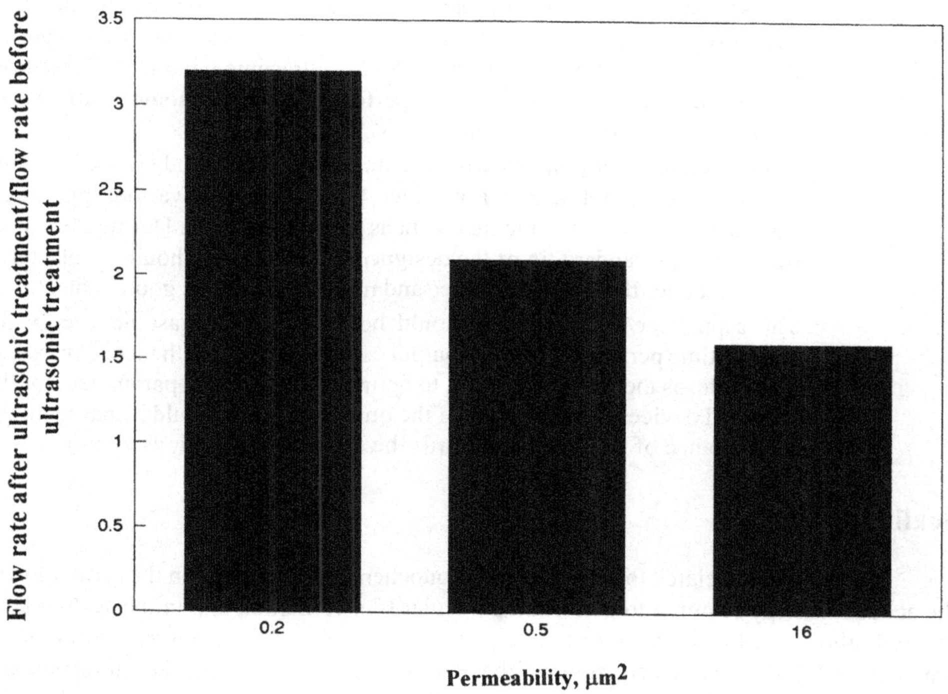

Figure 22. Effect of permeability on ultrasonic cleaning of asphaltenes.

for different time periods under different powers to optimize required parameters, under laboratory conditions. Recent investigations of emulsion flow through pipes indicated that a full scale pipeline should be used[(42)].

Viscosity Reduction

A quantitative analysis should be performed on the samples to find out the amount of asphaltenes present in the samples. During the ultrasonic treatment, special care should be taken to ensure proper temperature and pressure conditions. All the parameters should be re-optimized upon viscosity reduction. For different sets of experiments, operating conditions should be the same as those of cleaning efficiency experiments described below.

Cleaning Efficiency

For better understanding of the cleaning performance, a series of experiments should be carried out on collected samples using the designed devices. Reservoir conditions should be simulated and the experiments carried out in scaled bore holes.

i. Calculation of loss in weight fraction after the treatment
The collected crude oil and bitumen can be placed in scaled bore holes with water, kerosene, crude oil, motor oil, etc., separately. The samples should be irradiated using the designed high intensity ultrasonic processor for various time periods in continuous and pulse operation modes, separately. Subsequent to the treatment, the residual and cleaned out weights should be measured. Here, care should be taken to ensure the weight loss due to the escape of hydrocarbons from the system.

ii. Measurement of flow rates before and after ultrasonic treatment
   a. The scaled bore holes with asphaltene deposition on sides should be ultrasonically treated as explained earlier, in various media separately, and the flow rates be monitored before and after the ultrasonic treatment. These flow rates should be used to correlate the performance of ultrasonic treatment for improving reservoir performance.
   b. A long column of asphaltene/sand mixture should be placed vertically inside the scaled bore holes, and flow rates be measured at various pressures, subsequent to the ultrasonic treatment as explained earlier. During ultrasonic treatment the standard tip of the designed ultrasonicator should be placed in such a manner that the asphaltene/sand mixture maintains good contact.
   c. The asphaltene/sand sections should be exposed to ultrasonic energy for various time periods at different output control settings, with water, kerosene, crude, etc. as media, respectively, to optimize the essential parameters for the designed devices. The selection of the time and power should be based on the performance of the samples towards the ultrasonic conditions exposed.

Scaling Up

The problems related to scaling up of a sonochemical process lie in the estimation of the acoustic energy required to perform a particular transformation and in the scaling laws that will allow one to transform results to a large scale process[(64,65)]. Before scaling up an ultrasonic process, one must determine if the process is truly sonochemical or ultrasonically enhanced mechanical process. The latter process is relatively simple to scale up. On the other hand, if the process is sonochemical, one must provide sonication during the chemical reaction itself. This constitutes a transient chemical process which is very difficult to scale up properly[66]. In scaling up such a process, the following factors must be considered[64]

- viscosity and vapor pressure of the medium,
- presence of gases or solids particles in the medium,
- reaction temperature,
- overpressure,
- sonic power and frequency,
- permeability of the medium.

The physical properties of the reaction medium will dictate the magnitude of the power of the ultrasound necessary. For instance, higher viscosity media will require higher energy to produce cavitation effects. The effect of overpressure on a reaction has been investigated by Cum et al.[67]. Others[68,69] investigated the effect of reaction temperature and applied power. Also, frequency plays an important role in the ultrasonic process. In this regard, it is interesting to note that a much higher frequency can be used in the petroleum reservoir because of the underground nature of the process and less environmental impacts. Also, the size of the generator probe and the depth of the asphaltic damage needs to be addressed before a field device can be used. In general, all factors that affect cavitation should be optimized. One important aspect of ultrasonic treatment of asphaltic wellbore damage is the fact that the high pressure present in the formation may not be as detrimental for the process as a conventional ultrasonic reaction process.

IDENTIFICATION OF THE USE OF ULTRASOUND IN OTHER AREAS

Ultrasonic treatment of the wellbore can have effects on the near wellbore wettability and hence alter the relative permeability in a two-phase flow system. Asphaltene deposition has a strong role in wettability, the tendency of a fluid to adhere or adsorb to a solid surface in the presence of another immiscible fluid. Wettability is one of the major factors that determine the degree of oil recovery from a reservoir. A number of experimental observations in support of the strong role of asphaltene in altering the interfacial tensions and wettability is reported [70-71]. Kim et al.[70] presented a statistical mechanical theory of wettability and its reversal. They confirmed the role of asphaltenes in altering the interfacial properties of water and oil systems. They also proposed a probable reason behind the instability as the surfactant characteristic of asphaltene particles.

Reddi et al.[72] proposed enhancement of hydraulic conductivity of clayey sands using ultrasonic excitation. Their laboratory results encourage further studies on feasibility of ultrasonic enhancement of flow at contaminated sites. The use of these generators in other potential areas, such as in the remediation sites and ground water production opens up new areas of research. More recently, Venkitaraman et al.[73] reported a series of laboratory experiments that demonstrated the usefulness of the ultrasonic method for removal of near-wellbore damage caused by fines and mud solids. They used acoustic intensities of 20-250 watt/m^2 and frequencies of 20-80 kHz. The effective depth of treatment was found to be 2.5 inches. Among others, they also studied the effect of rock permeability on ultrasonic removal of fines. Along the same line, studies should be carried out to study the feasibility of using the devices for removing fines and paraffin in oil wells.

CONCLUSIONS

Ultrasonic treatment has potential in cleaning asphaltene precipitation. In an ultrasonic cleaning process, energy dissipation and viscosity reduction are considered to be the major factors. The increased flow rates due to asphaltene dispersion through ultrasonic treatment demonstrate that the ultrasonic treatment can be used to improve production in oil reservoirs that contain asphaltic crudes and have problems of precipitation near the wellbore. The ultrasonic technology can be useful where the production cannot be performed at low well head pressures and where deep choking process operations are not possible.

The prospect of using ultrasonic treatment for enhancing heavy oil upgrading is found to be promising. Several researchers have demonstrated that the treatment can be used in conjunction with other methods to enhance the effectiveness of the overall upgrading process. More experimental research is recommended in this area.

Ultrasonic treatment can also be used for facilitating asphaltic crude oil transportation. Preliminary experimental results show that the process can be as effective as currently used technologies for the reduction of heavy oil viscosity. The combination of ultrasonic treatment with emulsion transport technology holds a great promise in the future.

Finally, guidelines are provided for the design of an ultrasonic device for all three applications, namely, reservoir wellbore cleanup, heavy oil upgrading, and transportation. Scaling up of a truly sonochemical process poses great difficulties and should be handled with caution before recommending a field-scale application.

REFERENCES

1. Danesh, A., Krinis, D., Henderson, G. D., and Peden, J.M., 1988. Asphaltene deposition in miscible gas flooding of oil reservoirs. Chem. Eng. Res. Des., 66: 339-344.
2. Yen, T.F., 1972. Present status of the structure of petroleum heavy ends and its significance to various technical applications. Am. Chem. Soc. F., Reprint 17(4), p. 102-114.
3. Leontaritis, K.J., 1989. Asphaltene deposition: a comprehensive description of problem manifestations and modeling approaches. paper no. SPE 18892.
4. Dubey, S.T., and Waxman, M.H., 1989. Asphaltene adsorption and desorption from mineral surfaces. SPE paper no.18462.
5. Yen, T.F., 1995. Multiple structural orders of asphaltenes, *Asphaltenes and asphalts, 1*, T.F. Yen and G.V. Chilingarian (eds.), Elsevier, 459 pp.
6. Witherspoon, P.A., Ray, B.R., and Grim, R.E., 1957. A study of the colloidal characteristics of petroleum using an ultracentrifuge. J. Phys. Chem., 61: 1296.
7. Bunger, J.W., 1979. Chemistry of Asphaltenes - Summary of Symposium, Salt Lake City, pp. 1028.
8. Sachanen, A.N., 1945. The Chemical Constituents of Petroleum, Reinhold Publishing Corp., 1945.
9. Schwager, J. and Yen, T. F. 1978. Coal-liquification Products from Major Demonstration Processes, 1. Seaparation and Analysis. *Fuel*, 57, 100.
10. Anonymous, 1979. A Basic Asphalt Emulsion Manual. The Asphalt Institute. Manual Series No. 19 (MS-19).
11. Islam, M.R. and Chakma, A., 1989. Compositional dependence of viscosity of bitumens/heavy oils. CIM paper no. 89-40.
12. Islam, M.R., 1994. Role of asphaltenes on oil recovery and mathematical modeling of asphaltene properties., in *Asphaltenes and Asphalts, 1*, 249-298.
13. Islam, M.R. and Chakma, A., 1989. Mechanics of bubble flow in heavy oil reservoirs. SPE paper no. 20070.
14. Lian, H. J., Lin, J. R., and T. F. Yen, 1991. Peptization Studies of Asphaltene in Systems and Correlation by Solubility Parameter Spectra. in *Particle Technology and Surface Phenomenon in Minerals and Petroleum*, Plenum Press, New York.
15. Speight, J. G., 1981. Solvent effects in the Molecular Weights of Petroleum Asphaltenes. Preprints American Chemical Society, Div. Pet. Chem. pp. 25-30.
16. Koots, J. A. and Speight, J. G. 1975. Relation of Petrloleum Resins to Asphaltenes. Fuel 54, pp. 179-84.
17. Nellensteyn, F. I., 1938. The Colloidal Structure of Bitumens. The Science of Petroleum. A.E. Dunstan (ed.), Oxford U. Press, London, pp. 4, 1938.

18. Lin, J., Lian, H., Chen, J., and Yen, T. F. 1991. Morphological size of asphaltene micelles in asphalt and heavy residue. Particle Technology and Surface Phenomenon in Minerals and Petroleum, Plenum Press, New York.

19. Dwiggins, C.W., Jr., 1978, Study of the colloidal nature of petroleum with an automated Bonse-Hart X-ray small-angle scattering unit, *J. Appl. Crystallogr.*, vol. 11, 615-619.

20. Dwiggins, C.W., Jr., 1956. A small angle X-ray scattering study of the colloidal nature of petroleum. *J. Phys. Chem.*, vol.69, 3500-3506.

21. Klm, H. and Long, R.B., 1977. Characterization of heavy residuum by small-angle X-ray scattering technique. *Ind. Eng. Chem. Fundam.*, vol. 18, 60-63.

22. Monger, T.G. and Trujillo, D.E., 1988. Organic deposition during CO_2 and rich-gas flooding. SPE 18063.

23. Kawanaka, S., Park, S.J., and Masoori, G.A., 1988. The role of asphaltene deposition in EOR gas flooding: a predictive technique. SPE/DoE 17376.

24. Fuhr, B.J., Klein, L.L., Komishke, B.D., Reichert, C., and Ridley, R.K., 1989. Effects of diluents and carbon dioxide on asphaltene flocculation in heavy oil solutions. *Proc. 4th UNITAR/UNDP Conf. on Heavy Crude and Tar Sands*, 2: 637-648.

25. Monger, T.G. and Khakoo, A., 1981. The phase behavior of CO2-Appalachian oil systems. SPE paper no. 10269.

26. Mansoori, G.A. and Jiang, T.S., 1985. Asphaltene deposition and its role in EOR miscible gas flooding. *Proc. 3rd European Conf.*, Rome, Italy.

27. Thaver, R., Nicoll, D.C., and Dick, G., 1989, Asphaltene deposition in production facilities. SPE paper no. 10269.

28. Leontaritis, K.J. and Mansoori, G.A., 1988. Asphaltene deposition: a survey of field experiences and research approaches. *J. Pet. Eng. Sci.*, 1: 229-239.

29. Akbar, S.H. and Saleh, A.A., 1988. A comprehensive approach to solve asphaltene deposition problem in some deep wells. SPE paper no. 17965.

30. Hirschberg, A., deJong, L.N.J., Schipper, B.A., and Meijer, J.G., 1984. influence of temperature and pressure on asphaltene flocuulation. Soc. Pet. Eng. J., June, pp. 283.

31. Mansoori, G.A., Jiang, T.S., and Kawanka, S., 1987. Asphaltene deposition and its role in petroleum production and processing. Arabian J. Sci. Eng., vol. 13, no. 1, pp. 17-34.

32. Leontaritis, K.J., 1988 Asphaltene Deposition: A Thermodynamic-Colloidal Model, Ph.D. Dissertation, University of Illinois, Chicago, USA.

33. Lichaa, P.M., 1977. Asphaltene deposition problem in venezuela crudes - usage of asphaltene emulsion stability oil sands. CIM Special Volume, June, pp. 609-624.

34. Lichaa, P.M. and Herrera, L., 1975. Electrical and other effects related to prevention asphaltene deposition. SPE paper no. 5304.

35. Trbovich, M.G. and King, G.E., 1991. Asphaltene deposition removal: long lasting treatment with a co-solvent. SPE paper no. 21038.

36. Dickakian, Ghaz and Seay, S. 1988. Asphaltene precipitation primary crude exchanger fouling mechanism. Oil & Gas Journal, March 7 pp. 47-50.

37. Hasket, C.E. and Tartera, M. 1965. A practical solution to the problem of asphaltene deposits - Hassi Messaoud Field. Algeria. J. Pet. Tech., April, pp. 387-391.

38. Grosso, J.L., Briceno, M.I., Paterno, J., and Layrisse, I., 1984. Influence of crude oil and surfactant concentration in the rheology and flowing properties of heavy crude oil-in-water emulsions. 5th International Symposium on Surfactants in Solution, Bordeaux, France, July.

39. Layrisse, I., Rivero, M., Paterno, I., Guevara, E., Marin, S., and Gonzalez, J., 1985. Transportation of viscous crude oil-in-water emulsions through pipes. 3rd International Conference of Heavy Crude and Tar Sands, UNITAR, Long Beach, CA, U.S.A., July.

40. Layrisse, I., Rivas, H., Quintero, L., Rivero, M., and Martinez, A., 1989. Imulsion tecnhonology, the answer for the production, handling and transportation of extra heavy crudes and bitumens. *Proc. Fourth UNITAR/UNDP International Conference on Heavy Crude and Tar Sands*. vol. 5, 475-489.

41. Hardy, W.A., Sit, S.P., and Stockwell, A., 1989. Field trials of transoil technology for emulsion pipelining of bitumen. *Proc. Fourth UNITAR/UNDP International Conference on Heavy Crude and Tar Sands*. vol. 5, 490-499.

42. Nunez, G.A. and Rivas, H., 1991. Imulsion, A continuous evolving technology. *Fifth Unitar International Conference on Heavy Crude and Tar Sands*. vol. 4, 213-220.

43. Guevara, E., Zubillaga, V., Zagustin, K., and Zamora, G., 1991. Core annular flow: Preliminary operational experience and measurements in a 152 mm x 55 km pipeline. *Fifth Unitar International Conference on Heavy Crude and Tar Sands*. vol. 4, 237-243.

44. Angelidou, C. and Keshavarz, M.J., 1977. The removal of oil particles from water by flotation. *Ind. Eng. Chem. Procss Des. Dev.*, vol. 16(4), 436-441.
45. Pal, R. and Masliyah, J., 1989. Separation of oil from oil-in-water emulsions by a flotation column. *Proc. Fourth UNITAR/UNDP International Conference on Heavy Crude and Tar Sands*. vol. 5, 39-45.
46. Chakma, A., Islam, M.R. and F. Berruti, 1991. Asphaltene - viscosity relationship of processed and unprocessed bitumen. Asphaltene particles in fossil fuel exploration, recovery, refining and production processes. Editor M. K. Sharma, Plenium Publishers, New York, 1994.
47. Chakma, A., Chornet, E., Overend, R., and Dawson, W.H., 1989. Hydrovisbreaking of Athabasca bitumen in high shear jet reactor. paper persented at the AIChE Spring National Meeting, Houston, TX, U.S.A., April.
48. Islam, M.R. and Chakma, A., 1990. Modeling of visbreaking of bitumen in a jet reactor. *Chemical Engineering Science*. vol. 45, no. 8, 2769-2775.
49. Chakma, A. and Berruti, F. 1991. Ultrasonic visbreaking of athabasca bitumen. Heavy crude and sand tars. Proc. of the 5th UNITAR International Conference on heavy Crude and Tar Sands, Vol. 4, 101-104.
50. Price, G.J., 1992. Introduction to sonochemistry. *Current Trends in Sonochemistry*, G.J. Price (ed.), Royal Soc. Chem., 2-7.
51. Boudjouk, P., 1992. Organometallic processes promoted by ultrasound. *Current Trends in Sonochemistry*, G.J. Price (ed.), Royal Soc. Chem., 110-122.
52. Luche, J.L., 1992. New orientations in Sonochemistry. *Current Trends in Sonochemistry*, G.J. Price (ed.), Royal Soc. Chem., 34-49.
53. Mason, T.J., 1991. *Practical Sonochemistry, A users guide to applications in chemistry and chemical engineering*, Ellis Horwood, Chichester.
54. Malhotra, S.L., 1982. Ultrasonic degradation of hydroxypropyl cellulose solutions in water, ethanol, and tetrahydrofuran. *J. Macromol. Sci.-Chem.*, vol. A17(4), 601-636.
55. Kuo, J.F., Sadeghi, K., Jang, L.K., Sadeghi, M.A., and Yen, T.F., 1986. Enhancement of bitumen separation from tar sand by radicals in ultrasonic irradiation. *Applied Phys. Comm.*, vol. 6, 205-212.
56. Lee, A.S., Xu, X.W., and Yen, T.F., 1989. Upgrading of heavy crude at low-temperature and ambient atmosphere. *Proc. Fourth UNITAR/UNDP International Conference on Heavy Crude and Tar Sands*, vol. 5, 109-116.
57. Frederick, J.R., 1965. *Ultrasonic engineering*. Wiley, New York, p. 117.
58. Smordal, K. and Islam, M.R., 1993. Numerical simulation of a novel technique for upgrading heavy residuum. Fine Particle Society Annual Conference, Chicago, IL. July.
59. Gollapudi, U.K., Bang, S.S. Islam, M.R., "Use of Ultrasonic Treatment for Removal of Asphaltene Deposits During Petroleum Production", Proc. SPE 1994 Conf. Prod., New Orleans, Louisiana.
60. Schmid, G. and Rommel, O., 1939. *Z. Electrochem.*, vol. 45, 649.
61. Ono, S., 1940. *Rev. Phys. Chem. Japan*, vol. 14, 25.
62. Brett, H.W.W. and Jellinek, H.H.G., 1956. Degradation of long-chain molecules by ultrasonic waves. Part VI. Effect of pressure. *J. Polymer Science*, vol. XXI, 535-545.
63. Zhong, L. and Islam, M.R., 1995. Asphaltene propagation in porous media and capillary tubes. manuscript in preparation.
64. Mason, T.J. and Berlan, J., 1992. Ultrasound in industrial processes: the problems of scale-up. *Current Trends in Sonochemistry*, G.J. Price (ed.), 148-157.
65. Mason, T.J. and Lorimer, J.P., 1989. *Sonochemistry, Theory, Applications and Uses of Ultrasound in Chemistry*, Ellis Horwood.
66. Islam, M.R. and Farouq Ali, S.M., 1990. New scaling criteria for chemical flooding experiments. *J. Canadian Petroleum Technology*, vol. 29 (1), 29-36.
67. Cum, G., Gallo, R., Spadaro, A., and Galli, G., *J. Chem. Soc. Perkins Trans II*, 376.
68. de Sousa-Barboza, J.C. Petrier, J.L., and Reverdy, G., 1988. *J. Org. Chem.*, vol. 53, 1212.
69. Petrier, C., Jeunet, A., Luche, J.L., and Reverdy, G., 1992. *J. Amer. Soc.*, 114.
70. Kim, S. T., Boudh-Hir, M. E. and Mansoori, G. A. 1990. The role of asphaltenes in wettability reversal. Annual Technical Conference & Exhibition, SPE paper no. 20700.
71. Crocker, M. E. and Marchin, L. M. 1988. Wettability and adsorption characteristics of crude-oil asphaltene and polar fractions. Journal of Petroleum Technology, April, pp. 470-474.
72. Reddi, L. N., Berliner III, S. and Lee, K. Y. 1993. Feasibility of ultrasonic enhancement of flow in clayey sands. Journal of Environmental Engineering, Vol.119, No. 4, July/August, 1993.
73. Venkitaraman, A., Roberts, P.M., and Sharma, M.M., 1994. Ultrasonic removal of near-wellbore damage caused by fines and mud solids, SPE paper no. 27388.

Chapter VIII

NEW METHODS OF PETROLEUM SLUDGE DISPOSAL AND UTILIZATION

M. R. Islam

Department of Geological Engineering
South Dakota School of Mines and Technology
501 East St. Joseph Street
Rapid City, South Dakota 57701

INTRODUCTION

Disposal of the oil sludge is one of the fundamental problems of petroleum production. Increasingly stringent environmental control regulations, lack of final disposal sites, and high costs involved in disposal have resulted in limitations of oil sludge disposal options. The oil sludge is the primary unwanted by-product generated in the production of oil. All types of this by-product contain some sediments and water with different ratios of oil, solids and water in various states of emulsification. In some cases excessive levels of the by-product yield a product that is unsellable and untreatable.

The problem of sludge disposal is more acute in the tar sand industry. The sludge is generated as a result of mining the oil sands and commercially extracting the bitumen with a hot water process that uses sodium hydroxide that frees bitumen from the oil sand. Subsequently, bitumen is skimmed from the separator for which the control of fines becomes the most difficult problem[1]. As the tailings are discarded at the tailing ponds, the tailing water causes the fines to segregate from the tailing sand. Approximately, 50% of the tailings fines is trapped in between sand grains whereas the remainder flows into the tailings ponds and settles as a sludge [2]. More than 2.3 cubic meters of toxic sludge are accumulated in the current commercial mined oil sand plants for every cubic meter of synthetic crude oil (SCO) produced by the Canadian plants[3]. Figure 1 shows projected sludge accumulations due to SCO production in Canada. This astronomical volume will be discarded in ponds and will have to be reclaimed.

Recently, Storm et al. [4] have identified the nature of sludge during heavy oil conversion. The formation of sludge during visbreaking, or the catalytic hydro-conversion of residua creates problems for the efficiency of the process and generates wastes that are difficult to treat. It is believed that sludge forms because the asphaltenes flocculate during processing[4]. However, it has not been possible to link feedstock properties to the tendency to form sludge during processing of heavy oil.

Asphaltenes: Fundamentals and Applications, Edited by E. Y. Sheu and O. C. Mullins
Plenum Press, New York, 1995

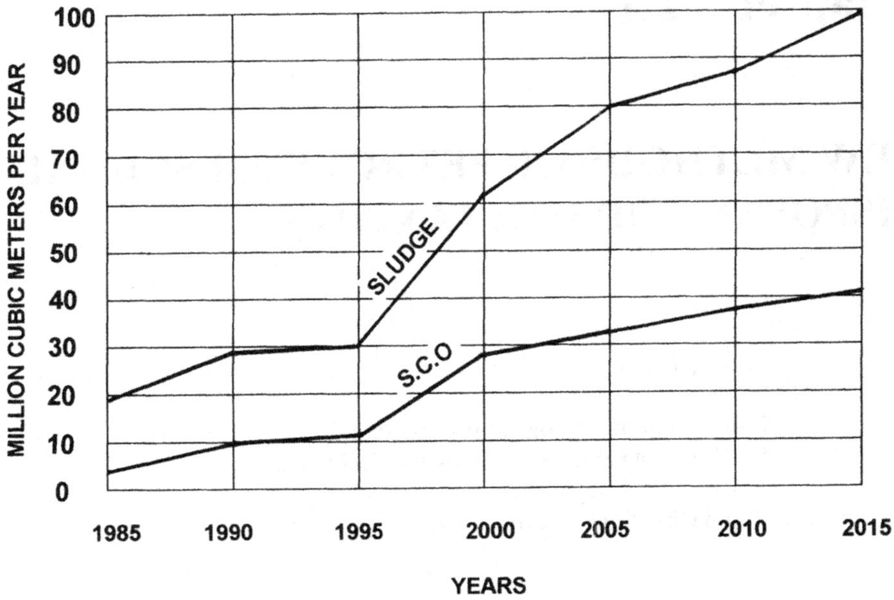

Figure 1. Projected sludge generation for Canada (from Ref. 3).

If the by-products are in the solid state, they can be biodegraded and the end products can be treated in a landfill or mixed with cement and used as a cementing material. If the by-products are in liquid state, they can be upgraded and used as asphaltic material. However, the problem is with the by-products containing solids, water and oil which cannot be biodegraded or upgraded. So, the solids and liquids should be separated for proper disposal.

The customary methods of waste oil management have primarily involved the storage of generated waste into convenient ponds and tanks from where the waste can be taken to designated waste sites. Often outside companies are asked to solidify the oil sludge and bury it at the cost of $900 per drum [5]. In these storage facilities, different kinds of pre-treatment of wastes are done to get valuable by-products, volume reduction, and acceptable toxicity level. All these disposal methods can affect air, land, and water environments [6]. This method of resource recovery has been largely used since the introduction of Resource Conservation and Recovery Act of 1976, and the 1984 Amendments to the Acts. Under the Act, disposal of bulk hazardous liquids in landfills is prohibited and they should be stored in properly constructed surface impoundments. Besides, the act will not allow storage of the oil sludge for a prolonged period of time [7].

Varieties of sources from oil refineries generate large volumes of different waste oil sludges, such as tank bottoms/cleaning sludge, dissolved and induced air flotation skimmings, and impoundment sludge. A cost-efficient means of treatment should be adopted to handle wide variations in oil, water, and solids.

Oleiphilic Sieve

This recently developed process has received some success in commercial applications[3]. When a suspension is passed through the Oleophilic sieve, water and solids pass through the apertures and bitumen is captured by the sieve surfaces. The sieve is mounted on a simple construction, a moving steel process belt with two zones, called the separation zone and the recovery zone. The separation zone captures bitumen from the suspension. The recovery zone, situated at the top, releases the bitumen. Even though separation can be done warm or cold, the bitumen product is warm and pumpable. This sieving technique makes use of the bitumen-to-surface adhesive forces combined with buoyancy forces. The combination of these forces being stronger than simple buoyancy forces of bitumen in water (used by conventional techniques), yields a process which is in general more effective than the conventional ones.

The AOSTRA Traciuk Process

The AOSTRA Traciuk Process (ATP) has been developed by the Alberta Oil Sands Technology and Research Authority and UMATAC Industrial Processor Division of UMA Engineering Ltd. (UMATAC)[8]. The process was originally developed as a method for simultaneously extracting and partially upgrading oil sands by thermal retorting of mined oil sands. However, the process has been used in pilot scale for converting petroleum waste into an oil-free, dry solids residue, a distillable oil, process water and flue gas. The heart of the process is a processor, which is a single, horizontal, rotating vessel in which separate, interconnected compartments are created by the geometry of the equipment. In these zones, the incoming feedstock is preheated by heat exchange with outgoing residual solids and the contained oil is then thermally cracked to generate distillable oil and coke, which is deposited on the solid matrix. The coke is subsequently burned off these solids and the resulting heat energy is used to drive the process. The final step involves heat recovery from the residual solids by heat exchange with fresh feed.

Emerging Technologies

The most desirable solutions to petroleum sludge remediation problems is in utilization of petroleum products. The utilization of asphaltic materials varies from the use of natural bituminous rocks for road construction[9] to the development of high performance paving materials[10,11]. The trend in the last decade has been in complete utilization (or zero waste) of the petroleum wastes. The solution becomes quite innovative if the solution offers good quality material construction. One of these solutions is the use of solid-liquid centrifugal separation. The other option is the use of the oil sludge as a cementing material. Recent studies have shown that the two options can be used in tandem for total remediation of the petroleum sludge. If the initial oil concentration in the sludge is high, high temperature centrifugation with biodegradable surfactant is recommended to lower the concentration to a reasonable value. The resulting solid extract can be mixed with cement and silica fume and examined for cementing properties. Evaluation of compressive strengths of the mortar cubes indicates that it is possible to produce cement from the oil sludge which will satisfy the strength requirements in the ASTM standard for masonry cement.

CASE STUDIES

The Grand Gulf nuclear station used a new disposal scheme based on separating solids and liquids using solid-liquid separator. A test run was made on one drum which has high levels of contaminants. First the oil was heated to 120° F and the oil was passed through the centrifuge. After passing the oil for four times all the water was removed with volume reduction to 50% and no suspended particles were seen in the oil [5].

Haley and Aldrich [12] recycled the contaminated soil into asphalt paving material. This contaminated soil was the residue found when oil storage tanks were removed from the site in Roxbury, Massachusetts. They also conducted studies to know the air quality during processing and construction. Their results showed no detectable levels of organic emissions.

In Casper, Wyoming [13] an oil-sludge pit was stabilized and solidified in less than two weeks by using fly ash and cement. The cost was \$13 per cubic yard of oil sludge. The area of pit was 7,000 cu yd and contained water based drilling fluids. This pit was used to cool drill bits and to clean the holes. In order to stabilize that pit, contractors used 820 tons of class C fly ash . Class C fly ash has good pozzolanic activity and forms good cementitious products. Large volumes of cement and fly ash were dumped into the pit by using pneumatic and mechanical methods. The mixture was thoroughly mixed and was allowed to set for 1-3 days for hardening. Along with fly ash, soil was added to the pit to absorb excess amount of water and to achieve hardening of the waste. By adding soil to the pit, the amount of fly ash and cement can be reduced by 25%. A five day compressive strength was conducted on the hardened waste and the strength was found to be 31.2 psi. After the waste was hardened, effectiveness of the treatment was tested. The results of these tests showed that the oil waste at the bottom of the pit was solidified and the soil lying under the pit was decontaminated. All these tests were done by Western Ash Co. of Denver, Colorado. They also conducted many tests to optimize the percent of locally available cementitious materials. According to their results, a minimum compressive strength of 20 psi was found to be sufficient to support 30 ft of overburden and the admixture was considered to be a good design mix [13].

In 1989, Kruyer [3] reported the use of Oleophilic sieve for treating bitumen sludges. The sieve was successfully tested with water-based emulsions, mined oil sand slurries, tailings, and tailing pond sludges. More than 1000 tons of toxic waste was processed. A high bitumen recovery was reported while energy and water requirements were considered to be minimal. Figure 2 shows pilot plant bitumen recovery efficiency from tailings pond sludge. This figure shows the effectiveness of the process in capturing bitumen. Commercial field-scale application of the process, however, is yet to be implemented.

Recent studies, based on the AOSTRA Traciuk Process provided some encouraging results[8]. A 105 tons per day continuous flow pilot plant has been operating successfully using this process. Even though successful technically, the process appears to be much more

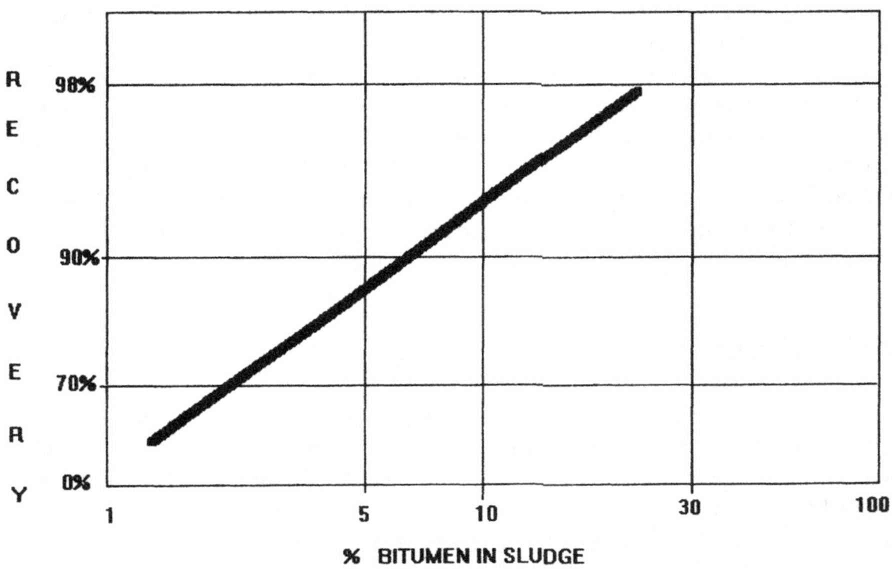

Figure 2. Pilot plant bitumen recovery with the Oleophilic Sieve (from Ref. 3).

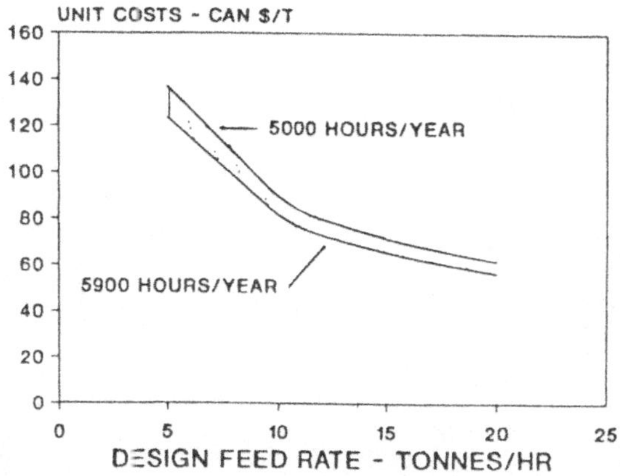

Figure 3. Treatment and disposal costs for an AOSTRA Traciuk Process Plant for Oilfield Wastes (from Ref. 8).

expensive than conventional techniques. Figure 3 shows the treatment and disposal costs for the process.

HEAVY HYDROCARBON RESIDUE AS A PAVING MATERIAL

Zanzotto et al. [10] developed a new method for utilization of hydrocracking residue of heavy oil as a paving material. This process holds a great promise because it eliminates waste while improving material quality. They first attempted to use the residue alone as a paving material, but tests indicated that the properties did not conform to the requirement of Alberta Transportation for premium grade asphalts. Consequently, the heavy oil residue was combined with different grades of paving materials. In a series of tests, heavy oil residues were combined in amount 3% to 30% with asphalt 85/100 Pen, 150/200 Pen and 200/300 Pen. The mixing of different compositions was carried out after a pre-heating at 160°C. The first test carried out was the consistency test that confirmed that there was no segregation

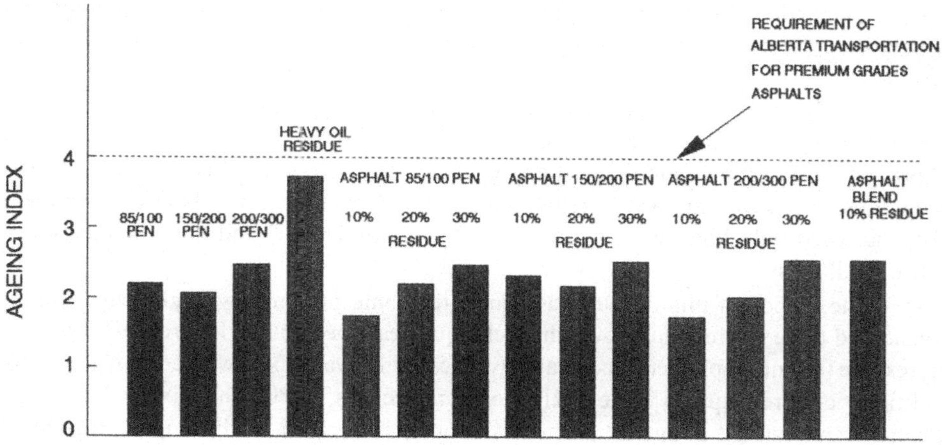

Figure 4. Ageing index of tested materials after measurement by the Thin Film Oven Test (from Ref. 10).

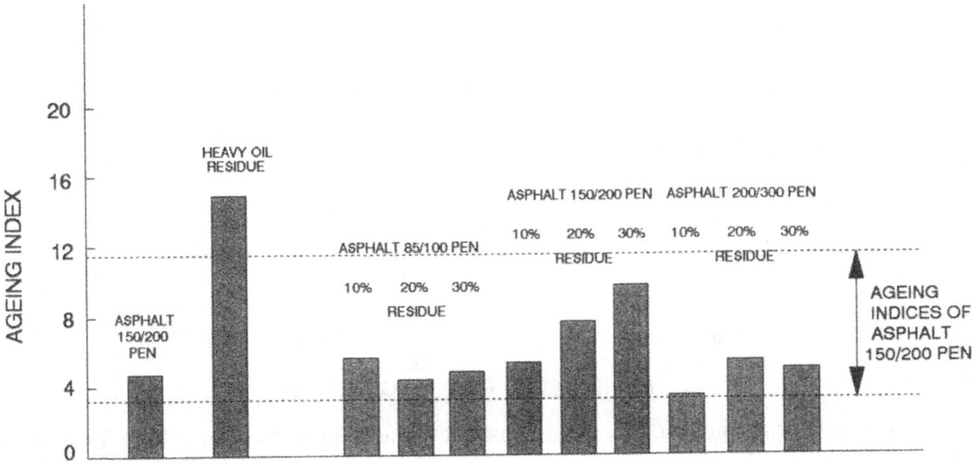

Figure 5. Ageing index of tested materials tested by the Microfilm durability test (from Ref. 10).

after cooling and bulk properties of the asphalt did not change. Subsequently, ageing, freeze-thaw and other tests were performed on the mixtures. Properties of these mixtures were compared with those of asphalt 150/200 Pen which is the most commonly used asphalt in the prairie provinces of Canada. The results show that properties of these compositions were substantially better than those of the most commonly used asphalts in the area. In most cases, it was found that the superiority of the new mixture holds for a heavy oil residue concentration of up to 30%. Figures 4 and 5 compare the ageing indices for different compositions. Figure 6 shows simulated distillation of the heavy oil residue as compared to common asphalts.

The Marshall Stability test (ASTM D1550) was performed for various composition of heavy oil residue along with asphalts and mineral aggregates. For this set of tests, three compositions, 10%, 20% and 30% were selected while the consistency of asphalt 150/200 Pen was chosen for comparison. Mineral aggregates contained approximately 36% limestone, 20% quartz, 16% sandstone, 11% chert, and 9% quartzite. A different set of tests was carried out using granite gneiss aggregates from Georgia, U.S.A. These aggregates are widely used for testing water damage to the paving material. Figure 7 shows some of these results.

In this study, it was concluded that the addition of hydrocracking residue to asphalt, the asphaltenes present in the residue act as an anti-stripping agent. This leads to drastic improvement in asphalt and mineral aggregates in terms of durability and other desirable properties. This invention was a distinct improvement over a previously developed method [14] that used only high temperature thermal cracking residue and the amount used was substantially less.

The author of this chapter has conducted some preliminary studies that might be considered to be improvement over the technique described above. In a recent work, heavy oil residue (mainly asphaltene and heavy hydrocarbons) was mixed with commercial asphalt and finely cut tire chips[15]. The mixing temperature was 180°C. The mixture later showed distinct improvement in ageing properties as well as durability. The improved properties were considered to be similar to the polymer-enhanced asphalts. Further work in this regard is in progress.

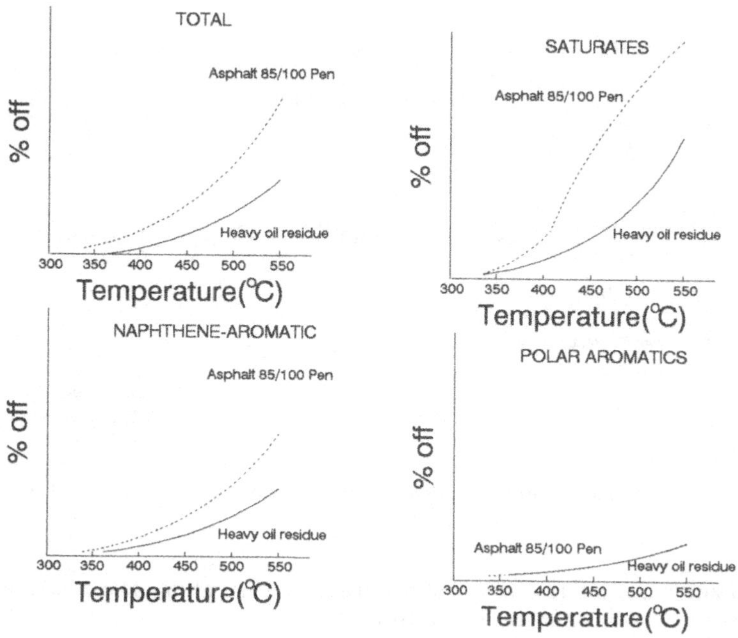

Figure 6. Gas Chromatography simulated distillation of the new and conventional mixtures (from Ref. 10).

A COMPLETE REMEDIATION METHOD

Oil sludge contains oil, emulsion, water and solids. The ratios of these components vary from one application to another and from tank to tank. The difficult part of any waste oil sludge is the oil-water-solids emulsion which is always present and which, in combination

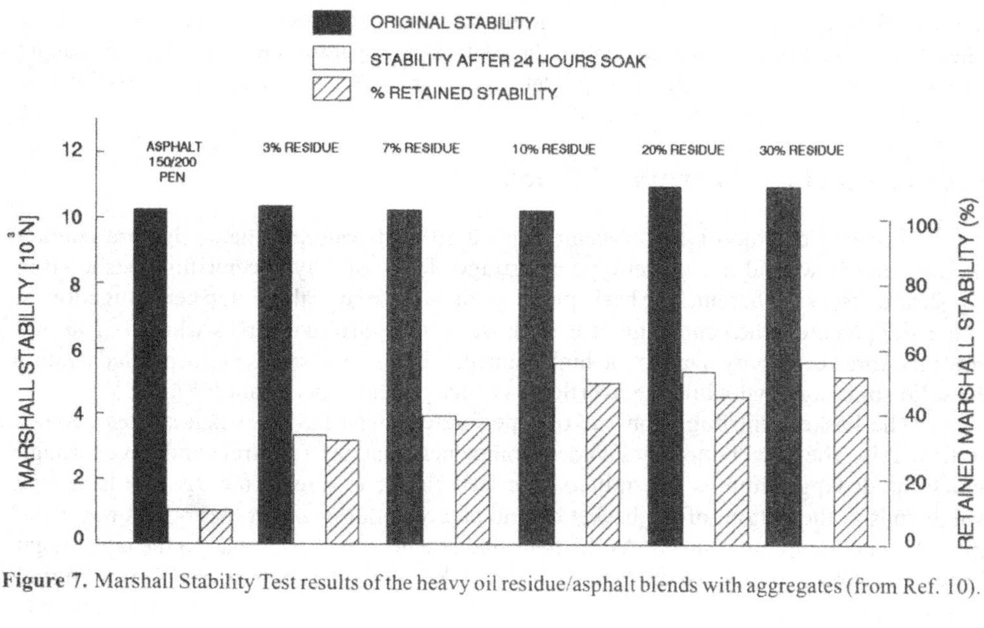

Figure 7. Marshall Stability Test results of the heavy oil residue/asphalt blends with aggregates (from Ref. 10).

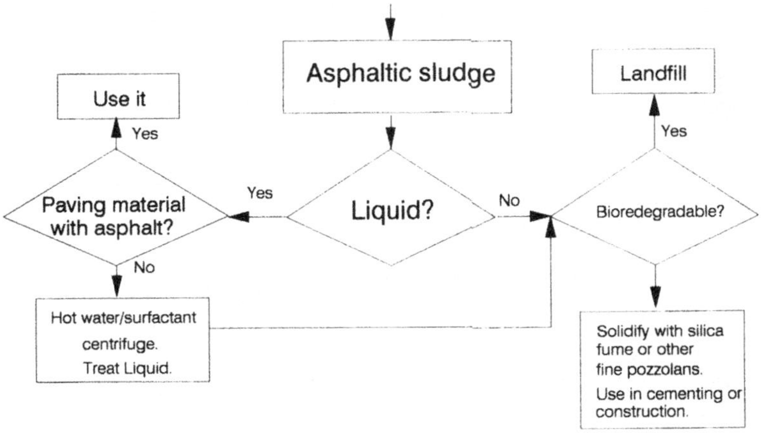

Figure 8. Flow chart for effective remediation of oil sludges (from Ref. 11).

with the oil and water, represents a distinct phase for separation. The emulsion phase is a stable composite which cannot be broken by centrifugal or filtration methods. Therefore, the first step is to de-stabilize the emulsion phase by heating which avoids the inter-droplet attraction due to their surface charges [16].

The overall methodology is shown in Fig. 8. Note that centrifugation is necessary only if the initial sludge state is liquid.

Heat Treatment (180°–200°F)

One of the most common treatments to break emulsions is heat treatment. Heating will reduce the viscosity of the emulsion which aids in the mixing of chemicals and sedimentation of the fine suspended particles. Heating produces a higher reduction of specific gravity of the oil component than in the water component, which causes an instability effect in the emulsion [16]. A natural gas heater constructed by L&R Drilling of South Dakota was used for this study. A biodegradable anionic surfactant SD 1000, as provided by Chevron Chemicals, was used in order to enhance the centrifugal separation process. This surfactant is thermally stable and is totally biodegradable [17]. The initial surfactant concentration was optimized before using in the field scale.

Description of the Centrifugal Process

The two common types of centrifuges used in dewatering the sludge are counter current solid bowl and the basket type centrifuge. The basket type centrifuge is shown in Fig. 9. and discussed herein. The basic principle of solids removal through centrifugation is, as the sludge enters the centrifuge, it is influenced by centrifugal forces which is 100-600 times the force of gravity. Due to the high centrifugal force, the solids are deposited against the solid spinning bowl while the overflow is a clear liquid supernatant [18].

The basket centrifuge consists of a perforated metal basket which rotates about a vertical axis. The bowl bottom has spider arrangement and has a central solid nave through which the driving torque is transmitted. The feed sludge is entered through the feed pipe which ends at the bottom of the basket behind an acceleration wheel that brings the speed up to the maximum bowl speed. As the feed sludge enters the basket due to the centrifugal force, the solids are collected on the side of the basket while the centrate overflow a lip ring

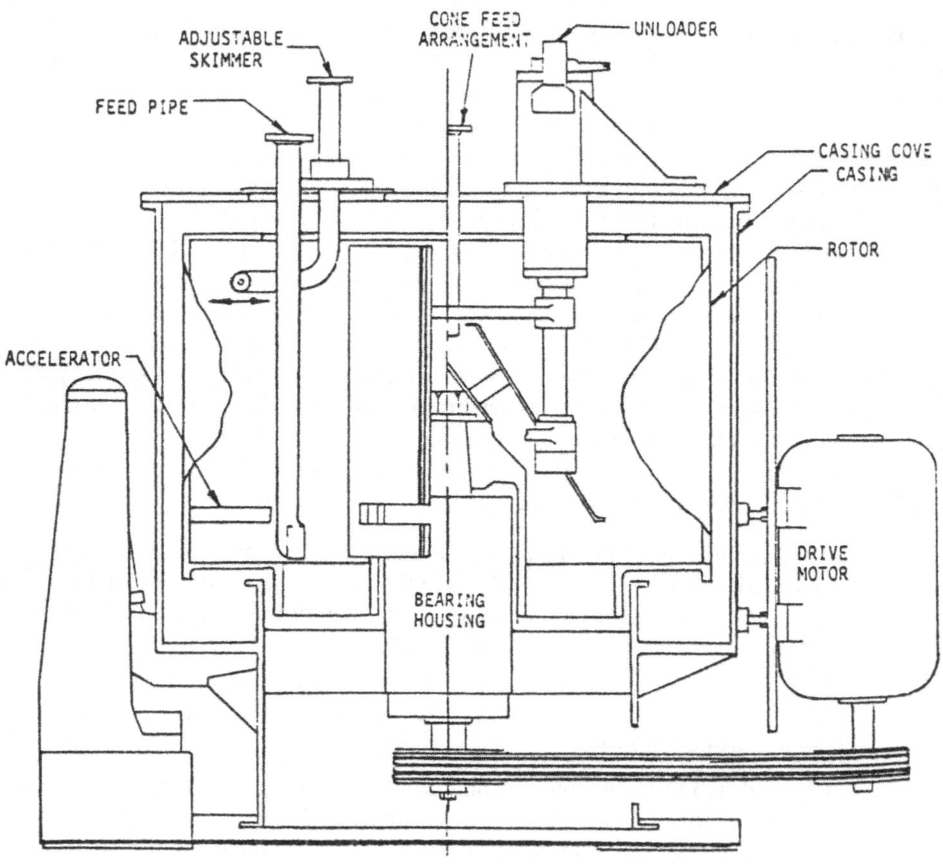

Figure 9. Schematic of the basket type centrifuge (from Encyclopedia of Chemical Processing and Design, 1987).

at the top of the basket. To retain the liquid and create the flow pattern, the solid portion of the bottom should have smaller internal diameter than that of the lip ring. When the solids attain certain thickness, the rotation of the bowl is decreased, and by means of a scrapping blade the solids are scrapped from the walls and discharged at the bottom of the basket. A solid-thickness detector can be used to detect the thickness of the solids. Soft cake may be removed by submerging a skimmer into the solid layer while the basket is spinning at full speed. Removal of solids can be enhanced by adding polyelectrolytes or other coagulants along with the feed sludge [18].

A centrifuge was constructed by L&R Drilling and was made available to the South Dakota School of Mines for research. This equipment now is being used commercially.

Principle behind Using Cement for the Disposal of Oil Sludge

The principle behind using the oil sludge as a cementing material is to decrease solubility and at the same time decrease the waste surface which is exposed to environment, and to simultaneously decrease the toxic nature of the sludge, so that it becomes safer to handle and to treat further [13].

The basic principle involved here is stabilization and solidification. Though these terms are interchangeable, they are two different processes. The main advantage of stabilization is that the solubility is decreased to immobilize the hazardous contaminants. By solidification we can produce a strong durable solid waste block which will be easy to dispose. By mixing the oil sludge with cement, one produces a solid mass and thereby achieve both stabilization and solidification. When cement is mixed with the sludge, the cement will maintain the sludge at high pH of 9-11, where most of the multivalent cations are immobilized to form insoluble hydroxides. These hydroxides will physically and chemically bind the metal ions. In some cases, the organic materials present in oil can interfere with hydration, i.e., the reaction between water and cement can decrease the hardened mass. Salts, such as zinc, copper and lead will retard the hardening of the cement mass and when this mass does not solidify properly it will lead to leaching of the waste and finally contaminating ground water and environment. When these cations are present in the oil sludge they can be pretreated with additives specifically chosen to immobilize those cations. Though anions are in general less toxic than cations, it is more difficult to bind them [13].

The following are some of the factors to be considered to achieve stabilization/solidification of the oil sludge or oil wastes [13]:

1. Leachability: Since the waste contains liquids and if proper solidification is not achieved, there is a risk for leaching of the waste. So, by conducting the leaching tests, one can determine the maximum concentration of the contaminants that ground water can remove from cement blocks.
2. Free-liquid content: According to EPA (Environmental Protection Agency) regulations no free water is allowed in the waste. The dead weight of the soil can drive the free water out of the waste.
3. Physical stability under buried conditions: If the waste is compressible, the waste may not withstand the construction equipment. Therefore, care should be taken to compact the waste properly and the waste should be impermeable. If the permeability is higher, there is a scope for the leaching of contaminants.
4. Reactivity and ignitability: Once the waste is stabilized/solidified, it should be nonreactive and nonignitable. But in some cases the waste has tendency to react with other wastes, synthetic or clay liners or absorbents. So tests should be conducted on solidified waste to know the reactivity and ignitability of the solidified waste.
5. Biodegradation: Biological activity can produce acid which will interfere with stabilized/solidified waste if the waste is not properly solidified and there is a danger for leaching of metals from the waste. So it is recommended to perform ASTM, tests G21 and G22 to know the biological effect on the waste.
6. Strength and durability: Strength of the final waste is one of the most important factors to be considered, depending upon the final usage of the waste. According to ASTM requirements, the final unconfined compressive strength of the final waste should range from 1 to 4000 psi. But 20 psi is enough to bury the waste. If the waste is used for backfill or foundation support, the minimum one day compressive strength should be 150 psi.

Durability is another important factor to be considered, because the waste is exposed to freeze-thaw and wet-dry cycles. If the waste is buried there will be very minor effect of temperature on the buried waste. But if the waste is used for construction, tests for freeze-thaw durability using ASTM D560 procedure and wet-dry durability using ASTM 559 procedure should be performed.

Table 1. Mixture proportions

Sample number	Cement percentage	Sludge percentage
1	100	0
2	80	20
3	85	15
4	90	10
5	95	5

Materials Used

Cement. Type I normal Portland cement satisfying the requirements of ASTM was used for all mixtures. The cement was produced by the South Dakota Cement plant in Rapid City, South Dakota.

Sand. 20/30 sand was used for all mixtures. The moisture content was calculated by oven drying known weight of sample for 24 hrs at 250°F. After 24 hrs the oven dried sample was weighed. The moisture content was calculated to be 2.6%.

Oil Sludge. The oil sludge used was supplied by Dakota Oil Producers, Pierre, South Dakota. This was essentially the solid extracts after centrifugation of the liquid sludge. The moisture content of the solid sludge was 4.9%. The liquid sludge was also provided by Dakota Oil Producers. This sludge contained crude oils from the New Castle area of Wyoming. Thirty percent of the sludge contained asphaltic materials.

Silica Fume. Silica fume is a by-product of the silicon and ferrosilicon industry. It has good pozzolanic activity and particle size is less than 0.15 micron. It is understood that silica fume is a rather expensive pozzolanic material. Fortunately, inexpensive substitutes of silica fume are under investigation [19].

Mixing Proportions. For six specimens of size 2 in. x 2 in. x 2 in., the proportions are: 1350 g of sand, 500 g of cement, and 242 g of water. The basic mix proportions are given in Tables 1 and 2.

Test Specimens

The specimens were cast in steel molds immediately after mixing. The specimens were covered with plastic sheets for 24 hours to prevent evaporation of water from the unhardened concrete. They were then demolded and placed in lime saturated water tanks with temperature of the water maintained at 75°F. The specimens for seven day and twenty

Table 2. Mixture proportions

Sample number	Cement percentage	Sludge percentage	Silica fume percentage
1	100	0	0
6	75	20	5
7	80	15	5
8	85	10	5
9	90	5	5

Table 3. Compressive strength measurement

Sample number	Seven day compressive strength in PSI	Twenty eight day compressive strength in PSI
1	4350	7187.5
2	1675	2687.5
3	1887.5	2950
4	2150	3425
5	2537.5	3937.5
6	1850	3125
7	2787.5	4475
8	3725	5225
9	3800	6087.5

eight day compressive strength were kept in water until they were tested. Prior to testing, the specimens were dried. The specimens were tested as per the ASTM C109 procedure. The loading had to be applied slowly because the maximum load was reached within 20 to 80 seconds. The machine should not be adjusted while the specimen is yielding before failure.

Calculation of Compressive Strength

The total maximum load is noted to calculate compressive strength. The compressive strength is measured in psi.

$$\text{Compressive Strength} = \frac{\text{Maximum load in pounds}}{\text{Area in square inches}}$$

The compressive strengths are calculated and tabulated in Table 3.

Centrifugation Results

A series of laboratory experiments was conducted to optimize the surfactant concentration for best separation of solid/liquid separation. For the oil sludge from the New Castle (WY) area, the optimum concentration was found to be 1% surfactant at a temperature of 95°C. The use of a higher temperature (higher than 95°C) was considered to be too expensive and was not deemed necessary because the separation was already enhanced considerably in the presence of the surfactant. Figure 10 shows the final concentration of petroleum in the solid residue as a function of surfactant concentration. This figure clearly shows the need for the use of surfactant without which the solid/liquid separation remains dismal. Also, as the surfactant concentration is increased beyond 1%, the crude oil concentration in the solid phase remains low. However, it becomes increasingly difficult to separate the oil phase from the aqueous phase because stable emulsions form at higher surfactant concentrations. Besides, at higher concentrations of surfactant the solid phase does not show much sign of a better clean up.

Solidification Results

Nine runs were conducted to determine the seven day and twenty eight day compressive strength. Cement, oil sludge and silica fume were used to prepare the specimens. Tables 1 and 2 give the sample numbers and sample proportions. In each run, 12 specimens were prepared, six for seven day and six for twenty eight day tests. The results of seven day and

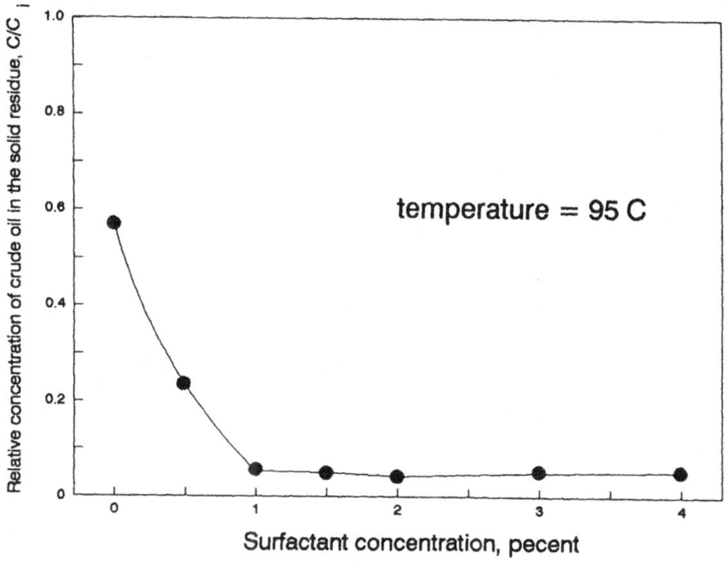

Figure 10. Effect of surfactant concentration on the performance of centrifuge (from Ref. 11).

twenty eight day compressive strength are tabulated in Table 3. Sample number 1 contains only cement and it is the control sample. Samples 2, 3, 4 and 5 contain different proportions of oil sludge and cement, samples 6, 7, 8 and 9 contains 5% silica fume along with different proportions of oil sludge and cement. The moisture content of oil sludge was 4.89% and sand was 2.6%. Type I normal Portland cement was used and the compressive strength determined in the laboratory for the control sample matched the specifications set by Rapid City cement plant, Rapid City, South Dakota. The control sample has seven day compressive strength of 4,350 psi and twenty eight day compressive strength of 7,187.5 psi. Figure 11 compares seven-day compressive strength values for different cement mixtures without the addition of silica fume. These results show that as the percentage of sludge decreases, the compressive strength increases. The same trend continues in the presence of 5% silica fume (Fig. 12). However, in the presence of silica fume, the compressive strength values are clearly

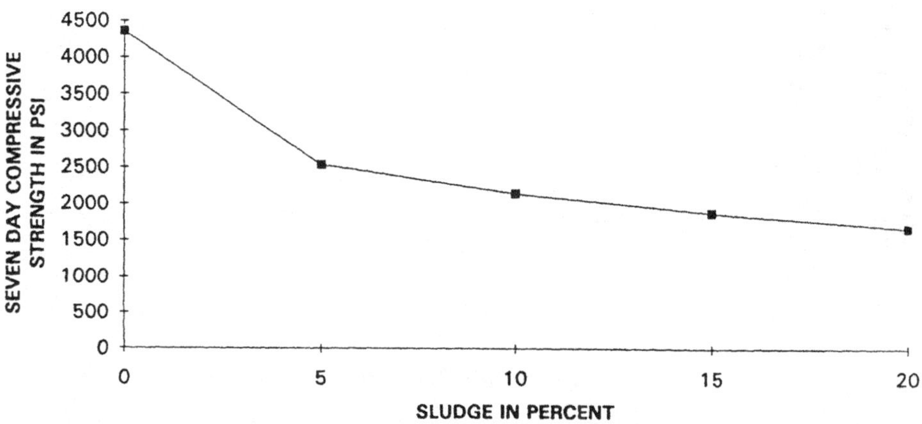

Figure 11. Seven-day compressive strength of cement sludge mixtures (from Ref. 11).

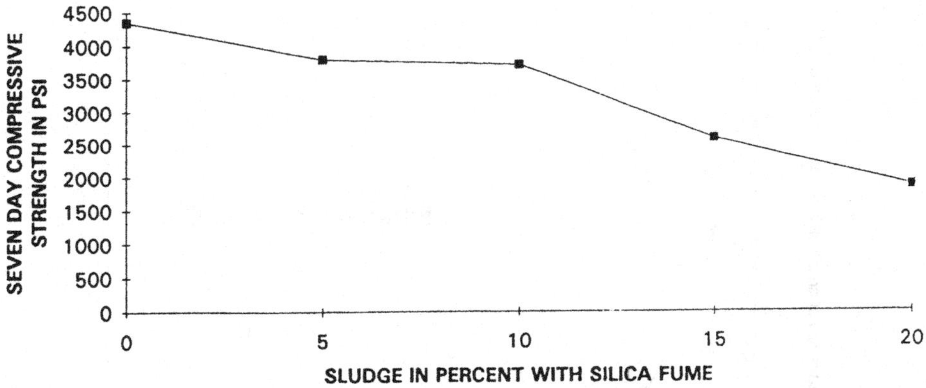

Figure 12. Seven-day compressive strength of cement sludge mixtures with 5% silica fume (from Ref. 11).

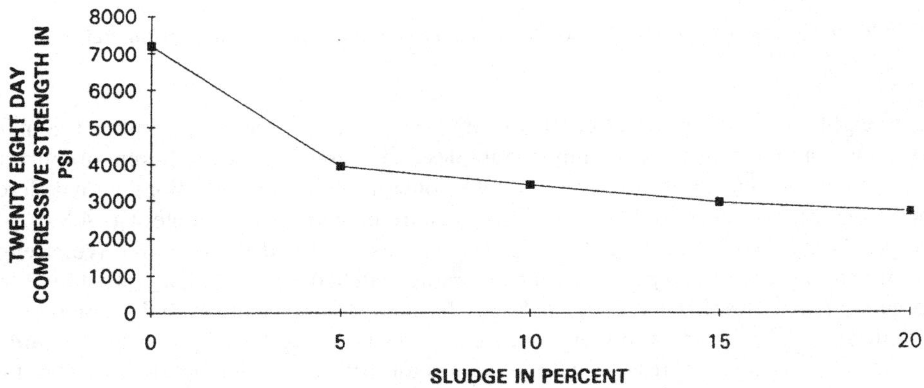

Figure 13. Twenty eight-day compressive strength of cement sludge mixtures (from Ref. 11).

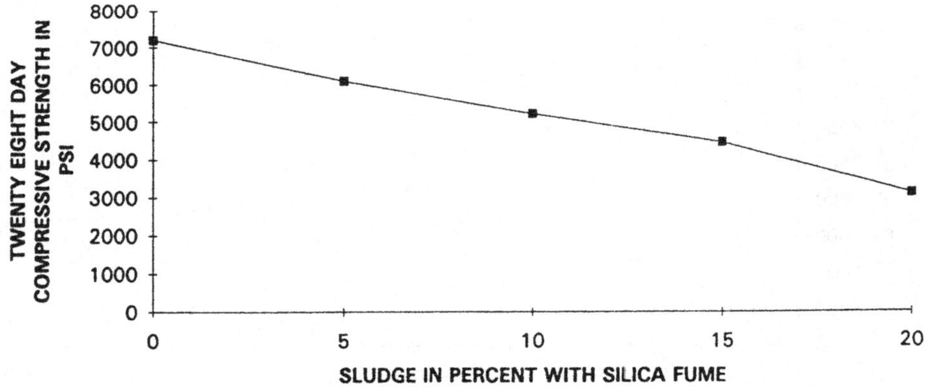

Figure 14. Twenty eight-day compressive strength of cement sludge mixtures with 5% silica fume (from Ref. 11).

higher. Figure 13 shows twenty-day compressive strength values in absence of silica fume. The same is shown in the presence of 5% silica fume in Figure 14. High compressive strength values are maintained for a sludge concentration as high as 10%. When Samples 2 and 6 are compared, they have the same percentage of sludge, but 0% silica fume in sample 2 and 5% silica fume in Sample 6. When their compressive strengths are compared, compressive strength is greater in the presence of silica fume for the same percentages of oil sludge. In order to determine if a given compressive strength is adequate, ASTM has a simple formula, given by:

$$\text{If } \frac{\text{Compressive Strength of Specimen with Additives}}{\text{Compressive Strength of Control Specimen}} * 100 = \begin{cases} \geq 70\text{—Good compressive strength} \\ \text{or} \\ \leq 70\text{—Not good compressive strength} \end{cases}$$

Table 3 shows sample number, sample proportions, compressive strength and optimum sample. It can be seen from Table 3 that Samples 8 and 9 provide adequate compressive strength values. So, the optimum proportions of cement, sludge and silica fume are 85%, 10%, and 5% to obtain sufficient compressive strength.

Other Properties

A series of permeability and durability tests were performed. For this, a recently developed local permeability measuring device, PDPK-200 of the Colorado School of Mines was used. This device is particularly suitable for measuring local permeability of an object of irregular geometry. No cylindrical core had to be prepared for conducting the permeability tests. Permeability values ranged from 120 to 180 mD for the concrete samples containing silica fume and the petroleum sludge. Permeability values ranged from 140 to 220 mD in the absence of petroleum sludge or silica fume. This observation demonstrated that the addition of silica fume and petroleum sludge in fact improve the permeability slightly. Durability of the proposed mixture was in the same range as the conventional concrete. Therefore, the new mixture at least met the requirement of the construction cement.

Effect of Asphaltene Concentration on Sludge Solidification

It is well known the presence of oil slows down the cementation process. While this aspect has been studied in detail, the effect of asphaltene and its possible role on cementation has been studied before. The author has started preliminary studies that appear to be encouraging [11]. A series of experiments was conducted to consolidate cement sludge with different amounts of asphaltene present in the mixture. The sludge concentration itself was maintained at 10%. Note that no other additive was added to the mixture. Figure 15 shows the 28-day compressive strength values for this set of experiments. None of them reaches the original compressive strength of concrete. However, one very interesting aspect is the possible role of asphaltene in improving the compressive strength. These results are rather preliminary and more research needs to be conducted, but they show that the presence of asphaltene can possibly improve the solidification process. The 7-day compressive strength values indicated that the curing of concrete is rather slow in the presence of asphaltene. However, the final compressive strength was distinctly superior for higher asphaltene concentration cases.

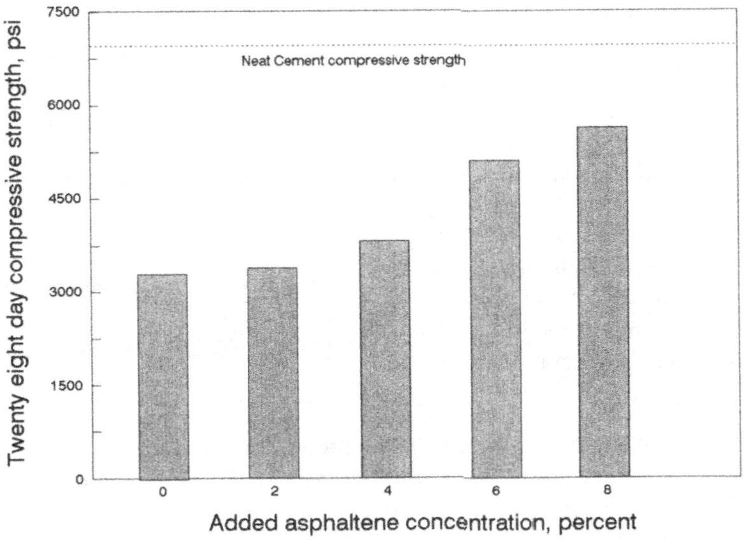

Figure 15. Effect of asphaltene concentration on solidified sludge strength.

CONCLUSIONS

Petroleum production is inherently related to the production of huge quantity of untreated sludge. Many techniques have been proposed for treating sludges of both heavy oil and light oil origins. Technologies in heavy oil recovery have focused on extraction of bitumen from the sludge. Unfortunately, many of these techniques are not yet commercially applied.

Only recently, techniques for total utilization of petroleum sludge have been developed. This involves the use of heavy oil residue as paving material. This patented technology has shown marked improvement in asphaltic properties when up to 30% of heavy oil residue was added to the conventional paving asphalt. The other method uses high temperature centrifugation in the presence of a biodegradable surfactant. The optimum concentration for surfactant was found to be 1% for the sludge studied. The solid residue was subjected to a series of laboratory tests. The results of these tests show that the use cement to stabilize oil sludge is an inexpensive method for disposal of the oil sludge. From the experimental results, one can conclude that the optimum amount of oil sludge and silica fume are a 10% and 5% to obtain an adequate compressive strength. The solid residue which cannot be disposed as per the new EPA regulation can be mixed with cement and silica fume in order to restore the original compressive strength. The mixture, therefore, can be used for oil well cementing or for any construction purpose.

Preliminar results indicate that asphaltenes play a favorable role in consolidation through cementing. However, more research needs to be done in this area.

REFERENCES

1. Takamura, K. and Wallace, D., 1986. The physical chemistry of the hot water process. paper 86-37-42, Annual Technical Meeting of the Petroleum Society of CIM, Calgary, Alberta, Canada, June.
2. Scott, J.D., 1974. Engineering contraints on environmental enhancement of oil sands tailings ponds. Symposium on Biolog. Issues in Energy Develop., Alberta Soc. of Prof. Biologists, Calgary, Alberta, Canada, April.

3. Kruyer, J., 1989. Reducing the sludge problem by co-processing sludge with oil sands ore. *Fourth UNITAR/UNDP International Conference on Heavy Crude and Tar Sands,* vol. 5, 19-35.
4. Storm, D.A., DeCanio, S.J., and Sheu, E.Y., 1994. Sludge formation during heavy oil conversion. *Asphaltene particles in fossil fuel exploration, recovery, refining, and production processes*, M.K. Sharma and T.F. Yen (eds.), Plenum Press, New York, 81-90.
5. Richard, D., 1986, "Centrifuge Reduces the Volume of Radioactive-Waste Oil", Electrical World, vol. 200, pp. 75.
6. Tay, J. H., and Show, K. Y., 1991, "Properties of Cement Made From Sludge", Journal of Environmental Engineering, vol. 117, No. 2, pp. 236-46.
7. Grasso, D., 1993, "Hazardous Waste Site Remediation Source Control", Lewis, Boca Raton, 1st ed.
8. Turner, L.R., 1991. Treatment of oil sands and heavy oil production wastes using the AOSTRA Traciuk Process. *Fifth UNITAR International Conference on Heavy Crude and Tar Sands,* vol. 4, 347-356.
9. Strelnikova, V.Y. and Kolesnikov, B.V., 1991. Natural bituminous rocks for motor road construction in Mongolian Republic. *Fifth UNITAR International Conference on Heavy Crude and Tar Sands,* vol. 1, 451-465.
10. Zanzotto, L., Foley, D.P., Jeffries, R.B., and Watson, R.D., 1990. Novel asphaltic composition. U.S. patent no. 4904305.
11. Kanakamedala, R.D. and Islam, M.R., 1995. A new method of petroleum sludge disposal and utilization", Sixth UNITAR International Conference on Heavy Crude and Tar Sands, in press.
12. Anonymous, 1988, "Oily Dirt Recycled for Parking Lot", Civil Engineering, vol. 58, pp. 12-13.
13. Bruce, A. S., Matthew, C. L., and Robert, L. S., 1990, 'Oilcrete", Civil Engineering, vol. 16, pp. 61-63.
14. Canadian patent 1221502, May, 1987.
15. Gollapudi, U.K. and Islam, M.R., "Reinforcement of Asphaltic Materials", Fine Particle Society Annual Conference, Chicago, IL, 1993.
16. Carl, P. S., Christopher, T. M., and James, D. W., 1987, 'Waste Oil Recovery", 42nd Purdue University Industrial Waste Conference Proceedings, pp. 457-63.
17. Sundaram, N.S., Sarwar, M., Bang, S.S., and Islam, M.R., 1994, "Biodegradation of Anionic Surfactants in the Presence of Petroleum Contaminants", *Chemosphere*, vol. 26 (6), 1253-1261.
18. Reynolds T. D., 1982, "Unit Operations and Processes in Environmental Engineering", Brooks/Cole Engineering Division, California.
19. Personal communication, Professor V. Ramakrishnan, South Dakota School of Mines and Technology, Rapid City, South Dakota, 1994.

INDEX

Made in the USA
Monee, IL
13 September 2026

62976349R00144